ELECTROMAGNETIC FIELDS AND WAVES

by
VLADIMIR ROJANSKY
Professor of Physics, Emeritus
Harvey Mudd College

Dover Publications, Inc.
New York

Copyright © 1971 by Vladimir Rojansky.
Copyright © 1979 by Dover Publications, Inc.
All rights reserved under Pan American and International Copyright Conventions.

Published in Canada by General Publishing Company, Ltd, 30 Lesmill Road, Don Mills, Toronto, Ontario.
Published in the United Kingdom by Constable and Company, Ltd.

This Dover edition, first published in 1979, is an unabridged and extensively corrected republication of the work originally published in 1971 by Prentice-Hall, Inc. The author has revised the Preface for the Dover edition.

International Standard Book Number: 0-486-63834-0
Library of Congress Catalog Card Number: 79-52648

Manufactured in the United States of America
Dover Publications, Inc.
31 East 2nd Street
Mineola, N.Y. 11501

To
J. H. Van Vleck
and to the memory of
W. E. Milne and B. H. Brown

Preface to the Dover Edition

This book will lead the reader through the rudiments of electric and magnetic fields to Maxwell's field equations for bodies at rest and to examples of electromagnetic waves. The prerequisites are an introductory course in physics and one in the calculus.

In 1955 I put away a nearly completed manuscript of a book on electricity and magnetism, and left teaching to join the TRW Space Technology Laboratories for ten years. There I worked side by side with other physicists and with engineers, mostly on projects—such as communication satellites—with which no one had any previous experience; and I came to see even better than before how important a knowledge of plain fundamentals is for the technical versatility of an engineer or scientist. As time went on, a general interest was growing in biophysics, geophysics, astrophysics, oceanography, and other studies involving this or that aspect of electricity and magnetism. Mathematics courses that used to be service courses for engineering and science students were becoming more abstract and less helpful for visualizing physical concepts. And so, when I unwrapped my old manuscript in 1965, I saw that it was out of step with the times. What was needed instead was a one-semester course devoted to a patient discussion of fundamentals—a base for more comprehensive books or for courses calling for a sound but not necessarily extensive background in electromagnetic fields and waves. This book was written to meet these needs.

The first four chapters should be rapid reading for a student who had a course in physics and one in the calculus; but they are also necessary reading—for one thing, they introduce much of the notation and language used later in the book.

What is the best vector field to take up first in this study? I have tried several possibilities in my courses and found that the choice of current density proved to be the most interesting and illuminating to the students. I have therefore adopted this choice in this book and have introduced it in Chapter 5 with a discussion of a steady flow of electric charge past a cylindrical hole in a conductor. This field is static, yet alive with moving electrons—and most students beginning this subject become particularly interested in situations in which "something is going on." Since current density has to do with an actual flow, it provides a vivid background for the flux integral and the concept of divergence. It also paves the way for the law of conservation of charge, which is the most concrete of the field equations involving both space and time derivatives.

The sequence of physical topics that leads in this book to Maxwell's equations is outlined in the chart that appears at the end of this preface, worded for compactness rather than precision. The discussion is confined to bodies at rest relative to the observer and to macroscopic phenomena, except for a few remarks on the microscopic viewpoint.

The physical discussion must of course have a substantial mathematical base, which I develop without pretense to rigor but fully enough, I hope, to bring out its vividness. In particular, the concepts of divergence, gradient, and curl are introduced as they become needed, and in settings that illustrate their physical meaning without much delay.

Most students who have not studied advanced calculus and come to a field theory for the first time, enter a world that is strange not only in geometrical and physical ideas, but also in language. Small misunderstandings at this stage can lead to serious misconceptions later, some so well hidden that they are hard for either the student or teacher to uncover. Therefore, I have made the presentation unhurried and explicit—to speed the reader's progress, to guard him against pitfalls, and to save lecture time for the instructor using this book in class. Certain topics, simple in retrospect, are not easy for most beginners to grasp without repeated exposure and without work on details that may help to reveal whether or not a topic is really understood. As the theory unfolds, the presentation remains unhurried; but it does of course become more compact.

After Maxwell's equations have been set up, I consider the radiation from a Hertzian dipole antenna, plane waves in nonconductors, and waves in rectangular guides. The dipole is studied first, because it provides a clear-cut source of the waves; and since the theory of plane waves does not help with the dipole problem, the reversal of the usual order does not make thi::gs any harder. The dipole radiation is worked out in Chapter 24 by hammer and tongs; in Chapter 27 with the aid of the magnetic vector potential; and in Appendix C on the basis of the Lorentz condition and in the complex-number notation.

Too many beginning students of this subject learn the mathematics but lose sight of the physics. Yet the world of ideas crystallized by Maxwell in his field equations is not just a formal world of mathematical symbols—it is the vivid world of Faraday. In his time, Maxwell resolved to read no mathematics concerning electricity until he had first read Faraday's papers. Today, a century later, Maxwell's advice to read Faraday is hardly practical. But one can still help to make the field equations come alive for the reader by showing him how the exacting formulas emerge from graphic though imperfect images. I have attempted to do this here and there in this book.

The exercises—most of them short—provide practice in the material already covered, tie up some of the loose ends in the discussion of the text, or clear the ground for the topics to come. The results of some exercises are used much later in the text; the reader who skips an exercise of this kind may have to return to it for a review.

Now a remark on terminology. If we are told that a voltmeter reads x volts, we cannot tell whether it reads the difference of potential across a capacitor, or the chemical emf of a battery, or the ir drop along a wire, or still something else. All we know is that a certain line integral is equal to x volts. It is convenient to have a uniform name for this integral. The logical thing to say is that voltmeters read voltages. Many beginners, however, think of voltage as a pressure, and this causes puzzlement when the reading of a voltmeter depends on the shape taken by its leads, as in electric fields that are induced by time-dependent magnetic fields and that I call "Faraday fields." The term "electromotive force" is freer from this difficulty; so, in this book, voltmeters read emf's, which in special cases are potential differences, ir drops, and so on.

Many people have helped to shape this book—my students, my former colleagues at Union College and later at the TRW Space Technology Laboratories, my present colleagues at Harvey Mudd College, friends elsewhere, and the anonymous reviewers of the manuscript. A persuasive article by Professor Francis W. Sears caused me to rewrite a good part of the chapter on Faraday's law. I am indebted to Dr. Robert M. Whitmer not only for many discussions, but also for reading the galleys and suggesting a number of significant improvements. My thanks for typing recent drafts of the manuscript go to Frances Menechios and Patricia Johnston. I am grateful to my wife Milla for typing the final draft and help with the proofs and the index.

The misprints and mistakes that have been found in the earlier edition of this book have been corrected in this one, and some new material has been added. I am greatly indebted to the editorial and production departments of Dover Publications, Inc., for making these improvements possible.

V. ROJANSKY

Claremont, California

Outline of the Discussion Leading to Maxwell's Equations

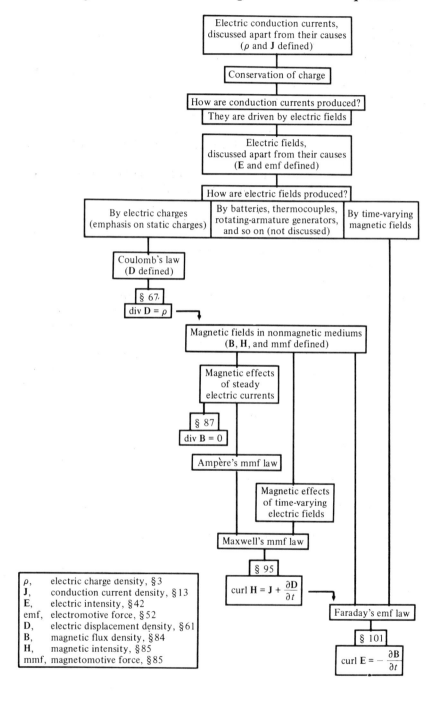

Contents

CHAPTER 1. Scalar Fields   1

     1. CIRCUIT QUANTITIES AND CIRCUIT EQUATIONS   2
     2. CONDUCTIVITY   8
     3. CHARGE DENSITY   11
     4. SCALAR FIELDS   18

CHAPTER 2. Mathematical Notes   23

     5. MISCELLANEOUS ITEMS   23
     6. TAYLOR'S EQUATION IN ONE DIMENSION   33
     7. TAYLOR'S EQUATION IN THREE DIMENSIONS   37

CHAPTER 3. Curves and Surfaces   42

     8. CURVES AND SURFACES   42
     9. SOLID ANGLES   48

CHAPTER 4. Vectors   56

     10. VECTORS   56
     11. CARTESIAN COMPONENTS   62
     12. NORMAL AND TANGENTIAL COMPONENTS   68

CHAPTER 5. Current and Current Density   72

     13. DEFINITIONS   72
     14. THE FLOW OF CHARGE PAST A CYLINDRICAL HOLE   78
     15. CURRENT AS AN INTEGRAL   84

Contents

CHAPTER 6. Vector Fields — 89

- 16. VECTOR FIELDS — 89
- 17. FIELD LINES, FLUX LINES, AND OTHER LINES — 94
- 18. UNIFORM VECTOR FIELDS — 100
- 19. THE FIELD $x\mathbf{1}_x$ — 103
- 20. CIRCULAR VECTOR FIELDS — 105
- 21. THE FIELD $y\mathbf{1}_x$ — 107

CHAPTER 7. The Flux Integral — 109

- 22. FLUX ACROSS A SURFACE — 109
- 23. FLUX AND FLUX LINES — 111
- 24. LINES OF CURRENT DENSITY — 112
- 25. GRAPHICAL ESTIMATION OF FLUX — 115
- 26. FLUX ACROSS A CLOSED SURFACE — 117

CHAPTER 8. Source Density and the Divergence — 121

- 27. SOURCES AND SINKS OF VECTOR FIELDS — 121
- 28. SOURCE DENSITY — 123
- 29. DIVERGENCE — 125
- 30. COORDINATE FORMULAS FOR THE DIVERGENCE — 126
- 31. THE DIVERGENCE THEOREM — 129
- 32. FLUX THROUGH A CONTOUR — 130

CHAPTER 9. Cylindrical and Spherical Coordinates — 134

- 33. THE VECTORS $\mathbf{1}_\rho$, $\mathbf{1}_\phi$, AND $\mathbf{1}_z$ — 134
- 34. TWO-DIMENSIONAL RADIAL VECTOR FIELDS — 137
- 35. TWO-DIMENSIONAL CIRCULAR VECTOR FIELDS — 142
- 36. THE VECTORS $\mathbf{1}_r$, $\mathbf{1}_\theta$, AND $\mathbf{1}_\phi$ — 144
- 37. SPHERICALLY SYMMETRIC VECTOR FIELDS — 147

CHAPTER 10. Conservation of Charge — 149

- 38. KIRCHHOFF'S CURRENT LAW — 149
- 39. LEAKAGE CURRENT IN A CABLE — 151
- 40. CONSERVATION OF CHARGE — 154

CHAPTER 11. Electric Intensity and the Laws of Coulomb, Ohm, and Joule — 158

- 41. ELECTRIC FIELDS — 159
- 42. ELECTRIC INTENSITY — 160
- 43. CHARGE-DRIVING FORCES — 164
- 44. OHM'S LAW — 166
- 45. JOULE'S LAW — 170

Contents xiii

CHAPTER 12. Electromotive Force   173

- 46. WORK   174
- 47. UNIFORM ELECTRIC FIELDS (WORK INDEPENDENT OF PATH)   176
- 48. THE INVERSE-SQUARE FIELD (WORK INDEPENDENT OF PATH)   179
- 49. ELECTROSTATIC FIELDS IN GENERAL (WORK INDEPENDENT OF PATH)   181
- 50. AN ELECTRIC FIELD PRODUCED MAGNETICALLY (WORK DEPENDS ON PATH)   182
- 51. CLOSED PATHS   183
- 52. ELECTROMOTIVE FORCE   185
- 53. D. C. VOLTMETERS   187

CHAPTER 13. Conservative Electric Fields   192

- 54. AN INTEGRAL TEST FOR CONSERVATIVE FIELDS   192
- 55. A DIFFERENTIAL TEST FOR CONSERVATIVE FIELDS   193
- 56. ELECTROSTATIC POTENTIAL   199
- 57. THE GRADIENT   208
- 58. COORDINATE FORMULAS FOR GRADIENTS   213

CHAPTER 14. Coulomb's Law in Rationalized Form   216

- 59. COULOMB'S LAW IN FREE SPACE   216
- 60. EXAMPLES OF ELECTROSTATIC FIELDS   218
- 61. ELECTRIC DISPLACEMENT DENSITY   229
- 62. DIELECTRICS   231

CHAPTER 15. Integral and Differential Forms of Coulomb's Law   238

- 63. GAUSS'S FORM OF COULOMB'S LAW   238
- 64. DELTA FUNCTIONS   244
- 65. THE DIVERGENCE OF A GRADIENT   249
- 66. POISSON'S AND LAPLACE'S FORMS OF COULOMB'S LAW   250
- 67. MAXWELL'S FORM OF COULOMB'S LAW   252

CHAPTER 16. Examples of Solutions of Laplace's Equation   253

- 68. HARMONIC FUNCTIONS OF PAIRS OF COORDINATE VARIABLES   254
- 69. THE POTENTIAL V IN A CURRENT-CARRYING COPPER BLOCK WITH A CYLINDRICAL HOLE   258
- 70. BOUNDARY CONDITIONS FOR $\mathbf{D}$ AND $\mathbf{E}$ 261
- 71. FIELDS INSIDE THE CYLINDRICAL HOLE AND CHARGES ON ITS SURFACE   264

xiv   Contents

CHAPTER 17. **Charged Conductors and Capacitors**   269

    72. CHARGED CONDUCTORS   269
    73. CAPACITORS   274

CHAPTER 18. **Electric Energy Density and Displacement Current**   280

    74. ELECTRIC ENERGY DENSITY   281
    75. TIME-VARYING VECTOR FIELDS   282
    76. ELECTRIC DISPLACEMENT CURRENT   283

CHAPTER 19. **Cross Products and Curls**   287

    77. TORQUES AND CROSS PRODUCTS OF VECTORS   287
    78. THE CURL OF A VECTOR FIELD   293
    79. DIRECTION LINES OF CURLS   295
    80. NORMAL COMPONENTS OF CURLS   296
    81. STOKES'S THEOREM   298
    82. CONSERVATIVE FIELDS   301

CHAPTER 20. **Static Magnetic Fields in Nonmagnetic Mediums**   304

    83. PRELIMINARIES   304
    84. MAGNETIC FLUX DENSITY   306
    85. MAGNETIC FIELD INTENSITY   308
    86. FORCES ON CURRENT-CARRYING WIRES   310
    87. THE $d\mathbf{B}$ FORMULA   313
    88. THE $d\mathbf{H}$ FORMULA   316
    89. THE NUMERICAL VALUE OF μ_0 320

CHAPTER 21. **Magnetomotive Force**   322

    90. MAGNETIC FIELDS AND SOLID ANGLES   322
    91. MAGNETOSTATIC POTENTIAL   325
    92. MAGNETOMOTIVE FORCE   327
    93. AMPERE'S MMF LAW IN INTEGRAL FORM   328
    94. AMPERE'S LAW IN DIFFERENTIAL FORM   333
    95. MAXWELL'S MMF LAW   334

CHAPTER 22. **Faraday's Law of Induction**   338

    96. FARADAY'S DISCOVERY   338
    97. THE $d\mathbf{E}$ FORMULA   341
    98. FARADAY'S LAW IN INTEGRAL FORM   344
    99. A. C. VOLTMETERS   349
    100. MAGNETIC ENERGY   353
    101. FARADAY'S LAW IN DIFFERENTIAL FORM   355

	102.	INDUCED EMF'S, CURRENTS, AND SURFACE CHARGES	356

CHAPTER 23. Maxwell's Equations, Wave Equations, and the Flow of Energy 364

- 103. MAXWELL'S EQUATIONS — 364
- 104. THE FIELDS **B**, **D**, **E**, AND **H** — 366
- 105. WAVE EQUATIONS — 368
- 106. THE FLOW OF ELECTROMAGNETIC ENERGY — 370

CHAPTER 24. Radiation from a Short Antenna 375

- 107. THE LOCAL FIELD — 375
- 108. THE COMPLETE FIELD — 379
- 109. THE DISTANT FIELD — 384

CHAPTER 25. Plane Electromagnetic Waves 388

- 110. PASSAGE TO PLANE WAVES — 388
- 111. RADIATION FROM A SHEET OF ALTERNATING CURRENT — 391
- 112. TERMINOLOGY AND NOTATION — 394
- 113. MORE BOUNDARY CONDITIONS — 397
- 114. REFLECTION FROM NONCONDUCTORS; NORMAL INCIDENCE — 398
- 115. REFLECTION FROM PERFECT CONDUCTORS; NORMAL INCIDENCE — 402
- 116. REFLECTION FROM PERFECT CONDUCTORS; OBLIQUE INCIDENCE — 407

CHAPTER 26. Rectangular Waveguides with Perfectly Conducting Walls 410

- 117. TE WAVES BETWEEN PARALLEL PERFECTLY CONDUCTING PLANES — 411
- 118. THE "CUTOFF" PHENOMENON — 418
- 119. TE WAVES IN RECTANGULAR GUIDES; THE DOMINANT MODE — 420
- 120. OTHER TYPES OF TE WAVES — 425

CHAPTER 27. Magnetic Vector Potential 429

- 121. FILAMENTARY CURRENTS — 429
- 122. FARADAY'S LAW — 434
- 123. RETARDED POTENTIALS FOR A SHORT ANTENNA — 435

xvi Contents

124.	THE ELECTROMAGNETIC FIELD OF A SHORT ANTENNA	437
125.	THE LORENTZ CONDITION	438

Appendices 442

A.	MATHEMATICAL FORMULAS	442
B.	THE MKS-GIORGI UNITS OF MEASUREMENT	449
C.	THE COMPLEX NUMBER SHORTHAND	453
D.	THE DEL NOTATION	457

Index 459

Note 1. In cross references, section numbers are indicated by superscripts: the symbol (1^8) denotes equation (1) of §8, the symbol Fig. 23^8 means that Fig. 23 appears in §8, and so on. Section numbers are printed at the right of page headings, so that the number at the upper right of a reference symbol is the number at the upper right of a page. References to Appendix A carry the superscript A.

Note 2. The marginal number at the left of an equation is its old number, or the number of another equation that might well be recalled.

CHAPTER 1

Scalar Fields

The theory of electromagnetic fields discussed in this book was developed in the 1860's by James Clerk Maxwell (1832-1879). It rests on four equations, called Maxwell's field equations and shown in the chart in the preface. These equations presumably describe all electromagnetic phenomena, except for quantum modifications in the atomic domain. They are built upon and include the law of interaction of static electric charges announced in 1785 by Charles Augustin de Coulomb (1736-1806), the law of interaction of steady currents formulated in 1822 by André Marie Ampère (1775-1836), and the law of induction of electric fields by time-varying magnetic fields discovered in 1831 by Michael Faraday (1791-1867). They also include the current law and the voltage law for electric circuits stated in 1848 by Gustav Robert Kirchhoff (1824-1887).

The roots of Maxwell's theory lie deep in Faraday's graphic concepts, which guided Maxwell in his work and which he eventually expressed in mathematical form. Some of these concepts, such as "lines of force," are still as useful today as they were to Faraday and Maxwell—for picturing electric and magnetic fields, for solving simple problems, and for making the implications of some of the mathematical formulas more vivid.

The reader has several tasks before him: to understand the physical content of Maxwell's equations, to learn the mathematical shorthand in which they are usually written, and to master the language of the theory—a language so blended of physics and mathematics that sometimes it is hard to tell which is which. We will lead up to Maxwell's equations by reviewing the more familiar laws of electricity and magnetism (such as the circuit laws illustrated for simple cases in §1), restating each in field-theoretic language, and rewriting each in field-theoretic symbols. As we do this, the reader will have to work in detail through quite a few unsophisticated preliminaries.

The portions of Maxwell's theory discussed in this book involve *scalar fields* and *vector fields*. Examples of scalar fields (conductivity and charge density) are given in §§2 and 3, without waiting for the mathematical cautions hinted at in Chapter 2. Scalar fields in general are taken up in §4, and vector fields in Chapter 6.

1. CIRCUIT QUANTITIES AND CIRCUIT EQUATIONS

Most of the topics presented in this section are presumably already familiar to the reader. We will nevertheless touch upon them, partly to have on hand certain equations for later reference, and partly because the concept of "field quantities" can be made clearer by contrasting it with the concept of "circuit quantities."

Units. The reader should gradually familiarize himself with Appendix B, which describes the MKS-Giorgi units of measurement used in this book, and with the conversion table on the inside front cover. For expository reasons, these units appear in the text in a different order from that in the appendix. For example, the definition of the ampere refers to forces between current-carrying wires. Therefore, although we will begin at once to call our unit of current the "ampere," the reader will have to wait until §89 before we make sure that this unit is in fact the ampere defined in the appendix. The MKS-Giorgi system is especially convenient for our purposes, but older systems —particularly the *cgs Gaussian* system, described in other books—have advantages in discussions of other aspects of electricity and magnetism.

Static charges. Conductors and insulators consist of positive electric charges (atomic nuclei) and negative charges (electrons), all in constant motion. Any extra charges placed on or in a body—say the extra electrons on the negative plate of a capacitor—are also in constant motion. The frequencies of the oscillation of all these charges are, however, very high—of the order of optical frequencies—so that in experiments not involving such frequencies only the time-averages of the fields of these charges come into play. It is, therefore, useful to introduce the term *static charges*, which stands for actual charges imagined to be held fixed in their average positions. The electric fields produced by static charges are called *electrostatic*.

Conduction currents. In metals, some of the electrons, called *conduction electrons*, are bound to the atomic nuclei only loosely and can be easily moved from place to place. A copper wire, for example, has one such electron per copper atom. If a piece of wire is connected across the terminals of a dry cell, an extra electric field is set up inside the wire in addition to the ever-present atomic electric fields, and the conduction electrons drift along the wire, amounting to a current. (We will not discuss "semiconductors," such as germanium, or "semimetals," such as arsenic, in which the situation is more complicated.)

The drift speed of the conduction electrons is much lower than the speed of electric signals along a wire (Exercise 2). The reason for this is the stop-and-go zigzag drift of the conduction electrons, caused by continual collisions with atoms. At each collision, some of the kinetic energy of a conduction electron is transferred to an atom and converted into vibrational energy of the atomic lattice, and thus into heat.

Despite their continual motion, the electrons that are tightly bound to atomic nuclei do not contribute to the conduction current flowing, say, through an ammeter—they move around the nuclei so fast and so close to the nuclei that from the large-scale viewpoint they can be regarded as standing still. Their motion is ignored in this book, which deals with large-scale (macroscopic) and not with small-scale (microscopic) phenomena. The motions of the tightly bound electrons cause magnetic effects, of course, effects that do not necessarily average out if neighboring atoms interact with one another sufficiently strongly. These effects must be considered in detail if one wants to account on the microscopic basis for, say, the magnetic permeability of a medium, which is itself a macroscopic quantity that describes an average effect.

We follow a standard convention and say that electric current is flowing[1] in a branch ab of a circuit from a to b if *in effect* a positive electric charge is moving in this branch from a to b.[2] For example, if the branch ab of a circuit is a copper wire, the statement that the current in this branch flows from a to b means *in fact* that conduction electrons move in this branch from b to a.

Circuit quantities. When we refer to such points as a and b in Fig. 1, we let the subscripts ab and (ab) mean, respectively, "from a to b" and "between a and b." The simpler circuit quantities are the resistance, say $R_{(ab)}$,

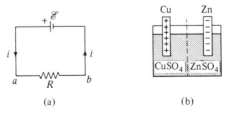

Fig. 1. A simple electric circuit and an example of a chemical cell activating it.

---

[1] We say "a current flows" when we mean that electric charge is flowing, just as we say "a river flows" when we mean that water is flowing.

[2] Positive charge is the kind of charge that collects on glass rubbed with silk. When electrons and protons were discovered, it was found that this kind of charge is carried by protons and that electrons are charged negatively.

4 Scalar Fields § 1

between the terminals a and b of a conductor (resistor); the potential drop, say V_{ab}, from a to b; the current, say i_{ab}, flowing from a to b; the capacitance between the terminals of a capacitor (condenser), and so on. In many conductors the current i_{ab} is directly proportional to V_{ab}; these conductors are said to satisfy a law formulated in 1826 by Georg Simon Ohm (1787–1854). In this book we consider only these "ohmic" conductors and take it for granted that in the equation

$$i_{ab} = \frac{V_{ab}}{R_{(ab)}} \qquad (1)$$

the resistance $R_{(ab)}$ is a *constant*; this equation then states *Ohm's law*. An example of a "nonohmic" conductor is thyrite, in which i_{ab} is roughly proportional to $(V_{ab})^3$, so that the resistance of a thyrite rod drops rapidly when the potential difference between its ends is increased.

Another way of writing (1) is

$$i_{ab} = G_{(ab)} V_{ab}; \qquad (2)$$

here $G_{(ab)}$ is the *conductance* of the conductor between the terminals a and b, defined as

$$G_{(ab)} = \frac{1}{R_{(ab)}}. \qquad (3)$$

We express resistance in ohms, and hence our unit for measuring conductance is the "reciprocal ohm," called the *mho*.

Equation (2) is usually written $i = GV$. We used double subscripts above to stress the fact that such circuit quantities as current, conductance, and potential drop each pertain not to a single point but to a *pair of terminals* (which are points in principle, but perhaps binding posts in practice). For example, let a and b be the terminals of a thick wire and let P be a point *inside* the wire. Then the question "What is the magnitude of the current flowing from a to b?" has a clear-cut meaning; the answer might be "five amperes." But the question "What is the magnitude of the current flowing *at* the point P?" is meaningless, because current is not the kind of quantity that can be evaluated at an individual point. (A proper question to ask is "What is the *current density* at P?")

Electromotive forces. The current i in Fig. 1 is caused by the chemical cell that has a positive terminal, marked plus, and a negative terminal. A simple cell is shown in Fig. 1(b), where the dashed line indicates a semi-permeable partition separating the $CuSO_4$ and the $ZnSO_4$ solutions, and where chemical action of the copper and the zinc ions keeps the copper plate charged positively relative to the zinc plate. Outside the cell, the conventional current flows "from plus to minus," but inside the cell it flows "from minus to plus." This means that the cell has a rather special property: it forces charges to move in an "unnatural" direction, and hence it does work on them. This property is described by saying that the cell is a source of an *electromotive force* (emf). The magnitude of this emf, say $\mathscr{E}$, is the work that the cell does per coulomb of positive charge that it moves from its negative to its positive plate. We express $\mathscr{E}$ in joules per coulomb or simply *volts*. The resistance R in the figure includes the internal resistance of the cell.

Given the values of $\mathscr{E}$ and R in the circuit of Fig. 1, the current i can be found by using *Kirchhoff's loop law*. In the case of constant currents in circuits consisting of resistances and sources of emf's, this law can be stated as follows: If we go around any closed loop of a circuit, the sum of the potential drops that we encounter along the resistances will be equal to the sum of the emf's included in the loop. Suppose, for example, that in Fig. 1 we go from a to b and then, through the cell, back to a. On the way from a to b we move "with the current," and hence encounter the potential drop V_{ab}; as we cross the cell "from minus to plus" we encounter the emf $\mathscr{E}$. Kirchhoff's law states in this case that $V_{ab} = \mathscr{E}$. By Ohm's law, the potential drop V_{ab} is the "iR drop" along the resistor. Accordingly, $iR = \mathscr{E}$ and $i = \mathscr{E}/R$.

Junction law. If a circuit comprises more than one loop, *Kirchhoff's junction law* comes into play: The sum of currents flowing toward a junction of wires is equal to the sum of the currents flowing away from this junction. Figure 2 shows three equivalent examples of two 1-ampere currents flowing toward a junction and one 2-ampere current leaving it.

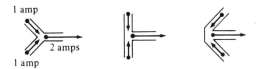

Fig. 2. Examples of two 1-ampere currents flowing toward a junction and one 2-ampere current leaving it.

Branch currents. The circuit shown in Fig. 3(a) has three branches: that containing the cell and its internal resistance r, that containing R_1, and that containing R_2. The corresponding currents are labeled $i, i_1,$ and i_2. The directions along the branches, marked by the arrowheads, are chosen arbitrarily, with the understanding that if i_1, say, actually flows in the direction opposite to that of the arrowhead, then the value of the number i_1 will prove to be negative. [In Fig. 3(a) the three branch currents obviously *do* flow as indicated by the arrowheads, and hence each of the three numbers $i, i_1,$ and i_2 is positive.] At the starred junction we have

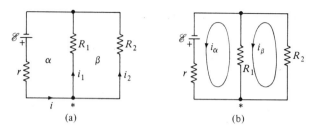

Fig. 3. Branch currents (i, i_1, i_2) and loop currents (i_α, i_β).

$$i = i_1 + i_2. \tag{4}$$

At the other junction we have the equation $i_1 + i_2 = i$, which adds nothing new.

The circuit has three loops (or meshes): that containing the cell and R_1, and marked α; that containing R_1 and R_2, and marked β; and that containing the cell and R_2—the "big" loop. If we start at the star and go counterclockwise around the first loop, we get

$$i_1 R_1 + ir = \mathscr{E}. \tag{5}$$

If we start at the star and go counterclockwise around the second loop, we get

$$i_2 R_2 - i_1 R_1 = 0. \tag{6}$$

The big loop contributes nothing new. We now have three equations—(4), (5), and (6)—ready to be solved for the three unknown currents. The solution for i proves to be

$$i = \frac{R_1 + R_2}{R_1 R_2 + r(R_1 + R_2)} \mathscr{E}. \tag{7}$$

Loop currents. To prepare for another method of computing the currents in Fig. 3, we imagine that the loops α and β carry the "loop currents" (or "mesh currents") i_α and i_β, pictured in Fig. 3(b), and we count up the potential drops and the emf's along these loops.

Starting at the star, we go around loop α. As we go along R_1, the net current flowing "with us" is $i_\alpha - i_\beta$, so the potential drop we encounter along R_1 is $(i_\alpha - i_\beta) R_1$. The potential drop along r is $i_\alpha r$, and consequently

$$(i_\alpha - i_\beta) R_1 + i_\alpha r = \mathscr{E}, \tag{8}$$

which can be written in the form (9). Similarly, loop β yields the equation

$i_\beta R_2 + (i_\beta - i_\alpha)R_1 = 0$, recorded below as (10). Thus the two loop currents satisfy the *two* equations

$$(R_1 + r)i_\alpha - R_1 i_\beta = \mathscr{E}, \tag{9}$$

$$-R_1 i_\alpha + (R_1 + R_2)i_\beta = 0, \tag{10}$$

which are, of course, consistent with the *three* equations (4), (5), and (6) of the branch-current method (Exercise 5).

The method of loop currents has two advantages: Kirchhoff's junction equations need not be written explicitly, and the coefficients in such sets of equations as (9) and (10) are easy to check. For example, the respective coefficients of i_α in (9) and of i_β in (10) are the total resistances of the corresponding loops; the coefficient of $-i_\beta$ in (9) and of $-i_\alpha$ in (10) is the resistance shared by the two loops.

EXERCISES

1. What does the statement "current is flowing from the terminal *a* to the terminal *b*" mean in the case of an electrolytic conductor?

2. A wire has *n* conduction electrons per unit volume and carries a current of magnitude *i*. Show that the drift speed of the conduction electrons is

$$v_d = \frac{i}{neA}. \tag{11}$$

   Here *e* is the magnitude of the electronic charge and *A* the cross-sectional area of the wire. Next, assume that a copper wire has 8.5×10^{22} atoms per cm^3 and 1 conduction electron per atom, and set $e = 1.6 \times 10^{-19}$ coulomb, $A = 1$ mm^2, and $i = 1$ ampere. Show that the drift speed of the conduction electrons is then only about 0.007 cm/sec.

3. What nontrivial information can be inferred from Fig. 108[38] about the signs of the numerical values of the currents i_a, i_b, and i_c in that figure?

4. Derive (7) from (4), (5), and (6).

5. Verify that the loop-current equations (9) and (10) give the right-hand side of (7) as the current flowing "from minus to plus" in the chemical cell of Fig. 3(b).

6. If you have met in your earlier work any circuit quantities that pertain to *four* terminals, name and define them.

2. CONDUCTIVITY

Such circuit quantities as $G_{(ab)}$, V_{ab}, and i_{ab} each pertain to a *pair of points*. By contrast, the basic quantities of the electromagnetic field theory—called *field quantities* or *point quantities*—each have a numerical value (and some also a direction) at every *single point* in space. For example, as we turn in this section from circuit theory to field theory, our attention will shift from the conductance $G_{(ab)}$ between the terminals of a conductor to the *conductivity*, say g, at any single point in space. But first some terminology and notation.

The unqualified term "space" means in this book the classical three-dimensional space in which we live. A "point in space" is a point located anywhere in this space, inside or outside of conductors or insulators, and thus not necessarily in "empty" or "free" space. A direction in space can be described, for example, with the help of two fixed points in space, or simply by pointing our hand. A direction relative to a given coordinate frame can be specified by listing the numerical values of the direction cosines. We must, however, stress the fact that the concept of a direction in space does not involve any coordinate frames; thus, the direction from a corner of this book to the point where the center of the moon appears to the reader to be at this instant is quite independent of any coordinate frame that anyone might wish to use.

We use the words "line" and "surface" in the usual way, but restrict the unqualified term "region" to a portion of *three-dimensional* space; accordingly, barring mathematical anomalies, a *line* has a *length*, a *surface* has an *area*, and a *region* has a *volume*. The concept of a line is very different from the concept of a length: a line is an infinite set of points, but the length of a line is a single number. The concepts of a surface and a region differ in a similar way from the concepts of an area and a volume. Therefore the reader should guard against using the word "area" interchangeably with "surface," or using "volume" interchangeably with "region." We usually write S for a particular surface (closed or open) and R for a particular region; and sometimes we write AB for a particular line (straight or curved) connecting the points A and B. We also write

> the line AB has the length l,
> the surface S has the area a,
> the region R has the volume v.

To stress the fact that a particular surface S is closed, we write (S) instead of S. Open and closed surfaces are contrasted in §8.

Conductivity. A substance or a body is called *homogeneous* if its properties are the same at every point in it; it is called *isotropic* if its properties at every point are the same in every direction. (Pure crystalline quartz is homogeneous but not isotropic.) This book deals only with isotropic substances.

A homogeneous conducting rod has the length l and the cross-sectional area A. Experiment shows that the conductance $G_{(ab)}$ between the ends a and b of such a rod is directly proportional to A and inversely proportional to l; that is,

$$G_{(ab)} = g\frac{A}{l}, \qquad (1)$$

where the constant of proportionality g is called the electric *conductivity* of the material of which the rod is made. The conductivity of a homogeneous material can thus be found by measuring the dimensions and the conductance of a cylindrical sample and then using the formula

$$g = \frac{l}{A}G_{(ab)}. \qquad (2)$$

We express conductance in mhos and length in meters, so that, according to (2), our unit for measuring conductivity is the "mho per meter." Several approximate conductivities are listed in Table 1. The conductivity *at a point*, say

TABLE 1. Electric Conductivities

Substance	g (mho/meter)
Silver	6.1×10^7
Copper	5.8×10^7
Iron	1.0×10^7
Sea water	3 to 4
Fresh water	10^{-2} to 10^{-3}
Wet ground	10^{-2} to 10^{-3}
Dry ground	10^{-4} to 10^{-5}
Bakelite	About 10^{-9}
Glass	About 10^{-12}
Sulfur	About 10^{-15}

P, in a homogeneous conductor is defined as the conductivity of the material of which the conductor is made. For example, at any point P in a copper wire we have $g = 5.8 \times 10^7$ mhos/meter. The reciprocal of the conductivity is called "specific resistance" or "resistivity." According to (1) and (3¹), the resistance between the ends of a homogeneous rod is[3]

$$R_{(ab)} = \frac{l}{gA}. \qquad (3)$$

[3] See Note 1 following the Table of Contents.

Next suppose that a cylindrical conductor is not homogeneous, but is made of an alloy whose composition varies from point to point. The quantity $lG_{(ab)}/A$ is then called the *average conductivity* of the material in the cylinder; that is,

$$g_{\text{average}} = \frac{l}{A} G_{(ab)}. \tag{4}$$

The conductivity *at a point*, say a point P, in a nonhomogeneous conductor is defined by a limiting process: A small "test cylinder," which includes P, is cut out of the conductor; its dimensions and its conductance are measured, and the average conductivity of the material in the neighborhood of P is then computed from (4). This process can in principle be repeated with smaller and smaller similarly shaped test cylinders, each including P. The series of average values of the conductivity in the neighborhood of P, obtained in this way, can be expected to approach a limit. The conductivity of the conductor *at* the point P is defined to be this limit.

Now, electric conductivity depends on the interplay of the atoms in the conductor, and therefore the hypothetical test cylinders, although small by ordinary laboratory standards, should remain large compared to atomic dimensions. The granular structure of matter precludes the possibility of letting the dimensions of the test cylinders tend to zero. Nevertheless, in some of our work (as in Exercise 1 below) we will allow certain volume elements to approach zero, in order that integration can be used instead of summation of finite series.[4]

The conductivity g of a nonhomogeneous conductor varies from point to point and is therefore a function of the coordinates used to identify the various points, say the cartesian coordinates x, y, and z. If a variable, say u, is a function of x, y, and z, it is customary in elementary work to write $u = f(x, y, z)$. Accordingly, to show that the conductivity g depends on x, y, and z, we might write $g = f(x, y, z)$. But this notation uses two letters, g and f, for dealing with a single physical quantity, and since in our work many letters have fixed significance, this notation would soon make us run out of suitable letters. Therefore, we will summarize the statement "g is a function of x, y, and z" by writing

$$g = g(x, y, z), \tag{5}$$

---

[4] If a length dl, an area da, or a volume dv is required to decrease to a size small compared to everyday standards yet large compared to atomic dimensions, we will, for simplicity of notation, write $dl \to 0$, $da \to 0$, and $dv \to 0$. Therefore, a limiting condition written in this book as $dx \to 0$ does not necessarily have its precise mathematical meaning. It should usually be apparent from the context whether such symbolism as "$dx \to 0$" is meant in its strict mathematical sense or in the sense imposed by the granular structure of matter and electric charge.

rather than by writing $g = f(x, y, z)$. Similarly, if g is a function of r, θ, and ϕ, we write $g = g(r, \theta, \phi)$.

At any point in a homogeneous conductor

$$g = g(x, y, z) = \text{constant}, \tag{6}$$

and at any point in a vacuum $g = 0$. Suppose, for example, that a copper sphere of radius r_0 is centered at the origin of a coordinate frame and all other bodies can be ignored. We can then describe the conductivity at any point P in space by writing

$$g = g(r, \theta, \phi) = \begin{cases} 5.8 \times 10^7 \text{ mhos/meter} & \text{if } r \leq r_0 \\ 0 & \text{if } r > r_0, \end{cases} \tag{7}$$

where r, θ, and ϕ are the polar coordinates of P. In this case g is actually independent of θ and ϕ; it is a continuous function of r (in fact, a constant) inside the sphere and also a continuous function of r (in fact, zero) outside the sphere; but it is discontinuous at the surface of the sphere, where it changes from one constant value to another.

EXERCISE

1. The conductivity of a nonhomogeneous rod (length l, cross-sectional area A) varies linearly from the value g_a at the end a to the value g_b at the end b; that is, at any point on the cross section of the rod lying at the distance x from a we have

$$g(x) = g_a + \frac{g_b - g_a}{l} x. \tag{8}$$

Divide the rod into discs of thickness dx, compute the resistance of a typical disc, and show by integration that the total resistance of the rod is

$$R = \frac{l}{(g_b - g_a)A} \ln \frac{g_b}{g_a}. \tag{9}$$

Here ln denotes the natural logarithm.

3. CHARGE DENSITY

When we speak of the electric charge located in a region, we mean the "net" charge, which is the algebraic sum of the total positive charge and the total negative charge. If a body contains as many electrons as protons, we say that it is uncharged. Our unit of charge is the *coulomb*; see page 450.

Substances having negligible conductivity and bodies made of these substances are called *insulators, nonconductors,* or *dielectrics.* The distinguishing property of a nonconducting substance is that an electric charge imbedded in it will remain where it was originally put, at least for a long time. All material insulators are at least slightly conducting. Therefore, the charges that one may place on them will in general slowly rearrange themselves, driven by the forces of their mutual attractions and repulsions. In particular, if charges are placed inside a body, these forces will presently cause the charges to cancel in the interior or to move outward, so that eventually the net charge (if any) will appear on the outer surface of the body. We will discuss this subject in §72 and in the meantime will consider hypothetical "perfect" insulators. For definiteness, we begin by speaking of dry wood, which we take to be a perfect insulator.

When dry wood is sawed, the dust particles usually become electrically charged, so that if sawdust is compressed into a compact solid body, this body is charged throughout its interior. If the volume of the body is v and the net charge in it is q, the average charge density in the body is q/v. If a small region R, which includes a point P, has the volume Δv and contains the charge Δq, the ratio $\Delta q/\Delta v$ is called the average charge density *near* the point P. Next imagine that the mathematical boundary (S) of R "shrinks" upon P without affecting the distribution of charge. We then have $\Delta v \to 0$, $\Delta q \to 0$, and call the limit of the ratio $\Delta q/\Delta v$ the *volume charge density,* or simply the "charge density," *at* the point P. Thus, by definition,

$$\text{volume charge density} = \lim_{\Delta v \to 0} \frac{\Delta q}{\Delta v}. \tag{1}$$

We express volume charge density in coulombs per cubic meter. The usual symbol for this density is ρ, but we will sometimes denote it by q_v, where the subscript stands for *volume.* (We will also write q_a for the charge per unit *area* of a surface and q_l for the charge per unit *length* of a rod.) Because of the granular structure of electric charge and because we are concerned with macroscopic rather than microscopic phenomena, the condition $\Delta v \to 0$ in (1) will mean in fact that Δv approaches, not zero, but a value small by laboratory standards, yet large compared to atomic dimensions.

Let a nonconducting sphere of radius r_0 be uniformly charged throughout its interior and carry the total charge q. We then have $\rho = q_v = 3q/4\pi r_0^3$ at every point in the sphere. In particular, if the sphere is centered on the origin and all other bodies and charges are ignored, the value of ρ at any point P in space is

$$\rho = q_v = \begin{cases} 3q/4\pi r_0^3 & \text{if } r < r_0 \\ 0 & \text{if } r > r_0, \end{cases} \tag{2}$$

where r is the distance from P to the center of the sphere.

If the value of q_v at every point in a region R is given, the total charge, say q, contained in R can be found by integration. Let dv be the volume of a sufficiently small portion of R that includes a point P and contains the charge dq. It then follows from (1) that

$$dq = q_v \, dv, \qquad (3)$$

where q_v is the charge density at P. Consequently,

$$q = \iiint_R q_v \, dv = \iiint_R \rho \, dv, \qquad (4)$$

where the integrals extend over the region R. If the charge distribution is discontinuous and must be regarded as a collection of point charges, the integral in (4) can be replaced by the sum of the point charges located in R, or handled as in §64.

A sinusoidally charged rod. If q_v varies from point to point in a body, it can be expressed as a function of the coordinates of the various points of the body. For instance, if cartesian coordinates are used, we have $q_v = f(x, y, z)$ or, as we write to save letters, $q_v = q_v(x, y, z)$. As an example of a nonuni-

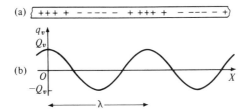

Fig. 4. A charged rod and a graph of its charge density.

formly charged body, consider a nonconducting rod whose axis lies along the x axis and whose interior is charged so that q_v does not depend on y and z but varies with x in a cosine fashion between the maximum of Q_v and the minimum of $-Q_v$ coulombs per cubic meter. This charge distribution is pictured by pluses and minuses in Fig. 4(a); its "shape," shown in Fig. 4(b), is given by the formula

$$q_v = Q_v \cos 2\pi \frac{x}{\lambda}, \qquad (5)$$

where λ is the distance between points at which q_v has its consecutive maxima, or "crests." Since (5) pertains only to the interior of the rod, we should more properly write

$$q_v = \begin{cases} Q_v \cos 2\pi \dfrac{x}{\lambda} & \text{inside the rod} \\ 0 & \text{outside the rod.} \end{cases} \qquad (6)$$

Functions of the form

$$f(x) = C \sin(ax + b), \qquad (7)$$

where C, a, and b are constants, are called *sinusoidal functions* of x. The function (5) is a special case of a sinusoidal function, obtained from (7) by writing $C = Q_v$, $a = 2\pi/\lambda$, and $b = \pi/2$. Hence one can say that the charge distribution in Fig. 4 is sinusoidal. (The seldom used term "cosinusoidal" would be more descriptive here.) The function (7) is said to be sinusoidal in x; the function $f(y) = C \sin(ay + b)$ is said to be sinusoidal in y; the function $f(x, y) = C \sin(a'x + a''y + b)$, where C, a', a'', and b are independent of x and y, is said to be sinusoidal in both x and y; and so on.

In a study of charged rods it is often necessary to consider the *linear charge density*, which we denote by q_l and express in coulombs per meter of length. Let dq be the net charge located between the cross sections of the rod at x and $x + dx$. If dx is small enough, the linear density q_l at the point x can be defined by the equation

$$dq = q_l\, dx, \qquad (8)$$

but a rigorous definition would involve a limiting process, as in the case of q_v. If the cross-sectional area of the rod of Fig. 4 is A, we have

$$dq = q_v\, dv = \left(Q_v \cos 2\pi \dfrac{x}{\lambda}\right)(A\, dx) = \left(AQ_v \cos 2\pi \dfrac{x}{\lambda}\right) dx, \qquad (9)$$

and comparison with (8) shows that

$$q_l = Q_l \cos 2\pi \dfrac{x}{\lambda}, \qquad (10)$$

where

$$Q_l = AQ_v. \qquad (11)$$

In this example, q_v is constant on any perpendicular cross section of the rod. An example in which q_v depends on the distance from the axis of the rod is given in Exercise 5.

A sliding charged rod. At the instant $t = 0$ the rod of Fig. 4 begins to slide along the x axis with a constant velocity v, dragging the charges with it. The charge density at any fixed point P on the x axis will then depend not only on the x coordinate of P but also on the time t. Our problem is to express q_l as a function of both x and t, if it is assumed that the rod is infinitely long.

Graphs of q_l for four instants of time are shown in Fig. 5 for the case when v is positive. (The rod slides toward the right.) It is, perhaps, apparent that, whether the rod slides to the right or to the left, the function q_l is *sinusoidal in*

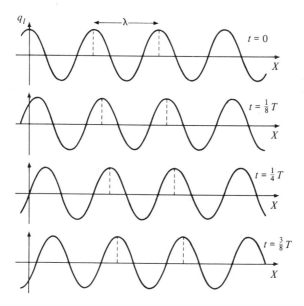

Fig. 5. Consecutive graphs of the charge density of a sliding rod.

x at any instant t and its "space period" is λ, which is the distance between successive crests of q_l. Similarly, at any fixed point P the function q_l is *sinusoidal in* t and its "time period," say T, is equal to the time needed to replace one crest of q_l by the next crest. Hence the formula for T is

$$T = \frac{\lambda}{|v|}, \tag{12}$$

where $|v|$ is the speed of the rod (the absolute value of the velocity). To illustrate the meaning of the vertical bars: $|2.5| = 2.5$ and $|-2.5| = 2.5$.

We will now verify that the required expression for q_l is

$$q_l = Q_l \cos \frac{2\pi}{T}\left(\frac{x}{v} - t\right). \tag{13}$$

Indeed, if the x in (13) is replaced by $x + \lambda$, the argument of the cosine increases by 2π if v is positive and by -2π if v is negative, so q_l is not changed in either case. Similarly, q_l does not change if t is replaced by $t + T$. Furthermore, the right-hand side of (13) has the correct amplitude, namely, Q_l. To complete the verification, we need only to check that (13) reduces to (5) when $t = 0$. If we set $t = 0$, the argument of the cosine in (13) becomes $\frac{2\pi|v|}{\lambda\,v}x$. The ratio $|v|/v$ is 1 if v is positive and -1 if v is negative, but the value of a cosine does not depend on the sign of its argument, and hence (13) does reduce to (5) in either case.

To get another check on (13), we ask how fast a moving point P' must travel along the x axis in order to have the charge density at P' remain constant. The obvious answer is that P' must move with the velocity v of the rod. To verify that (13) gives this answer, we denote the coordinate of P' by x and require that x depend on t in such a way that q_l stays constant. Now, in order that q_l may remain constant as time goes on, the argument of the cosine in (13) must be a constant. Therefore, since the factor $2\pi/T$ is constant, we have the condition

$$\frac{x}{v} - t = \text{constant.} \tag{14}$$

Since v is constant, it follows that

$$x = vt + \text{constant,} \tag{15}$$

and differentiation gives

$$\frac{dx}{dt} = v. \tag{16}$$

Thus (13) implies, quite correctly, that the velocity of the point P' must be equal to the velocity of the rod.

We have derived (13) for the artificial case of a sliding sinusoidally charged nonconducting rod. This formula is nevertheless of practical importance, because it applies equally well to a sinusoidal charge distribution sliding along a wire, and is, therefore, of interest for transmission lines and alternating currents.

Wave terminology. Let

$$f = A \cos \frac{2\pi}{T}\left(\frac{x}{v} - t\right), \tag{17}$$

where A and v are constants, and T is a positive constant. The function f is then called a one-dimensional sinusoidal *wave*. The absolute value $|A|$ of A is called the *amplitude* of this wave, although this term is sometimes used for A itself. The constant T is the time period or simply the *period* of this wave, and $1/T$ is the *frequency* of this wave. The constant v is called the *phase velocity* of the wave. The positive constant

$$\lambda = |v|T, \tag{18}$$

namely, speed times period, is called the space period or the *wavelength* of the wave. The charge density (13) is our first example of a wave.

A more compact form of (17) is

$$f = A \cos(\pm kx - \omega t), \tag{19}$$

where

$$\omega = \frac{2\pi}{T} \tag{20}$$

and

$$k = \frac{2\pi}{|v|T} = \frac{\omega}{|v|}. \tag{21}$$

The constant ω is sometimes called the *angular frequency* of the wave; its numerical value is positive and is equal to 2π times the frequency of the wave. The constant k is called the *propagation constant* of the wave. (In some books this name is used for our k divided by 2π.) Our respective units of measurement for ω and k are the "radian per second" and the "radian per meter." The direction of motion of a wave can be inferred from the sign written before the term kx in (19). For example, if $f = 5 \cos(7x - 20t)$, the velocity v in (17) is positive, and the wave moves to the *right* (in the $+x$ direction). But if $f = 5 \cos(-7x - 20t)$, the wave moves to the *left*.

The function (17) is a particularly simple solution of Equation (48⁵), which is the one-dimensional *wave equation*.

EXERCISES

1. A nonconducting sphere (radius r_0) is centered on the origin. A charge Q is distributed in it with a density proportional to the distance from the origin. Show that if all other bodies are ignored, the charge density at any point P in space is given by the formula

$$q_v = \begin{cases} \dfrac{Q}{\pi r_0^4} r & \text{if } r < r_0 \\ 0 & \text{if } r > r_0, \end{cases} \tag{22}$$

where r is the distance from P to the origin.

2. Show that in Exercise 1 the spherical region of radius $\frac{1}{2} r_0$, centered on the origin, contains the charge $\frac{1}{16}Q$. What is the radius of the spherical region (centered on the origin) containing the charge $\frac{1}{2}Q$?

3. Show that if $v \to 0$, then (13) reduces to (10) for every value of t.

4. A nonconducting rod, located as in Fig. 4, is charged so that $q_l = Q_l \sin(2\pi x/\lambda)$. When $t = 0$, it starts sliding along the x axis with the constant velocity v. Show that

$$q_l = Q_l \frac{v}{|v|} \sin \frac{2\pi}{T} \left(\frac{x}{v} - t \right). \tag{23}$$

5. A charged nonconducting rod has a circular cross section of radius ρ_0 and is coaxial with the z axis. Inside the rod, $q_v = c\rho^n$, where c and n are constants, and ρ is the distance from the z axis. Show that

$$q_l = \frac{2\pi c}{n+2} \rho_0^{n+2}. \tag{24}$$

4. SCALAR FIELDS

If a quantity can be described completely by a single number, it is called a *number*, a *numeric*, or a *scalar;* in particular, although a scalar can be positive, negative, or zero, it does not involve the idea of a direction in space. For example, if we are told that the mass of a body is, say, 10.0 pounds, nothing can be added to this statement to make the description of the mass of this body more complete within the implied accuracy; accordingly, mass is a scalar. Similarly, the net electric charge located in a body is a scalar, which can be positive, negative, or zero. Again, the x coordinate of the center of mass of a body relative to a given coordinate frame—say 5 feet—is a scalar.

If the numerical value of a scalar does not depend on the choice of the coordinate frame that one may be using, we call this scalar an *invariant scalar* or a *scalar invariant*. Accordingly, the mass of a body and the net electric charge located in a body are invariant scalars. On the other hand, the x coordinate of a point, given above as 5 feet, is not an invariant scalar, because it need not remain 5 feet if the coordinate frame is changed.

Throughout this book, the term "invariant" (either the noun or the adjective) is used in a very specific sense: it means that the quantity in question does not change when a coordinate frame is changed. An invariant quantity may vary with the time and may be different at different points in space, but it is not affected by the choice of a coordinate frame. Once a scalar has been identified as an invariant scalar, we will speak of it, for short, simply as a scalar; we will use similar abbreviated terminology with regard to invariant vectors.

Scalar quantities associated with individual points in space (inside or outside of material bodies) are called "scalar functions of position," "scalar point functions," or "scalar fields." A classical (nonrelativistic) definition reads: *A scalar field is a quantity that (at every instant t) has at every point in space a numerical value, which describes this quantity at this point (at the instant t) completely.* The word "completely" serves to distinguish scalar fields from such fields as vector fields, whose description includes not only numerical magnitudes but also directions in space. Some scalar fields of interest to us are invariant fields and can be described without referring to coordinate frames. But even then a coordinate frame—say a cartesian or a spherical polar frame—may be very useful for describing the field by mathematical formulas.

If the numerical value of a field is zero at an isolated point or at all points in a region, we say that the field vanishes there. If a field vanishes at all points outside a region, we say that it pervades only this region rather than all space. If a field remains constant at each point in space as time goes on, we call it *static*; otherwise we call it *time-varying* or *time-dependent*.

A familiar scalar field is temperature: at any instant t the temperature has a numerical value at every point P, and this value describes the temperature at P at this instant completely. For example, if the point P is located in pure water boiling at standard pressure, the temperature at P is 100°C; this single numerical value contains all the information that can be had about the temperature at P. Let one man describe the position of P in terms of cartesian coordinates pertaining to a certain coordinate frame, and let another man choose for this purpose another coordinate frame, say a frame having a different origin. The set of coordinates (x, y, z) identifying P for the first man will then differ from the set (x', y', z') identifying P for the second man. But the point P is the same point for both of them, and both will say that the temperature at P is 100°C. The temperature at any other point will, of course, also be independent of the choice of a coordinate frame. Consequently, temperature is an invariant scalar field.

If the Celsius[5] (centigrade) scale is used, the temperature at a point may be positive, negative, or zero. Since the words "positive" and "negative" suggest a direction, we must stress the fact that, when we say that a scalar field does not involve the idea of direction, we mean direction *in space* and not direction relative to the markings on the scale of a measuring device, such as a Celsius thermometer.

As long as we are ignoring nonisotropic substances, we can list the electric conductivity g as an invariant scalar field; this field can have only nonnegative values. (Its values can be only positive or zero.) At a point in empty space $g = 0$; at a point in copper $g = 5.8 \times 10^7$ mhos/meter; at a point in iron $g = 1.0 \times 10^7$ mhos/meter; and so on.

[5]Anders Celsius (1701-1744).

As another example, let us denote by P_0 a particular point in space and consider the quantity called "the distance from P_0." At any point P in space this quantity has a numerical value (equal to the length of the straight segment PP_0), which specifies this quantity at the point P completely. Hence "the distance from P_0" is a scalar field. This field can, of course, have only nonnegative values.

We denote scalar fields in general by f, so let us write f for "the distance from P_0." If we introduce a cartesian coordinate frame whose origin O is placed at P_0, this field can be described by the formula

$$f = \sqrt{x^2 + y^2 + z^2}. \tag{1}$$

Similarly, if we write f for the scalar field defined as "the square of the distance from the fixed point P_0" and use the same coordinate frame, we have

$$f = x^2 + y^2 + z^2. \tag{2}$$

If the origin O of the frame should be shifted through the distance x_0 in the $+x$ direction, the distance from P_0 to any point P would not be affected, but (2) would have to be replaced by

$$f = (x - x_0)^2 + y^2 + z^2, \tag{3}$$

and a similar change would have to be made in (1).

If f has the same numerical value at every point on a surface S, then S is called a *surface of constant f* or a *level surface of f*. In the case (2), for example, the surfaces of constant f are given by the equation

$$x^2 + y^2 + z^2 = c, \tag{4}$$

where c is an arbitrary positive constant. These surfaces are spherical and are

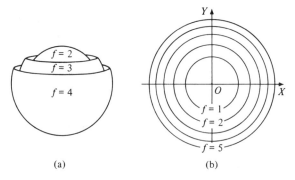

Fig. 6. Level surfaces of the field $f = x^2 + y^2 + z^2$ and their intersections with the xy plane.

centered on the origin. Parts of three of them are shown in Fig. 6(a), and the intersections of several of them with the xy plane in Fig. 6(b).

Next consider a scalar field f whose level surfaces are parallel planes, fixed in space and oriented as in Fig. 7(a). Suppose that the values of f on any of these planes increase by unity for steps of length d between the planes. This field can be described by setting up the coordinate frame shown in Fig. 7(b) and writing

$$f = \frac{y}{d}. \tag{5}$$

Note that (5) is not the definition of the field in question—it is merely an equation that describes in terms of coordinates the coordinate-free definition of f given above in words and illustrated in Fig. 7(a). If we should turn the coordinate frame in Fig. 7(b) so that the x axis replaces the y axis, we would have $f = x/d$ instead of (5). We are at liberty to turn the coordinate frame but not the planes of constant f, because, according to the description of f given above, the orientation of these planes is fixed in space.

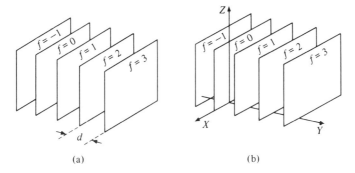

Fig. 7. Level surfaces of the field $f = y/d$.

If we want to emphasize the coordinates in terms of which a scalar field f has been expressed, we use such symbols as $f(x, y, z)$ or $f(r, \theta, \phi)$ if the field is static, and $f(x, y, z, t)$ or $f(r, \theta, \phi, t)$ if it varies with time. When we specify individual points in space by labeling them P, P', Q, and so on, we denote the numerical values of f at these points by $f(P)$, $f(P')$, $f(Q)$, and so forth.

We call a quantity such as (2) a "field" rather than a "function" because the word "field" fits in better with the language of the electromagnetic theory. But no harm is done by saying "scalar function" for "scalar field." A textbook should not be too informal, so we write, "Charge density and conductivity are scalar fields." The reader need not be so formal, and may say, "Charge density and conductivity are scalars"—provided he does not forget that each of these quantities has a numerical value at every point in space and is therefore a *field*.

EXERCISES

1. Give three physical examples of invariant scalar fields not mentioned above.

2. Consider the field $f = \sqrt{x^2 + y^2}$ in three-dimensional space and sketch the surfaces of constant f on which $f = 1, 2, 3$, and 4. Sketch also the intersections of these surfaces with the xy plane.

3. Do Exercise 2 for the field $f = x^2 + y^2$.

4. Suppose that the coordinate frame in Fig. 7(b) is turned about the x axis through 45 degrees, so that the y axis points upward and to the right. What equation will then replace (5)?

CHAPTER 2

Mathematical Notes

In this chapter we present, without proofs, certain mathematical topics, many of which are presumably already familiar to the reader.

Some formulas involve "coordinate variables," such as x, y, and z, that describe the positions of points relative to a specific coordinate frame; other formulas do not involve such variables and are called *coordinate-free*. The basic laws of electromagnetism (Coulomb's law, Ampère's law, Faraday's law, and so on) for mediums at rest relative to the observer can be written as coordinate-free equations. For this reason we stress in this book the mathematical expressions that are coordinate-free, and sometimes we even reverse the order in which mathematical concepts are usually introduced. For example, the differential of a function is defined in §7 without reference to coordinate frames, and the formulas for differentials in terms of partial derivatives with respect to coordinates are given later.

The letter P is used in this book only to label points, and the symbol $P(x, y, z)$ means that the cartesian coordinates of the point labeled P are x, y, and z; that is, this symbol does *not* stand for a function of x, y, and z. Similarly a Q with a list of coordinates after it means that the point labeled Q has these coordinates.

As before, we write $|n|$ for the absolute value of n. An unsigned square root of a positive number denotes the positive square root; thus $\sqrt{6.25} = 2.5$, $\sqrt{(-2.5)^2} = 2.5$, and $-\sqrt{6.25} = -2.5$.

5. MISCELLANEOUS ITEMS

Coordinate frames. Our basic coordinate frame is the right-handed cartesian frame XYZ of Fig. 8. The distinguishing property of a *right-handed* frame is this: If you grasp the z axis with the right hand, so that the thumb points in the positive z direction, and if you turn the frame about the z axis through 90 degrees in the sense indicated by the fingers, then the x axis will replace the y axis both in position and in direction. (Note the alphabetical order: x replaces y.) Left-handed frames are mirror images of right-handed frames; we will not use them.

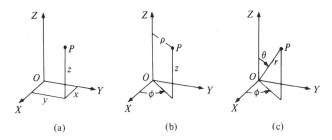

Figure 8

The cylindrical coordinates (ρ, ϕ, z) and the spherical coordinates (r, θ, ϕ) of a point P are defined in Fig. 8. Relations such as

$$\rho = \sqrt{x^2 + y^2} \tag{1}$$

and

$$r = \sqrt{x^2 + y^2 + z^2} \tag{2}$$

are listed in Appendix A. For compactness, we sometimes mix different kinds of coordinates, as when we write x/r for $x/\sqrt{x^2 + y^2 + z^2}$. We write ρ for the coordinate defined in (1) and also for the volume charge density defined in §3, but to avoid ambiguity we sometimes denote the latter by q_v.

Neighboring points. Let dl be the distance between the points P and Q involved in a computation in which $dl \to 0$. We then call P and Q "neighboring points," a term used in mathematics in a broader sense than we will need in this book. (In problems affected by the granular structure of matter, the condition $dl \to 0$ should be interpreted as in footnote 4[2].)

If a cartesian frame is used and the points in question are $P(x, y, z)$ and $Q(x + dx, y + dy, z + dz)$, then the equation

$$(dl)^2 = (dx)^2 + (dy)^2 + (dz)^2 \tag{3}$$

is exact for any values of dx, dy, and dz. But equations (13[A]) and (15[A]), which express dl in other coordinates, are approximate and become exact only if $dl \to 0$. By custom, we write ds for dl in some formulas.

Direction cosines. Let α_x, α_y, and α_z be the angles that a directed straight line makes with the positive directions of the x, y, and z axes of a cartesian frame. These angles are called the *direction angles* of this line relative

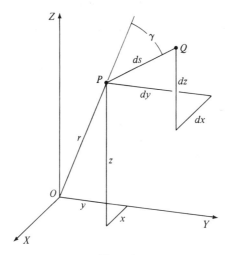

Figure 9

to this frame, and their cosines are called the *direction cosines* of the line relative to the frame. (If the direction of the straight line is reversed, each direction cosine reverses its algebraic sign.) For example, in Fig. 9 the coordinates of the points O and P are $(0, 0, 0)$ and (x, y, z). Accordingly, the direction cosines of the line OP (not PO) are

$$\cos \alpha_x = \frac{x}{r}, \quad \cos \alpha_y = \frac{y}{r}, \quad \cos \alpha_z = \frac{z}{r}, \tag{4}$$

where r is the distance from O to P.[1] The coordinates of Q are $(x + dx, y + dy, z + dz)$; consequently, if we denote the direction angles of the line PQ by β_x, β_y, and β_z, we have

$$\cos \beta_x = \frac{dx}{ds}, \quad \cos \beta_y = \frac{dy}{ds}, \quad \cos \beta_z = \frac{dz}{ds}, \tag{5}$$

where ds is the distance from P to Q.

Let the direction angles of one straight line be α_x, α_y, and α_z; let those of

---

[1] In more elementary expositions there is some advantage in denoting the coordinates of a particular point by particularized symbols, such as (x_0, y_0, z_0). In this book we usually denote the coordinates of a single arbitrarily chosen point by symbols such as (x, y, z), partly because they are simpler and partly because they stress the fact that the point has been chosen arbitrarily.

another straight line be β_x, β_y, and β_z; let γ be the angle between these lines. A theorem of analytic geometry then states that .

$$\cos \gamma = \cos \alpha_x \cos \beta_x + \cos \alpha_y \cos \beta_y + \cos \alpha_z \cos \beta_z. \tag{6}$$

For example, in view of (4) and (5), the cosine of the angle γ between the lines OP and PQ in Fig. 9 is

$$\cos \gamma = \frac{x}{r} \frac{dx}{ds} + \frac{y}{r} \frac{dy}{ds} + \frac{z}{r} \frac{dz}{ds}. \tag{7}$$

Partial derivatives. If f is a function of several independent variables, then the partial derivative of f with respect to any one of them is obtained by differentiating f with respect to this variable while treating the other independent variables as constants. For example, if the independent variables are x, y, and z, and if

$$f = \sqrt{x^2 + y^2}, \tag{8}$$

then

$$\frac{\partial f}{\partial x} = \frac{1}{2} \frac{1}{\sqrt{x^2 + y^2}} \cdot 2x = \frac{x}{\sqrt{x^2 + y^2}}, \tag{9}$$

$$\frac{\partial f}{\partial y} = \frac{1}{2} \frac{1}{\sqrt{x^2 + y^2}} \cdot 2y = \frac{y}{\sqrt{x^2 + y^2}}, \tag{10}$$

and

$$\frac{\partial f}{\partial z} = 0. \tag{11}$$

Writing (8) as

$$f = \rho, \tag{12}$$

we conclude that

$$\frac{\partial \rho}{\partial x} = \frac{x}{\rho}, \quad \frac{\partial \rho}{\partial y} = \frac{y}{\rho}, \quad \frac{\partial \rho}{\partial z} = 0. \tag{13}$$

Similarly, if we let

$$f = \sqrt{x^2 + y^2 + z^2} = r, \tag{14}$$

we find that

$$\frac{\partial r}{\partial x} = \frac{x}{r}, \quad \frac{\partial r}{\partial y} = \frac{y}{r}, \quad \frac{\partial r}{\partial z} = \frac{z}{r}. \tag{15}$$

Comparison with (4) shows that if r is taken to be a function of x, y, and z then the partial derivatives of r with respect to x, y, and z, evaluated at any point P, are equal to the direction cosines of the straight line drawn from the origin toward P.

Let f be a function of the variables u, v, and w, which are themselves functions of the variables α, β, and γ. If we should choose α, β, and γ as the independent variables, we have

$$\frac{\partial f}{\partial \alpha} = \frac{\partial f}{\partial u}\frac{\partial u}{\partial \alpha} + \frac{\partial f}{\partial v}\frac{\partial v}{\partial \alpha} + \frac{\partial f}{\partial w}\frac{\partial w}{\partial \alpha}, \tag{16}$$

with similar "chain rules" for $\partial f/\partial \beta$ and $\partial f/\partial \gamma$. For example, let f be an explicit function of the spherical coordinates (r, θ, ϕ), and let us choose as the independent variables the cartesian coordinates (x, y, z). Then

$$\frac{\partial f}{\partial x} = \frac{\partial f}{\partial r}\frac{\partial r}{\partial x} + \frac{\partial f}{\partial \theta}\frac{\partial \theta}{\partial x} + \frac{\partial f}{\partial \phi}\frac{\partial \phi}{\partial x}, \tag{17}$$

with similar formulas for $\partial f/\partial y$ and $\partial f/\partial z$. For instance, if

$$f = \frac{1}{\sqrt{x^2 + y^2 + z^2}} = \frac{1}{r}, \tag{18}$$

we can evalute $\partial f/\partial x$ in several ways. One way is to write

$$\frac{\partial}{\partial x}\left(\frac{1}{r}\right) = \frac{\partial}{\partial x}\frac{1}{\sqrt{x^2 + y^2 + z^2}}$$
$$= -\frac{1}{2}\frac{1}{(x^2 + y^2 + z^2)^{3/2}} \cdot 2x = -\frac{x}{r^3}. \tag{19}$$

Another way is based on (17): Since f does not depend on θ and ϕ, the derivatives $\partial f/\partial \theta$ and $\partial f/\partial \phi$ both vanish, and we have

$$\frac{\partial}{\partial x}\left(\frac{1}{r}\right) = \left[\frac{\partial}{\partial r}\left(\frac{1}{r}\right)\right]\frac{\partial r}{\partial x} = \left(-\frac{1}{r^2}\right)\frac{x}{r} = -\frac{x}{r^3}. \tag{20}$$

Partial derivatives of higher orders are defined by the equations

$$\frac{\partial^2 f}{\partial x^2} = \frac{\partial}{\partial x}\left(\frac{\partial f}{\partial x}\right), \quad \frac{\partial^2 f}{\partial x \partial y} = \frac{\partial}{\partial x}\left(\frac{\partial f}{\partial y}\right), \quad \frac{\partial^2 f}{\partial y \partial x} = \frac{\partial}{\partial y}\left(\frac{\partial f}{\partial x}\right), \tag{21}$$

and so on. Under certain continuity conditions, which are usually satisfied in physical problems, the order of partial differentiation is immaterial, so that, for example,

$$\frac{\partial^2 f}{\partial x \partial y} = \frac{\partial^2 f}{\partial y \partial x}. \tag{22}$$

The following special case will arise in our work: f is a function of a single variable u, which is itself a function of the coordinates x, y, and z, and the time t; that is,

$$f = f(u), \qquad u = u(x, y, z, t). \tag{23}$$

We then have the formulas

$$\frac{\partial f}{\partial x} = \frac{df}{du}\frac{\partial u}{\partial x}, \qquad \frac{\partial f}{\partial y} = \frac{df}{du}\frac{\partial u}{\partial y}, \quad \ldots, \tag{24}$$

which are special cases of (16). For example, let

$$f = \cos u, \qquad u = kx - \omega t, \tag{25}$$

where k and ω are constants. One way of computing $\partial f/\partial x$ is to eliminate u and write

$$\begin{aligned}\frac{\partial f}{\partial x} &= \frac{\partial}{\partial x}\cos(kx - \omega t) = -\sin(kx - \omega t)\frac{\partial}{\partial x}(kx - \omega t) \\ &= -k\sin(kx - \omega t),\end{aligned} \tag{26}$$

so that

$$\frac{\partial f}{\partial x} = -k\sin u. \tag{27}$$

To derive (27) with less writing, we note that

$$\frac{df}{du} = -\sin u, \qquad \frac{\partial u}{\partial x} = k, \tag{28}$$

and then use the first of Equations (24).

Directional derivatives. Let P and Q be two points, let the point Q' lie between P and Q on the straight line PQ, and let $f(P)$ and $f(Q')$ be the numerical values of a scalar field f at P and Q'. Next let Q' approach P along the line QP and write PQ' for the distance from P to Q'. The expression

$$\lim_{Q' \to P} \frac{f(Q') - f(P)}{PQ'} \tag{29}$$

is then called the *directional derivative of f taken at P toward Q* and is usually denoted by $\partial f/\partial s$. Note that the definition (29) is *coordinate-free*.

Consider, for example, the two-dimensional case shown in Fig. 10, where f is a function whose numerical value at any point is proportional to the distance

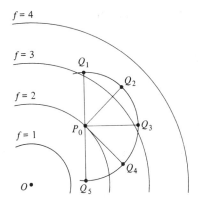

Figure 10

of this point from the point O. The circular lines of constant f are marked "$f = 1$," and so on. In this case the directional derivative, $\partial f/\partial s$, taken at P_0 toward Q_1 is positive; that taken at P_0 toward Q_5 is negative; and that taken at P_0 toward Q_4 is zero, because the line $P_0 Q_4$ is tangent at P_0 to a line of constant f. It is perhaps obvious that in this example the largest directional derivative of f at P_0 is that taken toward Q_2.

In terms of cartesian coordinates, the coordinate-free formula (29) for the directional derivative of f taken at P toward Q is

$$\frac{\partial f}{\partial x}\cos\alpha_x + \frac{\partial f}{\partial y}\cos\alpha_y + \frac{\partial f}{\partial z}\cos\alpha_z, \tag{30}$$

where the α's are the direction angles of the straight line PQ, and the derivatives are evaluated at P.

Notation for integrals. We denote integrals taken around a contour C, over a surface S—or (S) if the surface is closed—and over a region R by the symbols

$$\oint_C \cdots dl, \quad \iint_S \cdots da, \quad \iiint_R \cdots dv, \tag{31}$$

or simply

$$\oint \cdots dl, \quad \iint \cdots da, \quad \iiint \cdots dv. \tag{32}$$

As one reads a formula from left to right, the multiple integral signs tell the type of integral without delay. In equations of the form

$$\iiint_R \cdots dv = \iint_{(S)} \cdots da, \tag{33}$$

the closed surface (S) is the boundary of the region R. However, one may write (33) simply as

$$\iiint \cdots dv = \iint \cdots da. \tag{34}$$

Similarly, in equations of the form

$$\oint \cdots dl = \iint \cdots da, \tag{35}$$

the integral on the right is taken over an open surface that spans the contour used on the left.

Shrinking regions. Let a region R include a point P, and let Δv be the volume of R. Sometimes (as in the definition of charge density) we are concerned with the case when $\Delta v \to 0$. In §3 we did not go into mathematical details and simply said that the surface (S) of R shrinks upon P. In fact, our term "shrink" implies that $\Delta v \to 0$, that every point of (S) approaches P, that the area of (S) tends to zero, and that—to put it roughly—this surface does not become needle-shaped or crinkled. For example, the region involved in (1^3) is not allowed to be a shrinking needle-shaped region, whose width tends to zero but whose length remains constant. Similarly, (S) is not allowed to wrinkle as it contracts and to keep its area finite (there are anomalous mathematical possibilities in which this condition does not hold). In physical situations affected by the granular structure of matter, the process of shrinking must, of course, conform to the conditions stated in footnote 4[2].

The foregoing remarks apply in many places in this book, and the reader should keep them in mind. In particular, he should recall them when he comes to such equations as (5^{28}) and (1^{80}), which involve limits of ratios of vanishing integrals to vanishing volumes or areas.

Differentiation of integrals with respect to a parameter. We will be concerned here with integrals involving a parameter (say the time t) in the integrand but not in the limits.

If a and b are constants, the definite integral

$$\int_a^b f(x, t)\, dx \tag{36}$$

does not depend on x, but is usually a function of t. For example, if $f = (x + ct)^2$, $a = 0$, and $b = 1$, we have

$$\int_0^1 (x + ct)^2\, dx = \tfrac{1}{3} + ct + c^2 t^2, \tag{37}$$

so that

$$\frac{d}{dt}\int_0^1 (x+ct)^2\,dx = c + 2c^2 t. \tag{38}$$

Now, may we reverse the order of the integration and the differentiation in the left-hand side of (38) without affecting the result? More precisely: May we evaluate the left-hand side of (38) by integrating with respect to x the partial derivative of the integrand, taken with respect to t? The answer is yes; indeed,

$$\int_0^1 \frac{\partial}{\partial t}(x+ct)^2\,dx = \int_0^1 2c(x+ct)\,dx = c + 2c^2 t, \tag{39}$$

so that

$$\frac{d}{dt}\int_0^1 (x+ct)^2\,dx = \int_0^1 \frac{\partial}{\partial t}(x+ct)^2\,dx. \tag{40}$$

Equation (40) is not an accident, but is implied in a theorem stating that, if the interval of integration is kept fixed, then

$$\frac{d}{dt}\int_a^b f(x,t)\,dx = \int_a^b \frac{\partial}{\partial t}f(x,t)\,dx, \tag{41}$$

provided that f satisfies certain continuity conditions that are usually satisfied in physical problems. Under similar continuity conditions we also have the following relations, in which $f = f(x,y,z,t)$: If the surface S is kept fixed, then

$$\frac{d}{dt}\iint_S f\,da = \iint_S \frac{\partial f}{\partial t}\,da, \tag{42}$$

and if the region R is kept fixed, then

$$\frac{d}{dt}\iiint_R f\,dv = \iiint_R \frac{\partial f}{\partial t}\,dv. \tag{43}$$

Integrals that vanish over all domains. Suppose that, while trying to identify an unknown continuous function $f(x)$, we somehow succeed in proving that it satisfies the equation

$$\int_a^b f(x)\,dx = 0 \tag{44}$$

for *all* choices of the limits a and b. What conclusion can be drawn? The conclusion is that $f(x)$ is identically zero. For if $f(x)$ were positive for some value of x, say x_0, this function, being continuous, would be positive in some interval including x_0 and extending from, say, $x_0 - \epsilon$ to $x_0 + \delta$; if so, the condition

(44) could be violated by letting $a = x_0 - \epsilon$ and $b = x_0 + \delta$. If we permit $f(x)$ to be negative for some value of x, we also get a contradiction. The remaining possibility is that $f(x) = 0$ for *all* values of x.

Similarly, if a continuous function of the coordinates satisfies the stringent condition

$$\iiint_{\substack{\text{every}\\\text{region}}} f \, dv = 0, \tag{45}$$

then we must conclude that

$$f = 0 \tag{46}$$

at *every* point.

Inequalities. The inequality $b < a$ means that the number b is smaller than the number a; that is, the number $b - a$ is negative. The inequality $b \ll a$ means in this book that, in the given problem, b can be *ignored* compared to a; that is, the numbers $a + b$ and $a - b$ can each be replaced by a. Similarly, the condition $b^2 \ll a^2$ means that the numbers $a^2 + b^2$ and $a^2 - b^2$ can each be replaced by a^2. Note, however, that the condition $b^2 \ll a^2$ does not necessarily imply that $b \ll a$. For example, if $a = 1$ and $b = 0.03$, it may be permissible in a given problem to replace by 1 the numbers $1^2 + (0.03)^2$ and $1^2 - (0.03)^2$, namely, 1.0009 and 0.9991, but it is not necessarily permissible in the same problem to replace by 1 the numbers $1 + 0.03$ and $1 - 0.03$, namely, 1.03 and 0.97.

The relations $a > b$ and $c \gg d$ are equivalent, respectively, to $b < a$ and $d \ll c$.

EXERCISES

1. Let $u = u(r)$, recall (2), take x, y, and z as the independent variables, and show that

$$\frac{\partial u}{\partial x} = \frac{x}{r} \frac{du}{dr}, \quad \frac{\partial u}{\partial y} = \frac{y}{r} \frac{du}{dr}, \quad \frac{\partial u}{\partial z} = \frac{z}{r} \frac{du}{dr}. \tag{47}$$

2. Superimpose on Fig. 10 an XY frame (origin at O and the usual directions of the axes), interpret this figure as picturing the function $f = a\sqrt{x^2 + y^2}$ with a positive a, and show that the respective directional derivatives of f taken at P_0 toward Q_1, Q_2, Q_3, Q_4, and Q_5 are $a/\sqrt{2}$, a, $a/\sqrt{2}$, 0, and $-a/\sqrt{2}$.

3. What do the series (20^6) and (21^6) for $\cos x$ and $\sin x$ become if $x \ll 1$? If $x^2 \ll 1$? If $x^3 \ll 1$?

4. Show that the waves (19^3) and (23^3) satisfy the one-dimensional "wave equation"

$$\frac{\partial^2 f}{\partial x^2} = \frac{1}{v^2} \frac{\partial^2 f}{\partial t^2}. \tag{48}$$

6. TAYLOR'S EQUATION IN ONE DIMENSION

In this and the following section we recall the calculus equation that interconnects the values of a scalar field f at two different points in space.

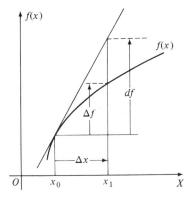

Fig. 11. A function $f(x)$ and its increment Δf and differential df associated with the step from x_0 to x_1.

Let $f = f(x)$, $f' = df/dx$, $f'' = d^2f/dx^2$, and so on. The symbol $f'(x_0)$ denotes the first derivative of f evaluated at $x = x_0$, the symbol $f''(x_0)$ denotes the second derivative, and so on. We assume that f and all of its derivatives are continuous in the interval under consideration. The main quantities of which we will speak are pictured in Fig. 11.

The quantity $x_1 - x_0$ is called the *increment of x* associated with the step from x_0 to x_1 and is denoted by Δx:

$$\Delta x = x_1 - x_0. \tag{1}$$

The quantity $f(x_1) - f(x_0)$ is called the *increment of f* associated with the step from x_0 to $x_0 + \Delta x$ and is denoted by Δf:

$$\Delta f = f(x_0 + \Delta x) - f(x_0). \tag{2}$$

The formula we want to recall is

$$f(x_0 + \Delta x) = f(x_0) + f'(x_0)\,\Delta x + \frac{1}{2!}f''(x_0)(\Delta x)^2 + \cdots. \tag{3}$$

The dots denote a series, usually infinite, whose terms involve the third and higher derivatives of f, evaluated at $x = x_0$ and multiplied by the respective factors $(\Delta x)^3$, $(\Delta x)^4$, and so on, and by the reciprocals of the corresponding factorials. Equation (3) is the one-dimensional version of *Taylor's series*, but it is not always written in the notation we are using here.[2] A special case is obtained by placing the origin in Fig. 11 at the point x_0 and writing x for x_1. Equation (3) then becomes

$$f(x) = f(0) + f'(0)x + \frac{1}{2!}f''(0)x^2 + \frac{1}{3!}f'''(0)x^3 + \cdots. \tag{4}$$

In view of (2) and (3),

$$\Delta f = f'(x_0)\,\Delta x + \frac{1}{2!}f''(x_0)(\Delta x)^2 + \cdots. \tag{5}$$

We must stress the fact that (5) holds for all values of Δx, small or large, provided that f and all its derivatives exist in the range of x under consideration and the series converges. The first term on the right-hand side of (5) is called the *first-order term* or the *linear term*. The other terms are called *terms of higher order*, so that we may write

$$\Delta f = f'(x_0)\,\Delta x + \text{terms of higher order}. \tag{6}$$

We now introduce the quantity denoted by df, defined by the equation

$$df = f'(x_0)\,\Delta x, \tag{7}$$

and called the *differential* of the dependent variable f, associated with the step from x_0 to $x_0 + \Delta x$. The graphical meaning of df is shown in Fig. 11. Since the size of the increment Δx is not restricted, the differential df may have any value, large or small. According to (6) and (7),

$$\Delta f = df + \text{terms of higher order}. \tag{8}$$

Finally, we introduce the symbol dx, defined by the equation

$$dx = \Delta x, \tag{9}$$

and called the *differential* of the independent variable x, associated with the step from x_0 to $x_0 + \Delta x$. We thus have two symbols, dx and Δx, for the same

---

[2] Brook Taylor (1685-1731). We use the term "Taylor's equation" for equations of the form (12). The reader should recall the related terms "Taylor's series," "Taylor's theorem," and "Taylor's formula," used for various purposes in books on calculus.

thing; the convenience of this arrangement shows up in the study of functions of functions of x. Since the value of Δx is not restricted, dx can have any value, small or large.

We now rewrite some of the preceding formulas without using primes; also, we replace Δx by the equivalent symbol dx and suppress the argument x_0 in symbols such as $f(x_0)$. Equations (7) and (8) then become

$$df = \frac{df}{dx} dx \qquad (10)$$

and

$$\Delta f = \frac{df}{dx} dx + \text{terms of higher order.} \qquad (11)$$

To abbreviate (3), we denote by P and Q the points on the x axis having the respective coordinates x_0 and $x_0 + \Delta x$ (or $x_0 + dx$, which is the same thing). The result is

$$f(Q) = f(P) + df + \text{terms of higher order} \qquad (12)$$

or, a little more explicitly,

$$f(Q) = f(P) + \frac{df}{dx} dx + \text{terms of higher order.} \qquad (13)$$

We refer to (12) as *Taylor's equation*.

So far we have stressed the fact that in all equations of this section the increment Δx of x can have any value, large or small. The final step of many calculations, however, is to make Δx tend to zero. Now, if Δx should tend to zero, the differential df will also tend to zero, and so will the sum of the higher-order terms in equations such as (11) and (12); furthermore, because of the factors $(\Delta x)^2$, $(\Delta x)^3$, and so on, the sum of the higher-order terms will tend to zero "faster" than df. Consequently, unless the contributions of the first-order quantities should cancel in a given problem, the contribution of the higher-order terms to the final result will, in general, be negligible. Accordingly, in computations in which Δx is small it is usually safe in physical problems to discard the higher-order terms, and to write df for Δf at the beginning of the work. The advantage of doing this lies in the fact that the expression for df is simpler than that for Δf, even when Δf is available in a form not involving any series. For example, if $f(x) = \sin x$, we have the exact formula (7), namely,

$$df = (\cos x_0) \Delta x, \qquad (14)$$

and also the exact but more complicated formula

$$\Delta f = \sin(x_0 + \Delta x) - \sin x_0$$
$$= \cos x_0 \sin \Delta x - (1 - \cos \Delta x) \sin x_0. \tag{15}$$

In both (14) and (15) one may write dx for Δx.

EXERCISES

1. Write Taylor's series in the form in which you first learned it and correlate the symbols with our symbols.
2. Let $f(x) = x^3$, $x_0 = 1$, and $\Delta x = 3$. Show that $\Delta f = 63$ and $df = 9$. Then show in detail that in this case the sum of the higher-order terms in (5) is 54, so that these terms account for the difference between Δf and df.
3. Let $f(x) = x^3$, $x_0 = 1$, and $\Delta x = 0.01$. Show that Δf is approximately 0.0303 and df is exactly 0.03. Then show, without going into details, that the sum of the higher-order terms in (5) has the correct order of magnitude to account for the difference between Δf and df.
4. Give an example in which $f(x_0)$, $f'(x_0)$, and Δx are positive but the sum of the higher-order terms in (3) is negative.
5. Derive the series

$$(1 + x)^n = 1 + nx + \frac{n(n-1)}{2!}x^2 + \cdots, \tag{16}$$

$$(1 + x)^{-1} = 1 - x + x^2 - \cdots, \tag{17}$$

$$\sqrt{1+x} = 1 + \frac{1}{2}x - \frac{1}{2}\left(\frac{1}{4}\right)x^2 + \frac{1}{2}\left(\frac{1 \cdot 3}{4 \cdot 6}\right)x^3 - \cdots, \tag{18}$$

$$\ln(1+x) = x - \frac{1}{2}x^2 + \frac{1}{3}x^3 - \cdots, \tag{19}$$

which converge if $-1 < x < 1$, and the series

$$\cos x = 1 - \frac{1}{2!}x^2 + \frac{1}{4!}x^4 - \cdots, \tag{20}$$

$$\sin x = x - \frac{1}{3!}x^3 + \frac{1}{5!}x^5 - \cdots, \tag{21}$$

$$e^x = 1 + x + \frac{1}{2!}x^2 + \frac{1}{3!}x^3 + \cdots, \tag{22}$$

which converge for all values of x. Here e is the base of the natural logarithms. The series (20), (21), and (22) are used in advanced mathematics as the *definitions* of $\sin x$, $\cos x$, and e^x for either real or complex values of x.

6. Check by trigonometry that the two forms of Δf given in (15) are equivalent. Then expand in powers of Δx the functions $\sin \Delta x$ and $\cos \Delta x$ occurring in the second form and verify that the result agrees with (5).

7. Suppose that the conductivtiy of the rod of Exercise 1[2] varies so slightly that $g_b = (1 - \epsilon)g_a$, where $\epsilon^2 \ll 1$. Use (19) to show that then 9[2] becomes $R = (1 - \frac{1}{2}\epsilon)l/Ag_a$. Next use (17) or some similar series to reduce this result to the symmetric form

$$R = \frac{2l}{(g_a + g_b)A}. \tag{23}$$

Finally, justify (23) on intuitive grounds, without using 9[2].

7. TAYLOR'S EQUATION IN THREE DIMENSIONS

Let $f(P)$ and $f(Q)$ be the values of a scalar function f at two arbitrarily chosen points P and Q in three-dimensional space. The "increment" Δf of f and the "differential" df of f associated with the step from P to Q are then defined by the equations

$$\Delta f = f(Q) - f(P) \tag{1}$$

and

$$df = \frac{\partial f}{\partial s} ds, \tag{2}$$

where $\partial f/\partial s$ is the directional derivative of f taken at P toward Q and ds is the distance from P to Q. The definitions (1) and (2) do not refer to any coordinate frames; they are coordinate-free. The increment Δf is the amount by which f actually changes when we go from P to Q. The differential df is the amount by which f would change when we go from P to Q if the space rate of change of f in the direction from P to Q were the same at all points on the straight line PQ as it is at P. (The step from P_0 to Q_4 in Fig. 10[5] illustrates a case in which Δf is positive but df is zero.)

Suppose now that a cartesian frame XYZ has been installed, that f has been expressed analytically as $f = f(x, y, z)$, and that the coordinates of P and Q are (x, y, z) and $(x + dx, y + dy, z + dz)$, as in Fig. 9[5]. One can then derive from (2) the following formula for the differential of f associated with the step from P to Q:

$$df = \frac{\partial f}{\partial x} dx + \frac{\partial f}{\partial y} dy + \frac{\partial f}{\partial z} dz, \tag{3}$$

where the partial derivatives are to be evaluated at P. One can also show that

$$\Delta f = df + \text{higher-order terms}, \tag{4}$$

where the higher-order terms include repeated partial derivatives, reciprocals of factorials, and such factors as $(dx)^2$, $dx\,dy$, $(dx)^2\,dz$, and so on. Combining (4) with (1), we get the equation

$$f(Q) = f(P) + df + \text{higher-order terms}, \qquad (5)$$

which we call *Taylor's equation* and which looks exactly like its one-dimensional version (12^6).

For the more general purposes of this book it is especially important to grasp the content of the coordinate-free definition (2) of df. But in detailed computations in particular problems it is usually convenient to introduce a coordinate frame and use such formulas as

$$f(Q) = f(P) + \frac{\partial f}{\partial x}dx + \frac{\partial f}{\partial y}dy + \frac{\partial f}{\partial z}dz + \cdots, \qquad (6)$$

$$f(Q) = f(P) + \frac{\partial f}{\partial \rho}d\rho + \frac{\partial f}{\partial \phi}d\phi + \frac{\partial f}{\partial z}dz + \cdots, \qquad (7)$$

and

$$f(Q) = f(P) + \frac{\partial f}{\partial r}dr + \frac{\partial f}{\partial \theta}d\theta + \frac{\partial f}{\partial \phi}d\phi + \cdots. \qquad (8)$$

Here the respective cartesian coordinates of P and Q are (x, y, z) and $(x + dx, y + dy, z + dz)$; their cylindrical coordinates are (ρ, ϕ, z) and $(\rho + d\rho, \phi + d\phi, z + dz)$; their spherical coordinates are (r, θ, ϕ) and $(r + dr, \theta + d\theta, \phi + d\phi)$.

Just as in the one-dimensional case, if we intend to make the distance PQ approach zero at the end of a computation (so that $\Delta f \to df \to 0$), it is usually safe in physical problems to write df for Δf at the beginning of the computation.

Example. To illustrate the computation of a differential, we consider the function

$$f(x, y, z) = \sqrt{x^2 + y^2 + z^2} = r, \qquad (9)$$

whose value at any point P is equal to the distance from P to the origin O. Since[3]

(15^5) $\qquad \dfrac{\partial r}{\partial x} = \dfrac{x}{r}, \qquad \dfrac{\partial r}{\partial y} = \dfrac{y}{r}, \qquad \dfrac{\partial r}{\partial z} = \dfrac{z}{r}, \qquad (10)$

Equation (3) becomes

[3] See Note 2 following the Table of Contents.

$$dr = \frac{x}{r} dx + \frac{y}{r} dy + \frac{z}{r} dz, \tag{11}$$

where we have written dr for df because in this example $f = r$.

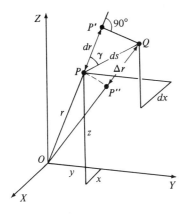

Figure 12

Equation (11) can be made more compact. The cosine of the angle γ between the lines OP and PQ in Fig. 12 is

(7⁵) $$\cos \gamma = \frac{x}{r} \frac{dx}{ds} + \frac{y}{r} \frac{dy}{ds} + \frac{z}{r} \frac{dz}{ds}, \tag{12}$$

where ds is the distance PQ. Therefore,

$$ds \cos \gamma = \frac{x}{r} dx + \frac{y}{r} dy + \frac{z}{r} dz \tag{13}$$

and

$$dr = ds \cos \gamma. \tag{14}$$

Thus dr is the projection PP' of ds upon the straight line drawn from O through P. (Remember that dr is not necessarily equal to Δr, which is the difference between the distances OQ and OP; in the figure, $OP'' = OP$, so $\Delta r = P''Q$.)

An approximate formula. A straight vertical antenna of length l, whose upper half is shown in Fig. 13, extends on the z axis from $z = -\frac{1}{2}l$ to $z = \frac{1}{2}l$. The distance s from an arbitrarily chosen point P to the upper end P' of the antenna is

$$s = \sqrt{r^2 + \tfrac{1}{4}l^2 - lr \cos \theta}. \tag{15}$$

We will suppose that the antenna is so short (or the point P so far) that

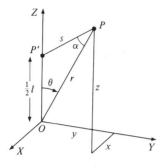

Figure 13

$$l^2 \ll r^2 \tag{16}$$

and will find the formula that can then be used instead of (15).

In view of (16), we have the approximate equation

$$s \approx \sqrt{r^2 - lr \cos \theta}, \tag{17}$$

so that

$$s \approx r\sqrt{1 - \frac{l}{r} \cos \theta}. \tag{18}$$

Now, according to (18^6),

$$\sqrt{1 - \frac{l}{r} \cos \theta} = 1 - \frac{l}{2r} \cos \theta - \frac{l^2}{8r^2} \cos^2 \theta - \cdots, \tag{19}$$

Therefore, remembering (16), we can replace (18) by

$$s \approx r\left(1 - \frac{l}{2r} \cos \theta\right). \tag{20}$$

We will rewrite (20) as

$$s = r\left(1 - \frac{l}{2r} \cos \theta\right) \tag{21}$$

and will keep in mind that (21) is an approximation.

EXERCISES

1. In Fig. 12, let $OP = 1$ meter, $PQ = 0.001$ meter, and $\gamma = 90$ degrees. Then show that, for the step from P to Q, Δr is about 5×10^{-7} meter and dr is exactly zero.

2. Consider the scalar fields $f = 1/r$ and $f = 1/r^2$, and show that

$$d\left(\frac{1}{r}\right) = -\frac{x}{r^3}\,dx - \frac{y}{r^3}\,dy - \frac{z}{r^3}\,dz = -\frac{ds}{r^2}\cos\gamma, \qquad (22)$$

and

$$d\left(\frac{1}{r^2}\right) = -\frac{2x}{r^4}\,dx - \frac{2y}{r^4}\,dy - \frac{2z}{r^4}\,dz = -2\frac{ds}{r^3}\cos\gamma, \qquad (23)$$

where the symbols ds and γ have the same meanings as in (14).

3. Show that, if $l^2 \ll r^2$, then the distances defined in Fig. 14 satisfy the following approximate equations:

$$s_1 = r - \frac{1}{2}l\cos\theta, \qquad s_2 = r + \frac{1}{2}l\cos\theta, \qquad (24)$$

$$\frac{1}{s_1} = \frac{1}{r} + \frac{l}{2r^2}\cos\theta, \qquad \frac{1}{s_2} = \frac{1}{r} - \frac{l}{2r^2}\cos\theta, \qquad (25)$$

$$\frac{1}{s_1^2} = \frac{1}{r^2} + \frac{l}{r^3}\cos\theta, \qquad \frac{1}{s_2^2} = \frac{1}{r^2} - \frac{l}{r^3}\cos\theta. \qquad (26)$$

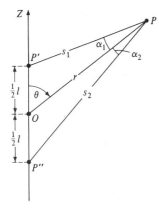

Figure 14

4. Show that, in Fig. 14,

$$\sin\alpha_1 = \frac{l}{2s_1}\sin\theta, \qquad \sin\alpha_2 = \frac{l}{2s_2}\sin\theta, \qquad (27)$$

and that, if $l^2 \ll r^2$, then $\cos\alpha_1 = 1$ and $\cos\alpha_2 = 1$.

5. In Fig. 13, draw a point Q lying at the distance $\frac{1}{2}l$ directly below P. The radial coordinate of Q is then equal to s. The difference between OP and s is, therefore, equal to the increment of r associated with the step from P to Q. Derive (20) on this basis.

CHAPTER 3

Curves and Surfaces

A student of the electromagnetic theory must learn to visualize and describe spatial relationships among various geometric elements. In this chapter we define some of these elements, outline the pertinent terminology, and discuss solid angles, which play important rôles in the theory of both the electric and the magnetic fields. We present even the simpler parts of this subject in detail, because, as we remarked in the preface, small misunderstandings early in this study may lead to serious misconceptions later.

Unless we specify otherwise, the curves and surfaces considered in this book are assumed to be stationary relative to the observer.

8. CURVES AND SURFACES

Curves and paths. A curve that does not intersect itself is called *simple*. If the ends of a simple curve are marked *a* and *b*, as in Fig. 15, a moving point can trace this curve by moving on it either from *a* to *b* or from *b* to *a*. Therefore, we say that a simple curve has *two senses*. It is often convenient to

Fig. 15. A curve and two paths.

single out one sense, to call it the *positive sense*, and to call the other sense *negative*. When this is done, the curve is called a *path*. We usually denote paths by Greek letters. The positive sense of a path can be indicated by an arrowhead, as in Fig. 15, where the paths α and β have opposite senses. If a point traces the path in the positive sense, we say that it moves *forward*; otherwise, we say that it moves *backward*.

If a simple path has no cusps, then it has, at every point P on it, a unique *direction*, namely, the direction in space in which a point moves at P when tracing the path forward. The term "sense" applies to the path as a whole, the term "direction" applies to the path at particular points on it. For example, the sense of the path α in Fig. 15 is fixed, but the direction of this path changes from point to point. The term "direction," however, can often be used instead of "sense" without ambiguity.

Contours. A closed line (in particular, a closed succession of straight lines or curves) is called a *contour*. A contour that does not intersect itself is called *simple*. Examples of simple contours are the boundary of a triangle, the shore line of a pond, a loop of thin wire whose ends have been soldered together, and the line formed by a twisted but unbroken thin rubber band. If a contour lies in a plane, we call it *flat*. We usually denote contours by the letter C or by a Greek letter in parentheses, say (γ).

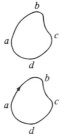

Fig. 16. An unsigned contour (a closed curve) and a signed contour (a closed path).

A contour, like a nonclosed curve, has *two senses*; for example, the contour at the top of Fig. 16 has the sense *abcda* (or *bcdab*, and so on) and also the sense *adcba*. If one of these senses has been designated as positive, we speak of the contour as a *signed contour* and say that it has a *signature*. The contour shown in the lower part of Fig. 16 is an example of a signed contour; as indicated by the arrowhead, its positive sense is *abcda*. An unsigned contour is a closed line; a signed contour is a closed path.

Flat surfaces. A surface that lies in a plane is called *flat*. One example is a flat rectangular surface, like a postcard of zero thickness. This surface has *one boundary* (consisting of four straight segments, called *edges*) and *two sides* (the picture side and the writing side of a picture postcard). This terminology is different from that of plane geometry; in particular, we say that a rectangular surface has two sides rather than four, so that the term "side" as we use it, is not synonymous with "edge." One can picture the two sides of a surface as having different colors. We denote surfaces by the letter S.

We will often designate one side of a surface as the *positive side* or the *front*, and the other side as the *negative side* or the *back*; and we will then speak of this surface as a *signed surface* and say that it has a *signature*.

A flat signed surface has at every point *two normals* (two perpendiculars) that protrude, so to speak, from its positive and negative sides. The normal protruding from the positive side is called the *positive normal* to the surface. One way of showing which side of a surface has been chosen as positive is to draw at some point on the surface a pointer in the direction of the positive normal. For example, in Fig. 17 we have chosen as positive the left-hand side

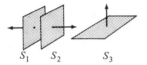

Figure 17

of S_1, the right-hand side of S_2, and the top side of S_3. We say that a flat surface *looks* or *faces* in the direction of its positive normal (the surface S_1 in Fig. 17 looks leftward).

If a flat surface has no holes in it, its boundary is a single contour. This contour *bounds* the surface, and the surface *spans* the contour.

The foregoing remarks apply equally well to flat surfaces having nonrectangular boundaries, such as a flat circular surface (like a coin of zero thickness, whose sides are "heads" and "tails"), a flat triangular surface (which, like a pennant, has three edges and two sides), or a flat surface with an irregular boundary.

The right-hand rule. If the contour C is the boundary of a simple surface S, we can decide arbitrarily which side of S to designate as positive *or* which sense of C to designate as positive, but we cannot make both decisions arbitrarily because of the following convention: *The signatures of a surface and of its boundary are connected by the right-hand rule.*

The right-hand rule can be stated in several equivalent ways. Here are two examples:

- Point the thumb of the right hand in the direction of a positive normal to S and curl the fingers around this normal; the fingers will then indicate the positive sense of C.

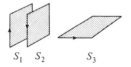

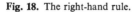

Fig. 18. The right-hand rule.   **Figure 19**

- Grasp a portion of C with the right hand so that the thumb indicates the positive sense of C; the fingers will then cross S from back to front.

The right-hand rule is illustrated in Fig. 18, where S is a flat circular surface.

We now have a second way of indicating the front of a surface in a diagram: We specify the positive sense of the boundary by an arrowhead and let the reader use the right-hand rule to infer the direction of the positive normal. For example, the signatures of the three surfaces in Fig. 19 are the same as those of the corresponding surfaces in Fig. 17.

"Across" and "through." The line ab in Fig. 20 goes from one side of the flat surface S to the other side and has one point in common with S: we say that it "crosses" S or "goes across" S. By contrast, the line cd, which lies in the plane of S, goes from edge to edge of S, and we say that it "traverses" S. This wording differs from that of elementary plane geometry, where what

Fig. 20. The line ab crosses (goes through) the surface S. The line cd traverses S.

we call an "edge" is called a "side." Therefore, we should stress the fact that by "across" we mean "from side to side" and *not* "from edge to edge." Our term "crossing a surface" is similar to "piercing a surface" or "going through a surface." To remind the reader of the meaning of the word "across" in discussions of surfaces, we sometimes write "across (through)" rather than simply "across." Strictly speaking, though, the word "through" should be reserved for dealing with holes or contours, as in saying "magnetic flux through a loop of wire." (Molecules diffuse *across* a membrane *through* the pores of the membrane.)

Curved surfaces. A surface that does not lie in a plane is called *curved*, even if it consists only of flat portions. A surface that does not intersect itself is called *simple*. We restrict ourselves to simple surfaces that have *two sides* and ignore such exceptions as the Möbius strip.[1] We call a surface *smooth* if one and only one tangent plane can be drawn at every point on it.

A surface S is called *open* or *nonclosed* if any two arbitrarily chosen points in space can be connected by a curve that does not cross S. Two examples are a flat rectangular surface of finite extent and a curved surface resembling the cloth of a butterfly net, pictured in Fig. 21.

---

[1] August Ferdinand Möbius (1790-1868). A Möbius strip is a surface that has only one side. To make a model of it, insert the end of a dress belt into the buckle after giving it a 180-degree twist. If you should proceed to paint a Möbius strip and continue until you return to where you started, no place will be left for a second color.

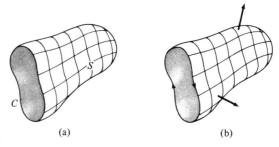

Fig. 21. An open surface, unsigned in (a) and signed in (b).

The positive sense of the contour C bounding an open curved surface S can be indicated by arrowheads. The positive side of S can be specified either by indicating the positive sense of C and referring to the right-hand rule, or by drawing one or more positive normals to S (that is, normals protruding from the positive sides of small and nearly flat portions into which S can be imagined to be subdivided). The reader should verify that in Fig. 21(b) the choice of the positive sense of C is consistent with the choice of the positive side of S.

Closed surfaces. A simple curved surface having a finite area is called *closed* if it divides all space into two regions—the *inner* and the *outer* region—in such a way that every curve connecting any point in one of these regions with any point in the other region crosses the surface an odd number of times (the identifying feature of the inner region is that the straight-line distance between any two points in this region has an upper limit). The two sides of a closed surface are called *inner* and *outer*, according to the following rule: If a moving point crosses S on its way from the inner to the outer region, it is said to cross S from the inner to the outer side.

Three examples of closed surfaces are the surface of a sphere, the surface of a doughnut, and the irregular surface of a smooth potato, pictured in Fig. 22. As these examples illustrate, *a closed surface has no boundary*, but a simple

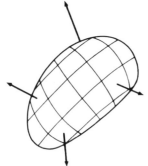

Fig. 22. A closed surface, resembling the surface of a smooth potato.

closed surface is itself the boundary of a portion of space, say a region R, called the inner region. According to a standard convention illustrated in Fig. 22, *the outer side of a closed surface is always taken to be the positive side.*

We denote both closed and open surfaces by S, but, as we remarked in §1, we replace S by (S) when we want to stress the fact that the surface in question is closed.

We will sometimes say that a point, say P, lies *just outside* a closed surface (S). This statement will mean that P lies outside (S), but at the end of a computation the distance between P and the nearest part of (S) will be made to approach zero without becoming zero. (The numbers in such a sequence as $1, \frac{1}{2}, \frac{1}{4}, \frac{1}{8}$, and so on, approach zero but do not include zero.) The term "just inside (S)" will have a similar meaning.

Connectivity. A region R is said to be *simply connected* if every contour lying in R can be spanned by a simple surface lying entirely in R. A cubical region is simply connected; a doughnut-shaped region is not.

Projections. Let a flat signed surface S_0, having the area da, be "projected" upon a signed plane S, as in Fig. 23, where the arrows show the direc-

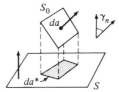

Fig. 23. A surface S_0 projected on a plane.

tions of the positive normals to S and S_0. The formula for the area, say da^*, of this "perpendicular" projection is taken to be

$$da^* = da \cos \gamma_n, \tag{1}$$

where γ_n is the angle between the positive normals to S and S_0. Note that if the plane S in Fig. 23 were facing downward, the angle γ_n would be obtuse and the area da^* would be negative.

EXERCISES

1. The lower contour in Fig. 16 can be described as having the "clockwise" sense, but only at the risk of an ambiguity. Does the earth move around the sun clockwise or counterclockwise?
2. Rephrase the right-hand rule in terms of a right-handed screw.
3. (a) Define a simply connected region R without referring to surfaces; instead, refer to contours lying in R and to the possibility of contracting each of them to a point. (b) Consider a simple open surface S and define what is meant by saying that "S has no holes in it."

9. SOLID ANGLES

This section is concerned with solid angles that are subtended by *surfaces* and enter in Gauss's restatement of Coulomb's law.[2] Solid angles subtended by *contours* play a part in the theory of magnetic fields produced by steady currents (§90). Our unit for measuring solid angles is called the *steradian*, but is seldom mentioned by name.

The area of the small signed surface S_0 in Fig. 24 is da; the point Q lies on S_0, at the distance r from P; the pointers drawn at Q show the respective directions of the positive normal to S_0 and the straight line PQ; the angle between

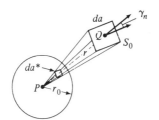

Fig. 24. A surface S_0 projected on a sphere.

these pointers is γ_n. If this angle is acute, as in the figure, we say that P sees the back of S_0. The figure also shows the projection of S_0 upon an imagined sphere of radius r_0 centered on P. We denote the area of the projection by da^* and take da^* to be *positive* if P sees the *back* of S_0 and negative if P sees the front of S_0. (This sign convention is consistent with Equation (1^8), which pertains to projections upon planes.) The solid angle subtended by S_0 at P, denoted below by $d\Omega$, is defined by the equation

$$d\Omega = \frac{da^*}{r_0^2}. \qquad (1)$$

As seen from Fig. 24, da^* is proportional to r_0^2; therefore, the choice of r_0 does not affect the value of $d\Omega$ and can be made arbitrarily.

The solid angle Ω subtended at P by a signed surface S of any size and shape is the sum of the solid angles subtended at P by the small flat "differential" portions into which S can be imagined to be subdivided. That is,

$$\Omega = \iint_S \frac{da^*}{r_0^2} \qquad (2)$$

or, since r_0 is a constant,

$$\Omega = \frac{1}{r_0^2} \iint_S da^*. \qquad (3)$$

---

[2] Karl Friedrich Gauss (1777-1855).

Here $\iint da^*$ is the sum, taken with due regard for signs, of the projections (upon the sphere of radius r_0) of the differential portions of S. We denote this sum by a^* and write (3) as

$$\Omega = \frac{a^*}{r_0^2}. \tag{4}$$

Example. Let a point P lie so far from a flat surface S that every linear dimension of S is negligible compared to the distance from P to any point on S. Let the radius r_0 of the imagined spherical surface drawn about P be so large that, except for higher-order corrections, the part of this surface lying near S can be taken as flat. In particular, consider Fig. 25, where the flat heart-shaped

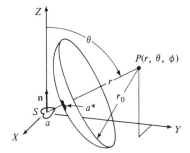

Figure 25

surface S (area a) lies in the xy plane around the origin and faces upward.[3] Finally, set r_0 equal to the radial coordinate r of P. It is then perhaps apparent from (1⁸) that, if P should lie far enough from S, the area a^* of the shaded projection in the figure would be $-a \cos \theta$. Accordingly, apart from higher-order corrections, the solid angle subtended by S at P is

$$\Omega = -a \frac{\cos \theta}{r^2} \qquad (P \text{ lies far from } S). \tag{5}$$

Second example. The flat circular disk-like surface S in Fig. 26 has the radius ρ_0 and faces upward; the (x, y, z) coordinates of the point P are $(0, 0, z > 0)$. We will compute the solid angle subtended at P by the disk and will use the abbreviation

$$s = \sqrt{\rho_0^2 + z^2}. \tag{6}$$

The projection of the disk upon the sphere shown in the figure is a spherical cap whose radius is $\rho_0 r_0 / s$. Since P sees the front of the disk, we reverse the sign of the standard formula for the area of such a cap and write

---

[3] Unless other directions are specified in our examples and exercises, a surface lying in the xy plane faces in the $+z$ direction, a surface lying in the yz plane faces in the $+x$ direction, and a surface lying in the zx plane faces in the $+y$ direction.

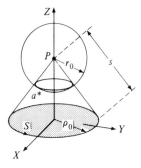

Figure 26

$$a^* = -2\pi r_0^2 \left(1 - \frac{z}{s}\right). \tag{7}$$

It now follows from (4) that

$$\Omega = -2\pi \left(1 - \frac{z}{s}\right) \qquad (z > 0). \tag{8}$$

The reader should show in a similar way (Exercise 7) that, if P lies below the disk, we have

$$\Omega = 2\pi \left(1 + \frac{z}{s}\right) \qquad (z < 0). \tag{9}$$

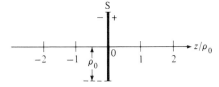

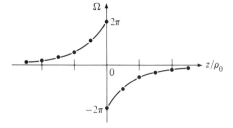

Fig. 27. The solid angle Ω subtended by the surface S (in Fig. 26) at the point P, plotted as a function of the position of P on the z axis.

Thus Ω would increase abruptly by 4π if P should cross the disk from front to back. A graph of Ω computed from (8) and (9) is shown in Fig. 27.

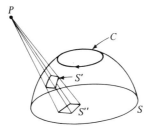

Figure 28

Curved surfaces. The contributions to Ω made by some parts of a *curved* surface S may cancel the contributions made by some other parts of S. The surface S in Fig. 28 is shaped as an inkwell and is signed as shown by the arrowhead on its boundary C. At the point P in the figure, the contribution, say $d\Omega'$, made by the element S' of S is negative, because P sees the front of S', but the contribution, say $d\Omega''$, made by the element S'' is positive, because P sees the back of S''. Furthermore, since the elements S' and S'' are cut out of S by the same cone emerging from P, the absolute values of $d\Omega'$ and $d\Omega''$ are equal. Consequently, $d\Omega' = -d\Omega''$, and the contributions made by S' and S'' to the total solid angle cancel each other. Inspection of Fig. 28 will, in fact, show that the solid angle subtended at P by the entire surface S is equal to the solid angle subtended at P by that part of S that is enclosed in the cone emerging from P and passing through the rim C into the inkwell.

Closed surfaces. It is perhaps obvious that if the surface in Fig. 28 did not have the opening at the top, the various $d\Omega$'s would cancel in pairs, and Ω would be zero. The reader should verify that this conclusion holds for a simple closed surface (S) of any shape if P lies *outside* it (Exercise 2). On the other hand, if P lies *inside* (S), then the area of the projection of (S) upon a sphere of radius r_0 centered on P is equal to the surface area of this sphere, and (4) gives $\Omega = a^*/r_0^2 = 4\pi r_0^2/r_0^2 = 4\pi$. To summarize:

$$\Omega = \begin{cases} 4\pi & \text{if } P \text{ lies inside } (S) \\ 0 & \text{if } P \text{ lies outside } (S). \end{cases} \tag{10}$$

Note that if P should move across (S) from front to back, Ω would increase by 4π. This discontinuity is similar to that implied in (8) and (9), and pictured in Fig. 27.

The case when P lies *on* the surface (S) is illustrated in Exercises 4 and 5.

Another formula for Ω. Suppose that the radius r_0 of the sphere in Fig. 24 is made equal to the distance r from P to Q. Since S_0 is small, the area da^* of its projection upon the new sphere is equal, apart from higher-order terms, to the area of its projection upon the plane tangent to this sphere at Q. In view of (1^8) we then have $da^* = da \cos \gamma_n$ and

$$d\Omega = \frac{\cos \gamma_n}{r^2} da, \tag{11}$$

so that (2) can be rewritten as

$$\Omega = \iint_S \frac{\cos \gamma_n}{r^2} da. \tag{12}$$

Equation (2) involves the projections of all the small portions of a surface S upon *the same* sphere of radius r_0, and hence the factor $1/r_0^2$ in (2) can be put outside the integral signs, as in (3). By contrast, in (12), different portions of S are projected upon *different* spheres, the distance r changes from point to point on S, and, therefore, the factor $1/r^2$ must be kept inside.

Combining (12) with (10), we find that, for a closed surface,

$$\iint_{(S)} \frac{\cos \gamma_n}{r^2} da = \begin{cases} 4\pi & \text{if } P \text{ lies inside } (S) \\ 0 & \text{if } P \text{ lies outside } (S). \end{cases} \tag{13}$$

To repeat: here P is an arbitrarily chosen point; (S) is a simple closed surface; da is the area of a differential element of this surface; r is the distance of this element from P; γ_n is the angle between the positive normal to this element and the straight line drawn from P to this element. Since (S) is simple and closed, its positive side is its outer side.

EXERCISES

1. Equation (10) implies that when you look at a soap bubble rising in the air, the solid angle subtended at your eye by the surface of the bubble is zero. Elucidate this statement, which may at first seem strange.

2. A point P lies outside a simple closed surface (S) of any shape. Convince yourself that any straight line that starts at P and extends to infinity will cross (S) from back to front as many times as from front to back. [Any point at which the line is tangent to (S) should be treated as a combination of back-and-forth crossings.] Consider in a similar way the case when P lies inside (S). Then replace the straight line by a narrow cone emerging from P and justify (10).

3. Show without integration that a face of a cube subtends the solid angle $2\pi/3$ at the center of the cube and the solid angle $\pi/6$ at a corner of the opposite face.

4. Let P be a point *on* a simple closed surface (S) at which (S) has only one tangent plane, and show that the entire surface (S) subtends at P the solid angle 2π.

5. Show that a cubical surface subtends at any one of its eight corners the solid angle $\frac{1}{2}\pi$. What would the solid angle be if the corner were rounded off, as on a playing die?

6. Use (12) to verify (8). (Divide the disk into concentric rings of radial thickness $d\rho$.)

7. Derive (9) and show that the equation

$$\frac{\partial\Omega}{\partial z} = \frac{2\pi\rho_0^2}{(\rho_0^2 + z^2)^{3/2}} \qquad \text{(at points on the } z \text{ axis)} \tag{14}$$

holds for both (8) and (9).

8. The surface S of Fig. 26 is reproduced in Fig. 29, where the straight line AB is parallel to the z axis; the distance between this line and the z axis is greater than ρ_0. A point Q moves on AB from a point far below the xy plane to a

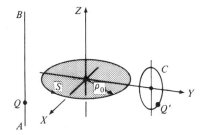

Figure 29

point far above. Sketch a graph of the solid angle subtended by S at Q. Note that Ω is now a continuous function of z, unlike the Ω plotted in Fig. 27. Repeat for the case when the distance from AB to the z axis is $\frac{1}{2}\rho_0$.

The contour C in Fig. 29 is a circle in the yz plane, centered on the y axis. The point Q' moves on C, describing a full circle. Sketch the solid angle subtended by S at Q'. Repeat for the case when C lies closer to the z axis and crosses S once.

Comment in a general way on the continuity or discontinuity of the solid angles subtended at the moving points Q and Q' by a *curved* surface S' that spans the same circular contour as S. Consider various forms of S', some of which are pierced by AB more than twice.

9. Show that if $z^3 \gg \rho_0^3$, then (8) and (9) reduce to (5) with the appropriate values of θ.

10. A rectangle lies in the xy plane, as in Fig. 30(a); its edge lengths are dx and dy; the coordinates of its center are $(x, y, z = 0)$. Show that, if dx and dy are small enough, the solid angle $d\Omega$ subtended by the rectangle at the point $P(x_0, y_0, z_0)$ is[3]

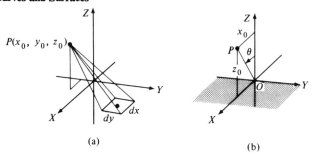

(a) (b)

Figure 30

$$d\Omega = -z_0 \frac{dx\,dy}{s^3}, \tag{15}$$

where

$$s = \sqrt{(x-x_0)^2 + (y-y_0)^2 + z_0^2}. \tag{16}$$

11. Use (15) to show that the solid angle subtended by the entire xy plane at the point P of Fig. 30(a) can be written as[3]

$$\Omega = -2[\tan^{-1}(\infty) - \tan^{-1}(-\infty)]. \tag{17}$$

(To simplify the integration without loss of generality, place P above the origin.)

Since an arctangent is a multivalued function of its argument, the formula (17) is ambiguous. Show, however, that, for the principal values of the arctangents, (17) becomes $\Omega = -2\pi$, and that this result agrees with that obtained from (8) by letting the radius of the disk in Fig. 26 increase without limit.

Next suppose that the xy plane is divided in half by a straight line passing through the origin O and show that, if P lies above O, then each half-plane subtends at P the solid angle $\Omega = -\pi$.

12. The coordinates of the point P in Fig. 30(b) are $(x_0 \geq 0, 0, z_0 \geq 0)$, so the colatitude of P, namely $\theta = \tan^{-1}(x_0/z_0)$, lies in the interval $0 \leq \theta \leq \frac{1}{2}\pi$. Show that the shaded half of the xy plane in the figure subtends at P the solid angle $\Omega = -2[\tan^{-1}(x_0/z_0) + \tan^{-1}(\infty)]$ and that, for the principal values of the arctangents, this angle is

$$\Omega = -2\theta - \pi. \tag{18}$$

Check (18) using the results of Exercise 11.

13. Take a clue from (18), and show without integration that the shaded half of the zx plane in Fig. 31 subtends at P the solid angle

$$\Omega = 2\phi - 2\pi \qquad (0 \leq \phi \leq 2\pi). \tag{19}$$

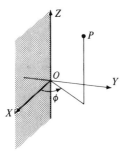

Figure 31

Then take δ to be a *positive* angle, small compared to $\frac{1}{2}\pi$; evaluate Ω for the following eight values of ϕ: $\delta, \frac{1}{2}\pi \pm \delta, \pi \pm \delta, \frac{3}{2}\pi \pm \delta$, and $2\pi - \delta$; finally, let $\delta \to 0$, and verify that (19) gives the values of Ω expected in the light of our earlier examples.

Note that (19) fails if $\phi = 2\pi + \delta$, even when $\delta \to 0$. Show, however, that the correct value of Ω for *any* value of ϕ can be obtained from the formula

$$\Omega = 2\phi - 2\pi + 4\pi n, \qquad n = 0, \pm 1, \pm 2, \cdots, \tag{20}$$

by choosing a proper value of n. [The incompleteness of (19) has to do with the fact that in deriving (18) we used only the principal values of the arctangents.]

14. The contours C and C' in Fig. 207(b) have the same sense and are connected by a ribbon-like surface, which we denote by the symbol S_{ribbon}. (The vectors $d\mathbf{s}$ and $d\mathbf{l}$ in the figure are not needed in the present exercise.)

    Imagine a surface S (say a soap film) spanning C, and a surface S' spanning C'. According to the arrowheads on the contours, the reader sees the fronts of S and S'. The signature of S_{ribbon} is shown by the vector $\mathbf{1}_n$. We use the symbols Ω, Ω', and Ω_{ribbon} for the respective solid angles subtended by S, S', and S_{ribbon} at the point P in Fig. 207(b).

    Note that if the signature of S is reversed, then the reversed surface, together with S' and S_{ribbon}, forms a drum-shaped *closed* surface. Use this fact and Equation (10) to show that

$$\Omega' = \Omega - \Omega_{\text{ribbon}}. \tag{21}$$

CHAPTER 4

Vectors

The simplest directed physical quantities are *vectors* and *vector fields*. This chapter is confined to vectors. An example of a vector field—current density—is analyzed in Chapter 5. Vector fields in general are discussed in Chapter 6.

10. VECTORS

A physical quantity is called a *vector quantity*, or simply a *vector*, if and only if

(a) it has a *numerical magnitude*,
(b) it has a *direction in space*,
(c) it obeys the *parallelogram rule* for addition.

Simple examples of vectors are the displacement, velocity, and acceleration of a particle.

As we remarked in §1, the concept of a direction in space does not involve any coordinate frames. Consequently, the three properties of vectors listed above imply that *the concept of a physical vector is coordinate-free*. Vectors can be defined mathematically by stating the rules for transforming vector components when a coordinate frame is turned about the origin; the statements (a), (b), and (c) given above are, however, more suitable for our immediate purposes.

A vector is pictured in diagrams by a *pointer* having the direction of the vector and a length equal to the magnitude of the vector in terms of a suitable scale unit. The end of the pointer marked by the arrowhead is its *terminus* or *tip*; the other end is its *origin* or *tail*. If a vector pertains to a point P, we put at P the tail of the pointer representing this vector.

Quantities that can be pictured by pointers do not necessarily obey the parallelogram rule. For example, constant electric currents in wires can be pictured as in Fig. 2[1], where the length of a pointer represents the magnitude of the corresponding current. But the pointers in that figure do not add by the parallelogram rule, and, therefore, currents are not vector quantities.

Complex numbers can be pictured by pointers and can be added by the parallelogram rule, but they do not have directions *in space* (the physical space in which we live), and consequently they do not conform to the definition of physical vectors stated above. (When the word "vector" is used for complex numbers in the theory of alternating currents, it is used in a mathematical sense.)

We denote vectors by bold-faced letters and their magnitudes by the corresponding light-faced letters; thus the magnitude of a vector **A** is denoted by A, although the symbol $|A|$ may be used instead. The magnitude of a vector is a nonnegative number, expressed in the appropriate units of measurement. A vector whose magnitude is zero (say the resultant of two equal and opposite forces) is called *zero* and is denoted by **0**.

Diagrams that show several kinds of vectors involve several kinds of scales. Imagine a ball revolving with a constant speed of 10 ft/sec in a circle whose radius is 5 feet, so that the magnitude a of the centripetal acceleration of the

Fig. 32. The velocity and acceleration of a ball moving in a circle, with a constant speed.

ball, namely v^2/r, is 20 ft/sec². The position, velocity, and acceleration of the ball at some particular instant of time can then be pictured as in Fig. 32, where the radius of the circle is 20 millimeters, the length of the pointer **v** is 20 mm, and the length of the pointer **a** is 10 mm. Consequently, the scales used in Fig. 32 are

distances:	1 mm represents $\frac{1}{4}$ ft,
speeds:	1 mm represents $\frac{1}{2}$ ft/sec,
accelerations:	1 mm represents 2 ft/sec².

Whatever the numerical values of r, v, and a may be, we may draw the circle and the pointers **v** and **a** of any size we please; every choice of sizes will involve three kinds of scales: one for distances, one for speeds, and one for accelerations.

In a diagram, we represent a vector **A** by a pointer marked **A**. The vector **A** (say the acceleration of a body) and the pointer **A** (say a chalk line on a blackboard) are very different things, but statements that distinguish explicitly between a vector and the pointer representing it are often cumbersome. Therefore, if a misunderstanding is not likely, we say "vector" for "pointer." When we say, for example, "the tip of the vector **A**," we mean the tip of the pointer representing the vector **A**.

If c is a *positive* number, the equation

$$\mathbf{A} = c\mathbf{B} \tag{1}$$

means that the direction of the vector **A** is the same as that of **B** and the magnitude of **A** is c times that of **B**. The equation $\mathbf{A} = \mathbf{B}$ means that **A** and **B** agree both in magnitude and in direction, but it does not necessarily mean that the tails of the pointers representing **A** and **B** lie at the same point. If c is *negative*, Equation (1) means that the direction of **A** is opposite to that of **B**, and the magnitude of **A** is $|c|$ times that of **B**.

Vector addition. The equation

$$\mathbf{A} = \mathbf{B} + \mathbf{C} \tag{2}$$

means that the pointer representing the vector **A** is the diagonal of the parallelogram based on the pointers representing the vectors **B** and **C**, as in Fig. 33. If $\mathbf{A} = \mathbf{B} + \mathbf{C}$, we say that the vector **A** is the *resultant* or *sum* of the vectors

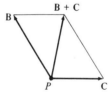

Fig. 33. The parallelogram rule for adding two vectors.

B and **C**. Equations (1) and (2) involve directed quantities and are geometric equations, even though they look like algebraic equations. The process of computing resultants of vectors is called *vector addition*.

Let **B**, **C**, and **D** be three vectors that do not necessarily lie in a plane, and let the symbols $(\mathbf{B} + \mathbf{C})$, $(\mathbf{C} + \mathbf{D})$, and $(\mathbf{B} + \mathbf{D})$ denote the respective resultants of these vectors taken in pairs; further, let the symbol $(\mathbf{B} + \mathbf{C}) + \mathbf{D}$ denote the resultant of the vectors $(\mathbf{B} + \mathbf{C})$ and **D**, and so on. It then follows from the parallelogram rule that the pointer representing the vector $(\mathbf{B} + \mathbf{C})$

+ **D** is the diagonal of a certain parallelepiped, and that the pointer representing the vector **B** + (**C** + **D**) is the *same* diagonal of the *same* parallelepiped, as is the pointer representing the vector (**B** + **D**) + **C**. Consequently, each of the three expressions (**B** + **C**) + **D**, **B** + (**C** + **D**), and (**B** + **D**) + **C** stands for the same vector. This vector is called the *resultant* or *sum* of **B**, **C**, and **D** and is denoted by the symbol **B** + **C** + **D**, so that

$$\mathbf{B} + \mathbf{C} + \mathbf{D} = (\mathbf{B} + \mathbf{C}) + \mathbf{D} = \mathbf{B} + (\mathbf{C} + \mathbf{D}) = (\mathbf{B} + \mathbf{D}) + \mathbf{C}. \quad (3)$$

The resultant of **B**, **C**, and **D** can be denoted by a single letter, say **A**, in which case we write **A** = **B** + **C** + **D**. Equations (3) tell us that this resultant can be found by three superficially different geometric constructions. The case of more than three vectors is similar. Equations (3) illustrate the fact that, when adding several vectors, we may associate them into pairs in any way we like. For this reason vector addition is said to be *associative*.

Dot products. The angle between two vectors is defined as the smallest angle through which one of the vectors can be turned so as to make its direction the same as that of the other vector. (If two vectors are oriented as the hands of a watch reading 7 o'clock, the angle between them is 150 deg.)

Let **A** and **B** be two vectors, γ be the angle between them, as in Fig. 34, and A and B be their respective magnitudes. The following three-factor product then often comes into play:

$$AB \cos \gamma. \quad (4)$$

For example, if **A** is a constant *force*, if **B** is a straight-line *displacement* of a particle on which this force is acting, and if γ is the angle between the displacement and the force, then $AB \cos \gamma$ is the expression for *work*. The product $AB \cos \gamma$ is an ordinary number (positive, negative, or zero) and has no direction in space. It is called the *scalar product* or the *dot product* of the vectors

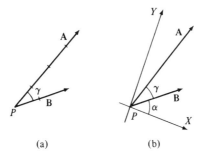

(a)   (b)

Fig. 34. In this figure, $A = 3.8$, $B = 2.2$, and $\gamma = 30$ deg; the angle α is arbitrary.

A and B, and is abbreviated by the symbol **A·B**, which is read "A dot B" (not "A times B"). The symbol **A·B** is thus defined as follows:

$$\mathbf{A} \cdot \mathbf{B} = AB \cos \gamma. \tag{5}$$

In Fig. 34, for example, we have $A = 3.8$, $B = 2.2$, and $\gamma = 30$ deg, so that $\mathbf{A} \cdot \mathbf{B} \approx 7.2$. We must stress the fact that the dot product **A·B** can be evaluated, as we have just done, without reference to any coordinate frames, such as the frame XY of Fig. 34(b).

The process of evaluating dot products is called *scalar multiplication* or *dot multiplication*. Multiplication of a vector by a number, vector addition, and dot multiplication are three of the processes of *vector algebra*, an algebra in which the words "addition" and "multiplication" may mean something very different from what they mean in elementary arithmetic and algebra.

Special cases of dot products are

$$\mathbf{A} \cdot \mathbf{B} = \begin{cases} AB & \text{if } \gamma = 0 \\ 0 & \text{if } \gamma = 90° \\ -AB & \text{if } \gamma = 180° \end{cases} \tag{6}$$

and

$$\mathbf{A} \cdot \mathbf{A} = A^2. \tag{7}$$

According to (5), we have $\mathbf{B} \cdot \mathbf{A} = BA \cos \gamma$, so that

$$\mathbf{A} \cdot \mathbf{B} = \mathbf{B} \cdot \mathbf{A}; \tag{8}$$

for this reason scalar multiplication is said to be *commutative*.

If the vectors **A** and **B** make the angle γ, the product $A \cos \gamma$ is called the *projection* of **A** upon **B** and is denoted by A_B; that is,

$$A_B = A \cos \gamma. \tag{9}$$

If γ is larger than 90 deg, the projection A_B is negative. (Remember that γ cannot exceed 180 deg.) The product $B \cos \gamma$ is called the projection of **B** upon **A** and is denoted by B_A; that is,

$$B_A = B \cos \gamma. \tag{10}$$

Comparison with (5) shows that

$$\mathbf{A} \cdot \mathbf{B} = AB_A = BA_B. \tag{11}$$

Let B_A, C_A, and $(B + C)_A$ be the respective projections of the vectors **B**, **C**,

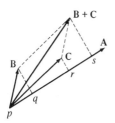

Figure 35

and **B** + **C** upon a vector **A**. These projections are shown in Fig. 35 for the case when the pointers **A**, **B**, and **C** have a common origin, lie in the plane of the page, and make acute angles with one another. In this figure $B_A = pq$, $C_A = pr$, and $(B + C)_A = ps$, so that, since $rs = pq$, we have

$$(B + C)_A = B_A + C_A. \tag{12}$$

Furthermore, it is perhaps apparent that (12) will hold even if the relative orientations of **A**, **B**, and **C** are not as simple as in Fig. 35. Multiplying (12) by A, we get the equation

$$A(B + C)_A = AB_A + AC_A, \tag{13}$$

which implies that

$$\mathbf{A} \cdot (\mathbf{B} + \mathbf{C}) = \mathbf{A} \cdot \mathbf{B} + \mathbf{A} \cdot \mathbf{C}. \tag{14}$$

Thus the factor **A** can be "distributed," so to speak, among the vectors appearing in the parentheses. For this reason, scalar multiplication is said to be *distributive*. Vector addition is associative, and scalar multiplication is distributive; therefore, when dealing with these operations, we may handle parentheses as in ordinary algebra. For example,

$$\mathbf{A} \cdot (\mathbf{B} + \mathbf{C} + \mathbf{D}) = \mathbf{A} \cdot \mathbf{B} + \mathbf{A} \cdot \mathbf{C} + \mathbf{A} \cdot \mathbf{D} \tag{15}$$

and

$$(\mathbf{A} + \mathbf{B}) \cdot (\mathbf{C} - \mathbf{D}) = \mathbf{A} \cdot \mathbf{C} - \mathbf{A} \cdot \mathbf{D} + \mathbf{B} \cdot \mathbf{C} - \mathbf{B} \cdot \mathbf{D}. \tag{16}$$

Terminology. If $\mathbf{A} = \mathbf{B} + \mathbf{C}$, then **A** is called the "resultant" or "sum" of **B** and **C**, but the converse terminology is not as uniform. Here are three versions:

B and **C** are a pair of *components* of **A**.
B and **C** are a pair of *vector components* of **A**.
B and **C** are a pair of *resolvents* of **A**.

The term "component" for a *vector* is usual in introductory physics books, but we will see in §11 that this terminology conflicts with the use of the same term in field theory for a scalar. The term "vector component" removes this ambiguity. Maxwell sometimes used the term "constituent." We use the term "resolvent" in the earlier part of this book. But after the reader learns to tell from the context whether a vector or an ordinary number is meant, he may find it most convenient to say "component" in either case—as we will do later in this book.

EXERCISES

1. State in words the geometric significance of the equations $\mathbf{A} = -\mathbf{B}$ and $\mathbf{A} = \mathbf{B} - \mathbf{C}$.

2. Write $|\mathbf{A} + \mathbf{B}|$ for the magnitude of the vector $\mathbf{A} + \mathbf{B}$ and show that

$$|\mathbf{A} + \mathbf{B}| = \sqrt{A^2 + B^2 + 2AB \cos \gamma}, \qquad (17)$$

where γ is the angle between $\mathbf{A}$ and $\mathbf{B}$.

3. The magnitude of $\mathbf{A}$ is 4; the magnitude of $\mathbf{B}$ is 5. (a) What is the angle between $\mathbf{A}$ and $\mathbf{B}$ if $\mathbf{A} \cdot \mathbf{B} = 20$? (b) If $\mathbf{A} \cdot \mathbf{B} = 10$? (c) If $\mathbf{A} \cdot \mathbf{B} = 0$? (d) If $\mathbf{A} \cdot \mathbf{B} = -10$? (e) If $\mathbf{A} \cdot \mathbf{B} = -20$?

4. If neither $\mathbf{A}$ nor $\mathbf{B}$ is zero and $\mathbf{A} \cdot (\mathbf{A} + \mathbf{B}) = A^2$, what geometric conclusion can be drawn?

5. If the magnitudes of $\mathbf{A}$ and $\mathbf{B}$ are equal but their directions are different, then the vectors $\mathbf{A} + \mathbf{B}$ and $\mathbf{A} - \mathbf{B}$ are mutually perpendicular. Verify this statement (a) by a geometric construction and (b) by evaluating the dot product of $\mathbf{A} + \mathbf{B}$ and $\mathbf{A} - \mathbf{B}$.

6. Suppose that $\mathbf{A} \cdot \mathbf{C} = \mathbf{B} \cdot \mathbf{C}$ for *every* choice of the vector $\mathbf{C}$ and show that then $\mathbf{A} = \mathbf{B}$.

7. The vectors $\mathbf{A}, \mathbf{B}_1, \mathbf{B}_2, \mathbf{B}_3,$ and $\mathbf{B}_4$ have a common origin. Show that if $\mathbf{A} \cdot \mathbf{B}_1 = \mathbf{A} \cdot \mathbf{B}_2 = \mathbf{A} \cdot \mathbf{B}_3 = \mathbf{A} \cdot \mathbf{B}_4$, then the tips of the four $\mathbf{B}$'s lie in a plane perpendicular to $\mathbf{A}$.

11. CARTESIAN COMPONENTS

In §10 we dealt with lengths, angles, and parallelograms, which are concepts belonging to "pure" or "synthetic" geometry. Now we will pass from that kind of geometry to "coordinate" or "analytic" geometry. This step is very useful, partly because it permits expressing *vectors* (which have directions in space) in terms of sets of *numbers* (which may be positive, negative, or zero, but have no directions *in space*).

Cartesian base-vectors. When we introduce a cartesian frame, say

XYZ, in a problem involving vectors, we also introduce *at every point in space* an imagined triplet of *base-vectors*, denoted in this book by $\mathbf{1}_x$, $\mathbf{1}_y$, and $\mathbf{1}_z$ ("one sub x," and so on). The magnitude of a base-vector is *unity* (not 1 meter, not 1 meter/second, not 1 newton, and so on, but simply 1). The vectors $\mathbf{1}_x$, $\mathbf{1}_y$, and $\mathbf{1}_z$ point, respectively, in the $+x$, $+y$, and $+z$ directions. The traditional symbols for these vectors are $(\mathbf{i}, \mathbf{j}, \mathbf{k})$; sometimes such symbols as $(\mathbf{e}_x, \mathbf{e}_y, \mathbf{e}_z)$ or $(\hat{x}, \hat{y}, \hat{z})$ are used for them; our notation is due to Skilling.[1] A vector whose magnitude is the "pure" number 1 is called a *unit vector*; the base-vectors $\mathbf{1}_x$, $\mathbf{1}_y$, $\mathbf{1}_z$ are examples of unit vectors.

Example: circular motion. Figure 36 is reproduced from Fig. 32, except for adding a coordinate frame and the base-vectors $\mathbf{1}_x$ and $\mathbf{1}_y$. The

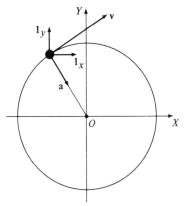

Fig. 36. The whirling ball of Fig. 32, a coordinate frame, and two base-vectors belonging to that frame.

length of these vectors is 8 mm, so that, so far as the base-vectors are concerned, 1 mm represents the "pure" number $\frac{1}{8}$ that has no units of measurement associated with it.

The velocity **v** is shown once again in Fig. 37. In Fig. 37(a) it is resolved into a vector parallel to the x axis and a vector parallel to the y axis. The magnitudes of these vectors are about 8.4 ft/sec and 5.4 ft/sec, so that **v** can be described as follows:

$$\mathbf{v} = (8.4 \text{ ft/sec}) \text{ rightward} + (5.4 \text{ ft/sec}) \text{ upward}. \tag{1}$$

The base-vectors $\mathbf{1}_x$ and $\mathbf{1}_y$ in Fig. 37(b) permit us to write (1) as

$$\mathbf{v} = (8.4 \text{ ft/sec}) \mathbf{1}_x + (5.4 \text{ ft/sec}) \mathbf{1}_y. \tag{2}$$

When a vector is expressed in terms of a set of base-vectors, as in (2), the

---

[1] H. H. Skilling, *Fundamentals of Electric Waves* (2nd Ed., New York, N. Y., John Wiley & Sons, 1948).

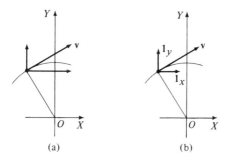

Figure 37

numerical coefficients of the base-vectors are called the *measure-numbers* or the *components* of the vector. In particular, the numerical coefficients of $\mathbf{1}_x$ and $\mathbf{1}_y$ in (2) are called the "x component" and the "y component" of $\mathbf{v}$, and are denoted by v_x and v_y. In this example,

$$v_x = 8.4 \text{ ft/sec}, \qquad v_y = 5.4 \text{ ft/sec}. \tag{3}$$

The units of measurement of the components of a vector are usually kept in mind, and an equation such as (2) is usually written as

$$\mathbf{v} = 8.4\mathbf{1}_x + 5.4\mathbf{1}_y. \tag{4}$$

The z axis in Fig. 37 is perpendicular to the page and hence the z component of $\mathbf{v}$ is zero. Therefore, the vector $\mathbf{v}$ can be described in full detail by writing

$$v_x = 8.4 \text{ ft/sec}, \qquad v_y = 5.4 \text{ ft/sec}, \qquad v_z = 0. \tag{5}$$

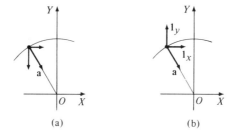

Figure 38

The acceleration $\mathbf{a}$, pictured in Fig. 32 and again in Fig. 38, can be treated similarly; the approximate result is

$$a_x = 10.8 \text{ ft/sec}^2, \qquad a_y = -16.8 \text{ ft/sec}^2, \qquad a_z = 0. \tag{6}$$

Note that a_y is an ordinary negative number and has no direction in space. The fact that it is negative does mean, however, that the vector -16.81_y points downward on the page, as in Fig. 38(a).

Second example: falling ball. If, unlike a base-vector, a given vector is a coordinate-free entity, its numerical description with the help of a coordinate frame will, in general, depend on the choice of the frame. To illustrate, we consider the gravitational force **f** acting on a freely falling ball and describe it first in terms of the frame S_1 of Fig. 39 and then in terms of the frame S_2 of that figure. Writing g for the magnitude of the acceleration of the ball, we get Table 2, which shows that the components of **f** relative to the frame S_1 do not all agree with the components of the same vector taken relative to the frame S_2.

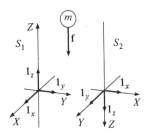

Fig. 39. Diagram for Table 2.

TABLE 2

	f_x	f_y	f_z
Frame S_1	0	0	$-mg$
Frame S_2	0	0	mg

Resolvents and components. Suppose that a vector **A** is resolved into three vectors that are, respectively, parallel or antiparallel to the x, y, and z axes of a cartesian coordinate frame. We then denote these vectors, as in Fig. 40(a), by $\mathbf{A}_x$, $\mathbf{A}_y$, and $\mathbf{A}_z$ (bold type), call them the *cartesian resolvents* of **A**, and write

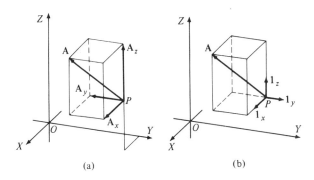

Figure 40

$$\mathbf{A} = \mathbf{A}_x + \mathbf{A}_y + \mathbf{A}_z. \tag{7}$$

Unless the vector $\mathbf{A}_x$ is zero, it is either parallel or antiparallel to the base-vector $\mathbf{1}_x$ and therefore has the form $A_x\mathbf{1}_x$, where A_x (light type) is either a positive or a negative number. All in all,

$$\mathbf{A}_x = A_x\mathbf{1}_x, \quad \mathbf{A}_y = A_y\mathbf{1}_y, \quad \mathbf{A}_z = A_z\mathbf{1}_z. \tag{8}$$

The numbers A_x, A_y, and A_z in (8) are called the *cartesian components* or *cartesian measure-numbers* of the vector $\mathbf{A}$. The cartesian "component form" of this vector, namely

$$\mathbf{A} = A_x\mathbf{1}_x + A_y\mathbf{1}_y + A_z\mathbf{1}_z, \tag{9}$$

can be written equally well as

$$\mathbf{A} = \mathbf{1}_xA_x + \mathbf{1}_yA_y + \mathbf{1}_zA_z. \tag{10}$$

In our two-dimensional example concerning the whirling ball of Fig. 32, we were given the scales used in drawing the figure and therefore we could find the numerical values of the components of $\mathbf{v}$ and $\mathbf{a}$ listed in (3) and (6). We need not describe the vector $\mathbf{A}$ of Fig. 40 in as much detail, but we should compare the two parts of the figure and note that the components A_x and A_z are positive, but A_y is negative.

We should stress the fact that in field-theoretic terminology *the components of a vector are ordinary numbers (positive, negative, or zero) and have no directions in space*. The directional features of the vector $\mathbf{A}$ are provided in (9) and (10) by the base-vectors $\mathbf{1}_x$, $\mathbf{1}_y$, and $\mathbf{1}_z$, which serve as *direction indicators*.

In the case of a vector $\mathbf{B}$, say, the notation is similar:

$$\mathbf{B} = \mathbf{B}_x + \mathbf{B}_y + \mathbf{B}_z \tag{11}$$

and

$$\mathbf{B} = B_x\mathbf{1}_x + B_y\mathbf{1}_y + B_z\mathbf{1}_z, \tag{12}$$

or $\mathbf{B} = \mathbf{1}_xB_x + \mathbf{1}_yB_y + \mathbf{1}_zB_z$. Here $\mathbf{B}_x$ is the x resolvent of $\mathbf{B}$, B_x is the x component of $\mathbf{B}$, and so on.

The vector $(-1)\mathbf{1}_x + (0)\mathbf{1}_y + (1)\mathbf{1}_z$ is written for short as $-\mathbf{1}_x + \mathbf{1}_z$; the notation for other vectors whose components include 1, 0, or -1 is similar.

When a geometric vector equation is rewritten in terms of components, it turns into three simultaneous numerical equations, which do not involve any quantities that have directions in space. For example, the geometric equation

$$\mathbf{A} = c\mathbf{B} \tag{13}$$

is equivalent to the three equations

$$A_x = cB_x, \quad A_y = cB_y, \quad A_z = cB_z, \tag{14}$$

which involve only numbers. Similarly, the geometric equation

$$\mathbf{A} = \mathbf{B} + \mathbf{C} \tag{15}$$

is equivalent to the three numerical equations

$$A_x = B_x + C_x, \quad A_y = B_y + C_y, \quad A_z = B_z + C_z. \tag{16}$$

The cartesian component formula for the magnitude A of a vector $\mathbf{A}$ is

$$A = \sqrt{A_x^2 + A_y^2 + A_z^2}. \tag{17}$$

Dot products. The vectors $\mathbf{1}_x$, $\mathbf{1}_y$, and $\mathbf{1}_z$ are mutually perpendicular, and the magnitude of each is unity; hence,

$$\mathbf{1}_x \cdot \mathbf{1}_x = 1, \quad \mathbf{1}_x \cdot \mathbf{1}_y = 0, \quad \mathbf{1}_x \cdot \mathbf{1}_z = 0, \tag{18a}$$

$$\mathbf{1}_y \cdot \mathbf{1}_x = 0, \quad \mathbf{1}_y \cdot \mathbf{1}_y = 1, \quad \mathbf{1}_y \cdot \mathbf{1}_z = 0, \tag{18b}$$

$$\mathbf{1}_z \cdot \mathbf{1}_x = 0, \quad \mathbf{1}_z \cdot \mathbf{1}_y = 0, \quad \mathbf{1}_z \cdot \mathbf{1}_z = 1. \tag{18c}$$

Now,

$$\mathbf{A} \cdot \mathbf{B} = (A_x \mathbf{1}_x + A_y \mathbf{1}_y + A_z \mathbf{1}_z) \cdot (B_x \mathbf{1}_x + B_y \mathbf{1}_y + B_z \mathbf{1}_z). \tag{19}$$

The right-hand side of (19) is the sum of nine products, such as $A_x B_x (\mathbf{1}_x \cdot \mathbf{1}_x)$, $A_x B_y (\mathbf{1}_x \cdot \mathbf{1}_y)$, and so on. In view of the multiplication table (18), six of these products vanish and (19) reduces to

$$\mathbf{A} \cdot \mathbf{B} = A_x B_x + A_y B_y + A_z B_z. \tag{20}$$

Another proof of this very useful formula is suggested in Exercise 6.

EXERCISES

1. Verify that if $\mathbf{A} = \mathbf{1}_x + 2\mathbf{1}_y + 3\mathbf{1}_z$ and $\mathbf{B} = 4\mathbf{1}_x - 3\mathbf{1}_z$, then $A = \sqrt{14}$, $B = 5$, $\mathbf{A} + \mathbf{B} = 5\mathbf{1}_x + 2\mathbf{1}_y$, $\mathbf{A} - \mathbf{B} = -3\mathbf{1}_x + 2\mathbf{1}_y + 6\mathbf{1}_z$, and $\mathbf{A} \cdot \mathbf{B} = -5$.

2. Show that the angle between the vectors $\mathbf{A}$ and $\mathbf{B}$ of Exercise 1 is about 105 deg.

3. Show analytically and check by a diagram that the vectors $2\mathbf{1}_x + 3\mathbf{1}_y$ and $3\mathbf{1}_x - 2\mathbf{1}_y$ are mutually perpendicular.

4. Give an example of two nonvanishing vectors that are perpendicular to the vector $\mathbf{1}_x - 2\mathbf{1}_z$ and to each other.

5. Verify (17), using (20) and (7[10]).

68   Vectors

6. Show that

$$A_x = A \cos \alpha_x, \quad A_y = A \cos \alpha_y, \quad A_z = A \cos \alpha_z, \tag{21}$$

where the α's are the direction angles of the vector **A**; then recall (6⁵) and derive (20) from equations of the form (21).

7. Show that

$$A_x = \mathbf{A} \cdot \mathbf{1}_x, \quad A_y = \mathbf{A} \cdot \mathbf{1}_y, \quad A_z = \mathbf{A} \cdot \mathbf{1}_z. \tag{22}$$

8. Let the symbol $d\mathbf{l}$ denote the "vector step" from a point P to a neighboring point Q, namely the displacement from P to Q, represented by a pointer whose tail is at P and tip at Q. Write (x, y, z) and $(x + dx, y + dy, z + dz)$ for the respective coordinates of P and Q, show that

(10ᴬ) $$d\mathbf{l} = \mathbf{1}_x\, dx + \mathbf{1}_y\, dy + \mathbf{1}_z\, dz, \tag{23}$$

and verify that (23) is consistent with (11ᴬ).

9. Express in terms of A, B, γ, and α the x and y components of the vectors **A** and **B** in Fig. 34(b). Then verify that (20) yields the same value of $\mathbf{A} \cdot \mathbf{B}$ as does (5¹⁰).

12. NORMAL AND TANGENTIAL COMPONENTS

Our notation for normal and tangential components of a vector **A** is as follows:

A_n = normal component of **A** relative to a *surface*,
A_t = tangential component of **A** relative to a *line*,

and

A_ν = normal component of **A** relative to a *line*,
A_τ = tangential component of **A** relative to a *surface*.

The symbols A_n and A_t will be needed often, but A_ν and A_τ only seldom.

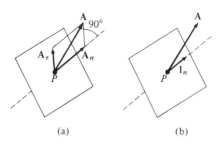

Fig. 41. The resolvent $\mathbf{A}_n$ of **A** is perpendicular to the plane; the resolvent $\mathbf{A}_\tau$ lies in the plane.

Sometimes (as when a body slides on an inclined plane) we are given a vector **A** and have to consider the "normal" resolvent and the "tangential" resolvent of this vector relative to a given plane. These resolvents, which we denote by $\mathbf{A}_n$ and $\mathbf{A}_t$, are defined in Fig. 41; we find $\mathbf{A}_n$ by projecting **A** upon a straight line (shown dotted) normal to the plane, and then find $\mathbf{A}_t$ by completing the parallelogram that has **A** as the diagonal and $\mathbf{A}_n$ as one edge. ($\mathbf{A}_n$ is perpendicular to the plane; $\mathbf{A}_t$ lies in the plane.)

Suppose now that one side of the plane has been chosen as the positive side, and let $\mathbf{1}_n$ be a unit vector having the direction of a positive normal to this plane. The normal resolvent of **A** will then be equal to the vector $\mathbf{1}_n$ multiplied by some number, say A_n, that is,

$$\mathbf{A}_n = A_n \mathbf{1}_n. \tag{1}$$

The number A_n defined by (1) is called the *normal component* of **A** relative to the given plane. An equivalent definition is

$$A_n = A \cos \gamma_n, \tag{2}$$

where A is the magnitude of **A** and γ_n is the angle between **A** and $\mathbf{1}_n$. If the positive side of the plane is chosen so that $\mathbf{1}_n$ points as in Fig. 41(b), then A_n is positive; but if the other side of the plane should be taken as positive, then **A** would not change, $\mathbf{1}_n$ would reverse its direction, and A_n would become negative. Thus the normal resolvent can be constructed whether or not the plane is signed; but the normal component can be evaluated only if the plane is signed. (Exercise 17[77] presents a systematic method for resolving a vector into two resolvents, one of which is parallel and the other perpendicular to a given vector.)

Sometimes (as when a horse walks alongside a track and pulls a cart moving on the track) we have to consider the normal resolvent and the tangential resolvent of a vector **A** relative to a given straight line. These resolvents, which we denote by $\mathbf{A}_v$ and $\mathbf{A}_t$, are defined in Fig. 42(a), where, for simplicity, the vector **A** and the straight line are understood to lie in the plane of the page.

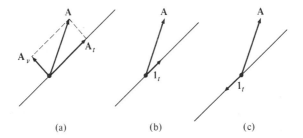

Fig. 42. Tangential and normal resolvents of a vector relative to a straight line.

Suppose now that one direction along the given line has been designated as positive, so that the line becomes a path. Let the symbol $\mathbf{1}_t$ denote the unit vector tangent to the path and pointing in the direction of the path. The tangential resolvent of $\mathbf{A}$ will then be proportional to $\mathbf{1}_t$; denoting the factor of proportionality by A_t, we then have

$$\mathbf{A}_t = A_t \mathbf{1}_t. \tag{3}$$

The number A_t defined by (3) is called the *tangential component* of $\mathbf{A}$ relative to the given path. An equivalent definition is

$$A_t = A \cos \gamma_t, \tag{4}$$

where A is the magnitude of $\mathbf{A}$ and γ_t is the angle between $\mathbf{A}$ and $\mathbf{1}_t$. The tangential component of a vector relative to a path is also called the *forward component* of the vector. If $\mathbf{A}$ is oriented as in Fig. 42(b) and the path is directed as shown by the vector $\mathbf{1}_t$, then A_t is positive, but if the direction of the path is reversed, as in Fig. 41(c), then A_t becomes negative (that is, the forward component of $\mathbf{A}$ is then negative).

The concepts of normal and tangential components relative to a given plane or line are *coordinate-free*; these components can be identified by purely geometric construction, without first setting up a coordinate frame.

We have confined ourselves above to normal and tangential components relative to a plane or a straight line. In most of our work we will deal, in much the same way, with normal components relative to elements of a curved surface, elements small enough to be regarded as flat; we will also deal with tangential components relative to segments of a curved line, segments short enough to be regarded as straight.

EXERCISES

1. Verify that Equations (1) and (2) are equivalent, that Equations (3) and (4) are equivalent, and that

$$A_n = \mathbf{A} \cdot \mathbf{1}_n, \qquad A_t = \mathbf{A} \cdot \mathbf{1}_t. \tag{5}$$

2. Given a vector $\mathbf{A}$ and a signed plane S, which is perpendicular to the x axis of a coordinate frame. Show that if S faces in the $+x$ direction, then $A_n = A_x$, and that otherwise $A_n = -A_x$.

3. A plane S faces away from the origin of the frame XYZ; its $x, y,$ and z intercepts are each equal to 1; the vector $\mathbf{1}_n$ points along the positive normal to S. Show that the cartesian components of $\mathbf{1}_n$ are each equal to $1/\sqrt{3}$.

4. The open hemispherical surface S in Fig. 43 is signed as shown. The "latitudes" of the points $a, b, c,$ and d are 0, 30, 60, and 90 deg. The four equal

vectors marked **A** point in the positive z direction. Compute the normal components and describe the tangential resolvents of the **A**'s relative to S. (At b, we have $A_n = 0.5$.)

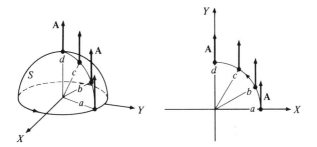

Figure 43   Figure 44

5. The path *abcd* in Fig. 44 is a quarter-circle in the xy plane; the points b and c divide it into thirds. Compute the tangential components and describe the normal resolvents of the **A**'s relative to the path *abcd*. (At c, we have $A_t = 0.5$.)

CHAPTER 5

Current and Current Density

We now come to the first topic listed in the chart in the preface: "Electric conduction currents, discussed apart from their causes." We restrict ourselves to steady currents carried by electrons in solid conductors and ignore the effects of the magnetic fields produced by these currents.

In field theory, conduction currents are defined less directly than in circuit theory. In fact, in field theory a conduction current is an integral of a quantity that is in some ways more fundamental, namely, a vector field called *current density* and denoted in this book by **J**. The definition of **J** is given in §13 and is followed by examples, centered on a steady flow of electric charge past a cylindrical hole in a conductor. The equations for this case are written without proof, which must wait until Chapter 16.

13. DEFINITIONS

Current. In circuit theory, the current i_{ab} flowing in the rod of Fig. 45 is defined as the time rate at which, in effect, positive charge enters the rod at point a and leaves it at point b. In field theory, however, we cannot always treat the terminals of conductors as mathematical points or ignore the cross-sectional dimensions of wires. Therefore, we will now define the current flowing in a conductor in terms of the flow of charge across (through) a physical

Figure 45

or imagined mathematical *surface*. The definition reads: *The current across a signed surface is the time rate at which, in effect, positive charge is crossing this surface from back to front* (for short: "the rate at which charge is crossing the surface"). Our unit for measuring current is the "coulomb per second," namely, the *ampere*.

Suppose that electrons are crossing (going through) a certain signed surface *from back to front* at the rate of 10^{19} electrons per second. In round numbers, the electronic charge is -1.6×10^{-19} coulombs, so that positive charge is in effect crossing this surface *from back to front* at the rate of -1.6 coulombs per second. Hence, the current flowing across this surface is -1.6 amperes. The current flowing across a given signed surface is described in complete detail by a single number (such as -1.6 amperes), and, therefore, current is a scalar quantity.[1] If a surface is not signed, the sign of the current flowing across it is indeterminate. The current flowing across a surface is, of course, not a field quantity, because it pertains to a surface rather than to a point.

In the wire of Fig. 46, positive charge is, in effect, flowing upward. (Electrons are flowing downward.) The two imagined surfaces S_1 and S_2 lie inside the wire. As shown by the arrowheads, S_1 faces upward and S_2 downward. Accordingly, the current flowing across S_1 is *positive*, and the current across S_2 is *negative*. The merits of this sign convention will appear when we come to Kirchhoff's current law.

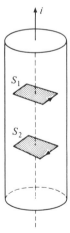

Figure 46

---

[1] In circuit theory it is useful to speak of the "direction" of a current in a wire, even though currents are not vectors but scalars (recall Fig. 2^1 and the top paragraph on page 57). The term "direction of the current," as used in the circuit theory, actually stands for "direction of the current density."

The direction of flow of charge. We will now consider an arbitrary point P inside a current-carrying conductor of an arbitrary shape, say the branching conductor of Fig. 47, and we will define the term "direction of the flow of charge at a point P."

Imagine placing at P a small flat signed mathematical "test surface" S having the area da; denote by di the current flowing across (through) S. The numerical value of di will then depend on the direction in which S is facing. In fact, for some orientations of S, positive charge will be crossing it in effect from back to front, and hence di will be positive; but for other orientations of S, positive charge will be crossing it in effect from front to back, and di will be negative. Now, it is perhaps obvious that there is one and only one special orientation of S for which the current di has a maximum (positive) value, say $(di)_{max}$. Accordingly, when $di = (di)_{max}$, the test surface S will be facing in a particular direction in space; *this direction we call the direction of the flow of charge at P*. In a metal, it is opposite to the average direction in which conduction electrons stream near P.

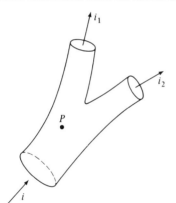
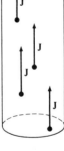

Figure 47　　　　　　　　Figure 48

Current density. Next we come to *current density*, which is a field quantity and which we denote by **J**. The current density **J** at any point P is a vector, and hence we must define its direction as well as its magnitude J, which we express in *amperes per square meter*.

The direction of **J** at P is defined to be the same as the direction of the flow of charge at P. For example, if, in effect, positive charge flows upward (electrons flow downward) in a long, vertical, uniform, and homogeneous wire, then the vector **J** at every point in the wire points upward, as in Fig. 48. But in a curved wire the direction of **J** changes from point to point.

The magnitude of **J** at P is defined with the help of a small imagined signed test surface S, having the area da, placed at P, and facing in the direction of **J**. We write $(di)_{max}$ for the current flowing across S when S is oriented in this way and define **J** as the limit of the ratio of the current $(di)_{max}$ to the area da, as da tends to zero. We will record this definition by writing $J = (di)_{max}/da$ or, which is the same thing,

$$(di)_{max} = J\,da, \tag{1}$$

and will keep in mind that $da \to 0$.

Current across a small surface. Suppose that the vector **J** pertaining to a point P is given and that an imagined small surface S is located at P in such a way that the unit vector $\mathbf{1}_n$, pointing along the positive normal to S, makes the angle γ_n with **J**, as in Fig. 49(a). To compute the current di

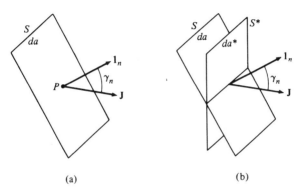

Fig. 49. Diagram leading to the equations $di = J\cos\gamma\,da = J_n\,da$.

flowing across S, we first project S upon a plane passing through P and normal to **J**, and construct the surface S^*, shown in Fig. 49(b). This surface faces in the direction of **J** and has the area

$$da^* = \cos\gamma_n\,da. \tag{2}$$

Since S^* faces in the direction of **J**, the current across it is simply $J\,da^*$. Further, it is perhaps obvious (especially if one looks at S and S^* edge-on) that, if S is small enough, the current di flowing across S is equal to the current flowing across S^*. Hence, $di = J\,da^*$ or, more explicitly,

$$di = J\cos\gamma_n\,da. \tag{3}$$

The product $J\cos\gamma_n$ is the normal component of **J** relative to S, so we can write (3) as

$$di = J_n\,da. \tag{4}$$

If S faces in the direction of **J**, then J_n takes on its maximum value, namely J, and (4) reduces to (1).

Parallelogram rule. To justify the statement that **J** is a *vector quantity* we must not only ascribe to it a magnitude and direction, which we have done, but must also prove that current densities add by the parallelogram rule. We will do this for a simple case, leaving generalizations to the reader (Exercise 3).

Imagine two streams of charges, each flowing in its own way in the same conductor. Suppose that the first stream alone would produce at the point P in Fig. 50(a) the current density $\mathbf{J}_1$, directed to the right. Suppose that the

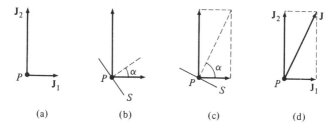

Fig. 50. Vector addition of current densities.

second stream alone would produce at P the current density $\mathbf{J}_2$, directed upward on the page. Our problem is to compute the total current density **J** produced at P jointly by the two streams. Let S (area da) be an imagined small flat surface, located at P, perpendicular to the page, and facing at some angle α to $\mathbf{J}_1$, as in Fig. 50(b). According to (3), the current across S produced by the first stream is $J_1 \cos \alpha \, da$, and the current across S produced by the second stream is $J_2 \cos(90° - \alpha) \, da = J_2 \sin \alpha \, da$, so the total current flowing across S is

$$di = (J_1 \cos \alpha + J_2 \sin \alpha) \, da. \tag{5}$$

By definition, the direction of the total current density **J** at P is the direction in which S will face when the current across it is a maximum. Differentiating the right-hand side of (5) with respect to α and equating the derivative to zero, we find that di is largest when

$$\tan \alpha = \frac{J_2}{J_1}, \tag{6}$$

that is, when S is oriented as in Fig. 50(c). Accordingly, **J** is directed along the diagonal of the parallelogram based on $\mathbf{J}_1$ and $\mathbf{J}_2$. To compute the maximum value of di we use (6) to eliminate α from (5) and find that

$$(di)_{\max} = \sqrt{J_1^2 + J_2^2}\, da. \tag{7}$$

Therefore, in view of (1),

$$J = \sqrt{J_1^2 + J_2^2}. \tag{8}$$

Consequently, **J** can be built from $\mathbf{J}_1$ and $\mathbf{J}_2$ by the parallelogram rule, as in Fig. 50(d). Thus, current density is indeed a vector quantity, and the relation among the current densities $\mathbf{J}_1$, $\mathbf{J}_2$, and **J** in Fig. 50 can be written simply as $\mathbf{J} = \mathbf{J}_1 + \mathbf{J}_2$.

Examples. An example of *uniform* current density is pictured in Fig. 48. In this case, **J** is the same at all points in the wire, the pointers are all parallel and have equal lengths, and the magnitude of **J** at any point in the wire is

$$J = \frac{i}{A}, \tag{9}$$

where A is the cross-sectional area of the wire and i the current in it. An example of *nonuniform* current density is given by the leakage current in the insulation of an armored cable, shown in cross section in Fig. 51(a). Suppose that, to test the insulation, we connect a battery between the steel armor and

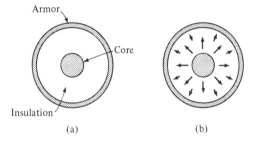

Fig. 51. A pointer map of the density of the leakage current in the insulation of an armored cable.

one end of the copper core (making the core positive relative to the armor), but do not connect the other end of the core electrically to anything. A small "leakage current" will then flow in the insulation from the core to the armor. If we consider a cross section of the cable under the simplifying assumptions listed in the footnote,[2] it is perhaps obvious that, as shown in Fig. 51(b), the

---

[2] These assumptions are: (a) the cable is long and straight, and its core, insulating sheath, and armor are all round and concentric; (b) the cross section in question lies far from the ends of the cable, where special end-effects take place; (c) the insulation is homogeneous; (d) all magnetic effects are negligible; (e) the potential drop *along the core* is negligible. (If the insulation were perfect, the circuit would be open, no current would flow in the core, and the potential drop along the core would be zero.)

vector **J** at every point in the insulation will point radially outward from the axis of the core, and its magnitude will decrease as we go (in the insulation) from the core to the armor. A formula for this **J** will be derived in §39.

Remarks. If electrons flow downward in a vertical wire, we may borrow a term from the circuit theory, and say, "The current in the rod flows upward," instead of saying, "The current density in the rod is directed upward." But the shorter statement is only an abbreviation for the longer one, and does not imply that current is a vector.

If the current flowing in a conductor is not constant in time, then both the magnitude and the direction of **J** at every point in the conductor will, in general, change with time.

Names, symbols, and definitions. The introduction of a physical quantity into a discussion involves three distinct steps: (a) naming this quantity, (b) specifying the symbol used for it, and (c) stating its definition. Thus the field introduced above is called *current density*. The symbol used for it in this book is **J**. The definition of **J** consists (as above) of two separate statements: one defines the direction of **J**; the other defines its magnitude. Ordinarily the definition of a physical quantity is "operational"; that is, it tells how the quantity in question can be measured, often (as above) under imagined highly idealized conditions. The reader should bear in mind the fact that naming a physical quantity and giving the symbol used for it are not equivalent to defining it.

EXERCISES

1. A thick wire carries a constant current of i amperes. Express this statement in field-theoretic terms.
2. Verify that $di = J_n \, da$ even when the angle γ_n in Fig, 49 is obtuse.
3. (a) Derive the parallelogram rule for the case when the vectors $\mathbf{J}_1$ and $\mathbf{J}_2$ in Fig. 50 are not perpendicular to each other. (b) Show that the parallelogram rule applies also when there are three streams of charges and the vectors $\mathbf{J}_1$, $\mathbf{J}_2$, and $\mathbf{J}_3$ are not coplanar.

14. THE FLOW OF CHARGE PAST A CYLINDRICAL HOLE

A large copper block connected across a battery carries a current whose density **J** is steady and uniform, and has the magnitude of J_0 amperes per square meter. Here "steady" means that **J** does not change with time, and "uniform" means that it does not change from place to place. A cylindrical hole of circular cross section (radius b) is then drilled through this block, at right angles to the original direction of **J**. When the switch is closed again and a steady state is reestablished, electric charge will detour around the hole, and **J** will no longer be uniform. Our problem is to consider the main features of the new **J**, which we will first describe by diagrams, based on formulas given later in this section. We assume throughout that the current in the block is so feeble that its magnetic effects can be ignored. We also assume that b is so small compared to the dimensions of the block that the presence of the hole does not affect **J** at points in the block lying far enough from the hole.

A pictorial description of **J**. The current density near the hole is pictured in Fig. 52 by pointers that represent **J** at a number of points in a plane perpendicular to the axis of the hole. The leads of the battery are presumably connected to perfectly conducting plates clamped to the right-hand and left-hand ends of the copper block; their polarity is shown in the figure by the

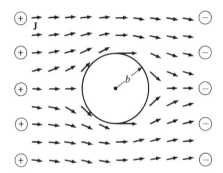

Fig. 52. A pointer map of the current density when electric charge flows past a cylindrical hole in a conductor.

plus and minus signs. We are assuming that at every point in the plane of Fig. 52 the flow of charge is directed in this plane, and that in every plane parallel to this plane the distribution of current density is the same as it is in this plane; a flow of this kind is called *two-dimensional*. The current density **J** is an example of a *vector field*; this term will be properly defined in the next chapter. Figure 52 is an example of a *pointer map* of a vector field. In this figure, the pointer pertaining to a point P (the pointer whose *tail* lies at P) represents the current density **J** at P; that is, this pointer has the direction of the flow of charge at P, and its length (in terms of a suitable scale unit) is equal to the time rate at which electric charge is crossing, per square meter, a small imagined surface located at P and facing in the direction of the pointer.

The main features of Fig. 52 are what one would expect intuitively. Far from the hole, where its effects are not important, the direction of **J** is nearly left to right, and its magnitude J is nearly constant. In the upper left-hand quarter of the figure, the pointers slant upward, because here the charges make an upward detour around the hole. Directly above and below the hole the pointers are longer (J and the drift speed v_d are larger); the charges that were originally aiming at the hole are deflected into these regions. Just to the left of the hole the pointers in the middle row are shorter (J and v_d are smaller) than elsewhere; the charges aiming at the hole have been slowed down. (In view of (11^1), $v_d = J/ne$.)

Surface charges. The charges approaching the hole begin to veer away from it at considerable distances from it. How do they know (if "know" is the right word) that an obstacle is standing in their way? Stated in the conventional terms (that is, on the assumption that the positive charges do the moving), the explanation is as follows. After the switch is first closed, positive charges begin to move in the copper block *from left to right*, as though the block had no hole in it. But the charges that run into the surface of the hole stop there, and hence a positive charge builds up on the left-hand half of this surface, as in Fig. 53. Also, the charges that leave the right-hand half of this surface are not

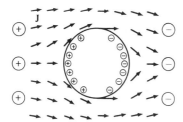

Fig. 53. The charges, shown here just inside the hole in the current-carrying conductor, actually settle *on the surface* of the hole.

replaced by other positive charges, and, therefore, this half of the surface becomes charged negatively. A positive charge that approaches the hole later on is repelled by the positive surface charges and attracted by the negative charges, but the repulsion is stronger, because the positive charges are nearer. Hence, the oncoming charges are in effect repelled by the hole and veer away from it. The surface charges continue to accumulate until a steady state has been reached, when the oncoming charges miss the hole altogether. Thus, one can say that the charges flowing toward the hole are avoiding not the hole itself, but the stationary charges that have collected on the surface of the hole during the initial surge of the charges. The density of the surface charges in Fig. 53 is discussed in §71.

Surface charges play a basic role even in the simplest electric circuits. For example, if conductors are connected in parallel, then, immediately after the switch is closed, surface charges become trapped or uncovered near the branching region and, by electrostatic forces, guide the further oncoming charges into the proper channels. Similarly, it is static surface charges that guide the moving charges around the bends and corners of a circuit such as a Wheatstone bridge.

In a *nonhomogeneous* conductor, some of the charges also settle *inside* the conductor after the closing of a switch and remain there even after the current becomes steady; to put it roughly, they balance out the inequalities of the interactions of the conduction electrons with the dissimilar atoms comprising a nonhomogeneous conductor. But if *magnetic fields are ignored* (see "Lorentz forces," page 374) and if the word "interior" excludes the surfaces of the conductor and of any cavities in it, then *the interior of a homogeneous conductor carrying a steady current remains uncharged.* To put it roughly again, although the conduction electrons keep streaming past the practically stationary atoms, the total number of electrons in any interior region of a *homogeneous* conductor remains equal to the number of protons located in this region. See Exercise 7[63].

A numerical description of **J**.   Figure 54 is an enlarged portion of Fig. 52. It includes a coordinate frame and the base-vectors $\mathbf{1}_x$ and $\mathbf{1}_y$ pertaining to the points at which **J** is shown. In Fig. 54, the length of the pointers representing $\mathbf{1}_x$ and $\mathbf{1}_y$ is made equal to the length of the pointers representing

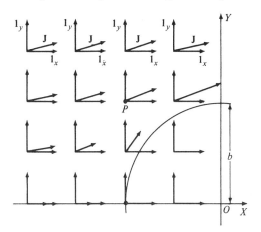

Fig. 54. An enlarged portion of Fig. 52, with the addition of a cartesian coordinate frame and a set of base-vectors associated with that frame.

J far from the hole, where the magnitude of **J** is J_0 amperes per square meter. With this choice of scale we can see, for example, that at the point whose coordinates are $x = -2b$ and $y = 3b/2$ we have $J_x = 0.9 J_0$ and $J_y = 0.2 J_0$, at least approximately. The reader should verify that Table 3 is at least roughly consistent with Fig. 54.

TABLE 3. Components of **J** at Several Points in Fig. 54

	$x = -\tfrac{3}{2}b$	$x = -b$	$x = -\tfrac{1}{2}b$
$y = b$	$J_x = 0.9 J_0$ $J_y = 0.3 J_0$ $J_z = 0.0$	$J_x = 1.0 J_0$ $J_y = 0.5 J_0$ $J_z = 0.0$	$J_x = 1.5 J_0$ $J_y = 0.6 J_0$ $J_z = 0.0$
$y = \tfrac{1}{2}b$	$J_x = 0.7 J_0$ $J_y = 0.2 J_0$ $J_z = 0.0$	$J_x = 0.5 J_0$ $J_y = 0.6 J_0$ $J_z = 0.0$	$J_x = 0.0$ $J_y = 0.0$ $J_z = 0.0$

An algebraic description of **J**. As illustrated by Table 3, the values of J_x and J_y at a point P in the copper block depend on the (x, y) coordinates of P. (The z coordinate is not involved, because the flow is two-dimensional.) In general, each component of **J** depends on each of the three coordinates of P; we record this fact by writing

$$J_x = J_x(x, y, z), \qquad J_y = J_y(x, y, z), \qquad J_z = J_z(x, y, z). \tag{1}$$

In our present problem, the cartesian components of **J** at a point P outside the hole prove to be

$$J_x = J_0 - J_0 b^2 \frac{x^2 - y^2}{(x^2 + y^2)^2}, \tag{2}$$

$$J_y = -2 J_0 b^2 \frac{xy}{(x^2 + y^2)^2}, \tag{3}$$

$$J_z = 0; \tag{4}$$

here b is the radius of the hole, J_0 is the magnitude of **J** far from the hole, and x and y are the coordinates of P. We will prove these equations from first principles in §69, and in the meantime we will ask the reader to take them for granted and use them in further computations. Figures 52, 53, and 54 were drawn from these equations.

Equations (2), (3), and (4) pertain to an idealized conducting block of infinite size. They are, nevertheless, of interest in practice, because, if the radius of the hole is small compared to the lateral dimensions of a conducting block, they give good approximations for the components of **J** near the hole.

The various partial derivatives of the components of **J** will be of interest to us. In the present example they are

$$\frac{\partial J_x}{\partial x} = 2 J_0 b^2 \frac{x(x^2 - 3y^2)}{(x^2 + y^2)^3}, \tag{5}$$

$$\frac{\partial J_x}{\partial y} = 2J_0 b^2 \frac{y(3x^2 - y^2)}{(x^2 + y^2)^3}, \tag{6}$$

$$\frac{\partial J_y}{\partial x} = 2J_0 b^2 \frac{y(3x^2 - y^2)}{(x^2 + y^2)^3}, \tag{7}$$

$$\frac{\partial J_y}{\partial y} = -2J_0 b^2 \frac{x(x^2 - 3y^2)}{(x^2 + y^2)^3}, \tag{8}$$

and

$$\frac{\partial J_x}{\partial z} = \frac{\partial J_y}{\partial z} = \frac{\partial J_z}{\partial x} = \frac{\partial J_z}{\partial y} = \frac{\partial J_z}{\partial z} = 0. \tag{9}$$

Some of the properties of these derivatives are peculiar to the special case with which we are dealing; for instance, Equations (9) hold because our example is two-dimensional. But the relations

$$\frac{\partial J_z}{\partial y} = \frac{\partial J_y}{\partial z}, \quad \frac{\partial J_x}{\partial z} = \frac{\partial J_z}{\partial x}, \quad \frac{\partial J_y}{\partial x} = \frac{\partial J_x}{\partial y}, \tag{10}$$

and

$$\frac{\partial J_x}{\partial x} + \frac{\partial J_y}{\partial y} + \frac{\partial J_z}{\partial z} = 0, \tag{11}$$

which hold among the derivatives are more fundamental; we will meet them in other problems. In fact, we will find that (11) corresponds in the electromagnetic field theory to Kirchhoff's current law for d.c. circuits, and that the three equations (10) have to do with Kirchhoff's voltage law.

EXERCISES

1. Verify that Table 3 is consistent with (2), (3), and (4).
2. Convince yourself on intuitive grounds that, at points outside the hole, the current density pictured in Fig. 52 should have the properties listed below, and verify that (2) and (3) provide for these properties.

    (a) J_x is never negative; at points lying sufficiently far from the hole in any direction it approaches the constant J_0.
    (b) J_y is zero at points on the x axis or the y axis.
    (c) J_y is positive in the even quadrants and negative in the odd quadrants; at points lying sufficiently far from the hole in any direction it approaches zero.
    (d) If two points are symmetric with respect to the x axis or the y axis, the values of J_x at these points are equal, and the values of J_y have equal magnitudes but opposite signs.
    (e) If two points are symmetric with respect to the axis of the hole, the **J**'s at these points are equal.

3. Use (2) and (3) to show that at points on the y axis $J_x > J_0$. Explain this inequality on intuitive grounds.

4. Use (2) and (3) to show that at the top and at the bottom of the cross section of the hole (where $x = 0$ and $y = \pm b$) we have $J_x = 2J_0$ and $J_y = 0$.

5. Find the locus of the points which lie in the xy plane at finite distances from the hole and at which $J_x = J_0$.

6. Verify Equations (5) through (11).

7. Note the point P in Fig. 54, at which $x = -b$ and $y = b$, so that $J_x = J_0$ and $J_y = \frac{1}{2}J_0$ at P. Introduce a new coordinate frame, say $X'Y'$, by turning the frame XY clockwise through 45 deg about the origin O; note that the new coordinates of P are $x' = -\sqrt{2}\,b$ and $y' = 0$, and denote the new basevectors by $\mathbf{1}_{x'}$ and $\mathbf{1}_{y'}$. Show that the new components of $\mathbf{J}$ at P are $J_{x'} = J_0/\sqrt{8}$ and $J_{y'} = 3J_0/\sqrt{8}$. As a partial check, verify that the value of J is not affected by the change of frame.

8. Use (1^A) and (2^A) to show that, in terms of cylindrical coordinates, Equations (2) and (3) are

$$J_x = J_0 - J_0 b^2 \frac{\cos 2\phi}{\rho^2}, \qquad J_y = -J_0 b^2 \frac{\sin 2\phi}{\rho^2}. \tag{12}$$

9. Verify that Equations (12) provide for the features of $\mathbf{J}$ listed in Exercise 2.

10. Show that at points on the surface of the hole the vector $\mathbf{J}$ is tangent to this surface (unless it vanishes).

11. The radius of the circular cross section of a conducting rod is ρ_0 meters. The rod is infinitely long, and its axis runs along the z axis of a coordinate frame. The rod carries a steady and uniformly distributed upward current of I amperes, as in Fig. 48[13]. Verify that if all other bodies can be ignored, the current density at any point P is

$$\mathbf{J} = \begin{cases} \dfrac{I}{\pi \rho_0^2} \mathbf{1}_z & \text{if } \rho \leq \rho_0 \\ 0 & \text{if } \rho > \rho_0, \end{cases} \tag{13}$$

where ρ is the distance from P to the z axis.

15. CURRENT AS AN INTEGRAL

From the standpoint of the electromagnetic field theory, current density (which is a field quantity) is a basic quantity and current (which is not a field quantity) is a derived quantity. In this section we show how the current flowing across (through) a surface can be computed from the current density.

Let **J** be the current density at an arbitrarily chosen point P in a current-carrying conductor. The current di flowing across a small signed surface located at P is

(4¹³) $$di = J_n \, da, \qquad (1)$$

where da is the area of this surface and J_n is the normal component of **J** relative to it. Consider now a signed simple surface S of any size and shape, and imagine it to be subdivided into small portions. The total current i flowing across the whole surface is the sum of the small currents flowing across the small portions of it. Each of these currents has the form (1), and hence

$$i = \iint_S J_n \, da. \qquad (2)$$

In words: *The current flowing across a surface S is equal to the integral (taken over S) of the normal component of **J** relative to S.*

If the current across one part of S is positive and that across the remainder of S is negative, then the current i across the whole surface S may be positive, negative, or zero. To stress this fact, one may speak of i as the *net current* across S.

Example.[3] The surface S_1 shown edge-on in Fig. 55 is an imagined rectangular surface, lying in the yz plane and facing toward the right. Its near edge cd runs along the y axis and has the length l. The point c lies on the surface of the hole. The depth of S_1 is h. (That is, $z = -h$ at all points on its far edge.) We will compute the current i flowing across (through) S_1.

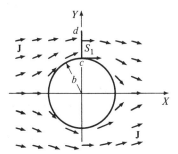

Fig. 55. The imagined surface S_1 (see footnote 4 on p. 87).

Let the point $P(0, y, z)$ lie on S_1. Since S_1 faces toward the right, we have

$$J_n = J_x. \qquad (3)$$

[3] This example and all the exercises at the end of this section deal with the flow of charge past a cylindrical hole.

In the present example, the general formula for J_x at a point with the coordinates (x, y, z) is

$$J_x = J_0 - J_0 b^2 \frac{x^2 - y^2}{(x^2 + y^2)^2}. \tag{2^{14}}\tag{4}$$

Therefore, since $x = 0$ at all points on S_1, the normal component of **J** at P is

$$J_n = J_0\left(1 + \frac{b^2}{y^2}\right). \tag{5}$$

According to (1) and (5), the current flowing across a small rectangular portion of S_1, located at P and having the edges dy and dz, is

$$di = J_n\, da = J_0\left(1 + \frac{b^2}{y^2}\right) dy\, dz. \tag{6}$$

Consequently,

$$i = \iint_{S_1} J_n\, da = \int_{-h}^{0}\!\int_{b}^{b+l} J_0\left(1 + \frac{b^2}{y^2}\right) dy\, dz, \tag{7}$$

and the final result is

$$i = J_0\, lh\left(1 + \frac{b}{b+l}\right). \tag{8}$$

In this example, J_n is independent of z, and hence i can be computed by a single rather than a double integration; to do this, we would subdivide S_1 into strips of width dy and depth h.

EXERCISES

1. The area of the surface S_1 in Fig. 55 is lh. Explain on intuitive grounds why the current (8) is larger than $J_0 lh$ and why it is nearly equal to $J_0 lh$ if l is much larger than b.

2. If l is much smaller than b, the current (8) is nearly equal to $2 J_0 lh$. Correlate this fact with a result of Exercise 4^{14}.

3. The surface S_1 in Fig. 55 is moved upward on the page until c lies at the distance s from the surface of the hole. Show that then

$$i = J_0 lh\left[1 + \frac{b^2}{(b+s)(b+l+s)}\right]. \tag{9}$$

Check this result for several simple limiting cases.

4. Show that the (net) current flowing across the surface S_2 in Fig. 56 is zero.

5. The point P lies on the surface S_3 in Fig. 56. Show that, at P, the normal component of J relative to S_3 is

$$J_n = 2J_0 b^3 \frac{x}{(x^2 + b^2)^2}, \tag{10}$$

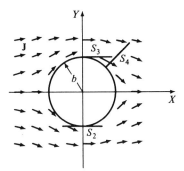

Fig. 56. The imagined surfaces S_2, S_3, and S_4 (see footnote 4 below).

where x is the x coordinate of P. Then show that the current flowing across S_3 is

$$i = J_0 lh \frac{bl}{b^2 + l^2}. \tag{11}$$

6. If l is much larger than b, the current (11) is nearly equal to $J_0 bh$. Explain on intuitive grounds.

7. Show that, in the case of the surface S_4 in Fig. 56, we have

$$J_n = \frac{1}{\sqrt{2}} J_0 \left(1 + \frac{b^2}{2x^2}\right) \tag{12}$$

and

$$i = J_0 lh \frac{2b + l}{\sqrt{2}\ (b + l)}. \tag{13}$$

Check Equation (13) for the limiting cases when l is very large or very small compared to b.

8. The surface S_5 in Fig. 57 is one-half of the surface of a circular cylinder of radius ρ_0 and depth h, coaxial with the hole; its straight edges lie in the yz

---

[4] In the exercises about the rectangular surfaces shown edge-on in Fig. 56, write l for width and h for depth, as was done in the text for S_1. S_2 is tangent to the surface of the hole, is symmetric with respect to the yz plane, and faces the bottom of the page. S_3 has one edge in the yz plane (at the surface of the hole) and faces the bottom of the page. S_4 faces toward the southeast and lies in a plane containing the z axis and making the angle of 45 deg with the x axis; one edge of S_4 lies at the surface of the hole.

plane. Show that at the point P we have $J_n = J_x \cos \phi + J_y \sin \phi$ and that, in view of (12^{14}),

$$J_n = J_0 \left(1 - \frac{b^2}{\rho_0^2}\right) \cos \phi. \tag{14}$$

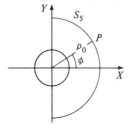

Figure 57

Then show that the current across S_5 is

$$i = 2J_0 \rho_0 h \left(1 - \frac{b^2}{\rho_0^2}\right). \tag{15}$$

9. Write $\rho_0 = b + l$ and show that (15) then reduces to twice the value of i given by (8). Explain the factor 2 on intuitive grounds.

CHAPTER 6

Vector Fields

From our first example of a vector field—the current density **J**—we now turn to vector fields in general. Section 16 is a collection of definitions, symbols, and formulas; we omit the proofs, which are much like those of Chapter 4. In §17 we tell how to picture vector fields by "field lines," such as "flux lines." The rest of this chapter describes by pictures and formulas a few simple types of vector fields that come up in many physical problems. Each type is introduced by remarks on a physical situation in which it can arise. These remarks, however, are not explanations but only hints, intended merely to help the reader to make some connection between our formulas and the physical fields that he has met in his earlier work. Each of these fields can be described in words or diagrams without referring to any coordinate frames. The introduction of a coordinate frame does, however, usually help in computations.

16. VECTOR FIELDS

A *vector field* is a quantity that has at every point in space the three properties discussed in §10, namely magnitude, direction in space, and the parallelogram rule for addition. If the magnitude of a vector field is zero at an isolated point or throughout a region, this field is said to vanish there.

We usually denote vector fields in general by **F** and their magnitudes by F. The vector pertaining to a point P is then denoted by $\mathbf{F}(P)$. The vector pertaining to a point whose cartesian coordinates are (x, y, z) is denoted by $\mathbf{F}(x, y, z)$ if the field is static and by $\mathbf{F}(x, y, z, t)$ if the field is time-dependent; the notation is similar in other coordinates.

The magnitude and the direction of a vector field **F** at a point P can be pictured by a pointer that has its tail at P, has the direction of **F** at P, and has a length which, in terms of a suitable scale unit, is equal to the numerical value of F at P. A *pointer map* of a vector field F is a diagram showing the pointers representing F at several points. Figure 52[14] is such a map.

The branch of mathematics called the *algebra of vector fields* is based on the following rule: Consider one point in space at a time and apply at this point the rules of the *algebra of individual vectors*, described in Chapter 4. For example, the equation

$$\mathbf{F} = \mathbf{G} + \mathbf{H} \qquad (1)$$

(where F, G, and H are vector fields) means that the pointer representing the field F at any point P is the diagonal of the parallelogram based on the pointers representing the fields G and H at P. The field F in (1) is called the "sum" or the "resultant" of the fields G and H. Conversely, G and H are a pair of "resolvents" or "vector components" or—for short and at the risk of an ambiguity—a pair of "components" of the field F.

The equation

$$\mathbf{G} = f\mathbf{F} \qquad (2)$$

(where G and F are vector fields and f is a scalar field) states the following: (a) at any point in space where f is positive, the direction of G is the same as that of F and the magnitude of G is f times that of F; (b) at any point where f is negative, the direction of G is opposite to that of F and the magnitude of G is $|f|$ times that of F; (c) at any point where $f = 0$, the field G vanishes.

The definition of the dot product $\mathbf{F} \cdot \mathbf{G}$ is

$$\mathbf{F} \cdot \mathbf{G} = FG \cos \gamma. \qquad (3)$$

In words: The dot product of two vector fields, say F and G, at any point P is the product of three purely numerical factors: the magnitude of F at P, the magnitude of G at P, and the cosine of the angle γ between F and G at P. This product does not depend on any coordinate frame, and, therefore, *the dot product of two invariant vector fields is itself an invariant scalar field.*

In the light of §10, it is perhaps obvious that addition of vector fields is associative and dot multiplication is both commutative and distributive.

Cartesian components. If a cartesian coordinate frame is given, we may construct at any point P the vectors $\mathbf{1}_x$, $\mathbf{1}_y$, and $\mathbf{1}_z$, oriented as in Fig. 58 and called the *cartesian base-vectors* pertaining to the given frame. Any vector F can then be written as

$$\mathbf{F} = F_x \mathbf{1}_x + F_y \mathbf{1}_y + F_z \mathbf{1}_z, \qquad (4)$$

where F_x, F_y, and F_z are numerical coefficients, called the *cartesian components*

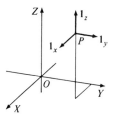

Fig. 58. The cartesian base-vectors at a point P.

of **F** relative to the given frame. In general, each component is a numerical function of each of the three coordinates (x, y, z) of P; that is, in general,

$$F_x = F_x(x, y, z), \quad F_y = F_y(x, y, z), \quad F_z = F_z(x, y, z). \tag{5}$$

The F's in (5) may depend on the choice of the coordinate frame (recall Exercise 7^{14}); consequently, they are scalars but *not* invariant scalars. Equations (2^{14}) through (4^{14}) are examples of explicit equations of type (5).

In terms of cartesian components, the magnitude or "strength" F of a vector field **F** at a point P is

$$F = \sqrt{F_x^2 + F_y^2 + F_z^2}. \tag{6}$$

In general, F varies from point to point, so

$$F = F(x, y, z). \tag{7}$$

However, as illustrated in Exercise 7^{14}, the value of F does not depend on the choice of the coordinate frame, and consequently *the magnitude of an invariant vector field is itself an invariant scalar field.* For example, it is perhaps obvious that the magnitude of the current density at any point in Fig. 52^{14} does not depend on any coordinate frame that might be added to that figure.

A geometric equation connecting vector fields can be replaced by three numerical equations connecting the cartesian components of these fields. For example, if **F** is the sum of the fields **G** and **H**, so that the geometric equation connecting these fields is

$$\mathbf{F} = \mathbf{G} + \mathbf{H}, \tag{8}$$

then the numerical equations

$$F_x = G_x + H_x, \quad F_y = G_y + H_y, \quad F_z = G_z + H_z \tag{9}$$

hold at every point. Again, if

$$\mathbf{G} = f\mathbf{F}, \tag{10}$$

where f is a scalar function of position, then

92   Vector Fields   § 16

$$G_x = fF_x, \quad G_y = fF_y, \quad G_z = fF_z \tag{11}$$

at every point.

The cartesian component formula for the dot product of two vector fields, namely,

$$\mathbf{F} \cdot \mathbf{G} = F_x G_x + F_y G_y + F_z G_z, \tag{12}$$

follows from (20¹¹).

Let $f(P)$ and $f(Q)$ be the values of an algebraic function f at the neighboring points $P(x, y, z)$ and $Q(x + dx, y + dy, z + dz)$. In view of (3⁷) and (5⁷) we then have

$$f(Q) = f(P) + \frac{\partial f}{\partial x} dx + \frac{\partial f}{\partial y} dy + \frac{\partial f}{\partial z} dz + \text{higher-order terms.} \tag{13}$$

Applying (13) to the functions F_x, F_y, and F_z, we get

$$F_x(Q) = F_x(P) + \frac{\partial F_x}{\partial x} dx + \frac{\partial F_x}{\partial y} dy + \frac{\partial F_x}{\partial z} dz + \ldots, \tag{14}$$

$$F_y(Q) = F_y(P) + \frac{\partial F_y}{\partial x} dx + \frac{\partial F_y}{\partial y} dy + \frac{\partial F_y}{\partial z} dz + \ldots, \tag{15}$$

$$F_z(Q) = F_z(P) + \frac{\partial F_z}{\partial x} dx + \frac{\partial F_z}{\partial y} dy + \frac{\partial F_z}{\partial z} dz + \ldots. \tag{16}$$

Here $F_x(P)$ is the value of F_x at the point P, $F_x(Q)$ is the value of F_x at the point Q, and so on. The partial derivatives are to be evaluated at P. The dots denote terms of higher order.

Tangential and normal components.   Figure 59 shows a point P on a path ab, the tangential unit vector $\mathbf{1}_t$ erected at P in the direction of the path,

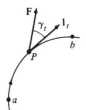

Fig. 59. Diagram leading to the equation $F_t = F \cos \gamma_t$.

the pointer $\mathbf{F}$ representing an arbitrary vector field $\mathbf{F}$ at P, and the angle γ_t between $\mathbf{F}$ and $\mathbf{1}_t$. The product $F \cos \gamma_t$ is denoted by F_t and is called the *tangential component* or the *forward component* of the field $\mathbf{F}$ at P relative to the path ab; that is,

(4¹²) $$F_t = F \cos \gamma_t \qquad (17)$$

or, which is the same thing,

(5¹²) $$F_t = \mathbf{F} \cdot \mathbf{1}_t. \qquad (18)$$

In general, both the direction and the magnitude of a vector field change from point to point on the path and, if the path is not straight, so does the direction of the vector $\mathbf{1}_t$. Consequently, the value of F_t usually changes along the path.

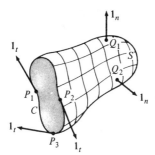

Figure 60

(For example, if the path is the contour C in Fig. 60 and if the field $\mathbf{F}$ should be directed upward on the page, the values of F_t at the points P_1 and P_2 would have opposite signs.)

Let P lie on a signed surface S, let $\mathbf{1}_n$ be the unit vector erected at P and pointing along the positive normal to S at P, and let γ_n be the angle between $\mathbf{1}_n$ and the pointer representing a vector field $\mathbf{F}$ at P. The product $F \cos \gamma_n$ is then denoted by F_n and called the *normal component* of $\mathbf{F}$ at P relative to S; that is,

(2¹²) $$F_n = F \cos \gamma_n \qquad (19)$$

or

(5¹²) $$F_n = \mathbf{F} \cdot \mathbf{1}_n. \qquad (20)$$

The value of the normal component of a given vector field relative to a given surface S usually varies from point to point on S. (For example, if S is the surface pictured in Fig. 60 and $\mathbf{F}$ should be directed upward on the page, then the values of F_n at the points Q_1 and Q_2 would have opposite signs.)

We reserve the symbols F_τ and F_ν for the tangential component relative to a surface and the normal component relative to a path.

EXERCISES

1. Find the locus of the points in the xy plane at which the vectors

$$(x - y)\mathbf{1}_x + (x + y)\mathbf{1}_y \quad \text{and} \quad (x + y)\mathbf{1}_x + (x - y)\mathbf{1}_y$$

   are mutually perpendicular.

2. A large current-carrying metal block contains a small *spherical cavity* (say an air bubble) of radius r_0, centered at the origin O of a cartesian coordinate frame. Far from the cavity, the direction of the current density is practically the $+z$ direction, and its magnitude is practically a constant, say J_0 amperes per square meter. The point $P(x, y, z)$ lies outside the cavity; its distance from O is r. As we will see in Chapter 16, the components of $\mathbf{J}$ at P then prove to be

$$J_x = -\frac{3}{2} J_0 r_0^3 \frac{xz}{r^5}, \tag{21}$$

$$J_y = -\frac{3}{2} J_0 r_0^3 \frac{yz}{r^5}, \tag{22}$$

$$J_z = J_0 - \frac{1}{2} J_0 r_0^3 \frac{(2z^2 - x^2 - y^2)}{r^5}. \tag{23}$$

   List a few features that you would intuitively expect the current density to have in this case and verify that Equations (21) to (23) include them. (Example: At all points in the xy plane outside the cavity, the charge should flow in the $+z$ direction.)

3. The "equator" of the cavity of Exercise 2 is defined by the equations $x^2 + y^2 = r_0^2$ and $z = 0$. Show that, at points on the equator $J_z = 1.5 J_0$. Explain on intuitive grounds why this value of J_z is larger than J_0 but smaller than the value of J_x found in Exercise 4[14].

4. Show that Equations (21) to (23) are consistent with (10^{14}) and (11^{14}).

5. In Exercise 2, why don't the oncoming charges run into the surface of the cavity? Or do they?

6. Show that

$$F_x = \mathbf{F} \cdot \mathbf{1}_x, \quad F_y = \mathbf{F} \cdot \mathbf{1}_y, \quad F_z = \mathbf{F} \cdot \mathbf{1}_z, \tag{24}$$

   and that any vector field, say $\mathbf{G}$, can be written as

$$\mathbf{G} = (\mathbf{G} \cdot \mathbf{1}_x)\mathbf{1}_x + (\mathbf{G} \cdot \mathbf{1}_y)\mathbf{1}_y + (\mathbf{G} \cdot \mathbf{1}_z)\mathbf{1}_z. \tag{25}$$

17. FIELD LINES, FLUX LINES, AND OTHER LINES

The static electric field of an electric dipole can be pictured as in Fig. 61. The imagined "field lines" begin on positive and end on negative charges; the direc-

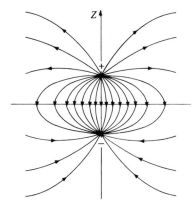

Fig. 61. The electric field of a dipole.

tion of a line at any point P is the same as the direction of the electric field at P, and the density of the lines at P is presumably equal to the strength of the electric field at P. Here "density" means the number of field lines per square meter that cross (go through) an imagined small flat surface placed at P and facing in the direction of the electric field at P.

Figure 61 is an example of a *line map*. Figure 62, which is equivalent to the pointer map of Fig. 52[14], is a line map of the current density **J** near a cylindrical hole in a conductor; the equations of the lines in Fig. 62 are derived below.

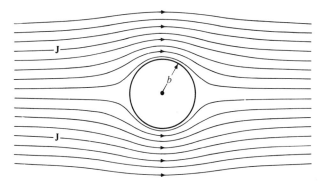

Fig. 62. A line map of the current density pictured in Fig. 52[14] by pointers.

95

Our discussion of field lines will be vague here and there. Some of the vagueness should wear off as we work out examples, but some of it is inherent in the very concept of field lines. We will, nevertheless, discuss this concept at some length, for several reasons. It provides vivid pictures of physical vector fields and should enable the reader, as it enabled Faraday, to grasp intuitively some properties of these fields. It will also lead us, as it led Maxwell, to mathematical concepts that are free from any vagueness and are essential for an understanding of electromagnetic waves. The results of experiments on the quantization of magnetic flux fit in well with the concept of field lines.[1] Some day this concept may help in the understanding of why electric charge is quantized.[2]

We call the field lines of specific vector fields—say **F**, **G**, **H**, and so on—"lines of **F**," "lines of **G**," and so forth. The rules for drawing or imagining a line map of a given physical vector field, say **F**, are

(a) The direction of a line of **F** at any point P on this line should be the direction of the field **F** at P.
(b) The density of the lines of **F** at any point P should be equal, as nearly as one can make it, to the magnitude F of **F** at P.

These rules are not altogether precise, so line maps of the same vector field, drawn by different persons, need not agree in all detail. In practice, line density is made not equal to but proportional to F.

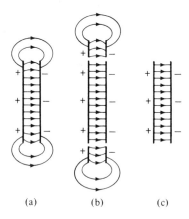

Fig. 63. The electric field of a charged parallel-plate capacitor with and without guard rings; also, an idealized map of this field.

[1] B. S. Deaver and W. M. Fairbank, "Experimental Evidence for Quantized Flux in Superconducting Cylinders," *Physical Review Letters*, Vol. 7, 1961, p. 43. R. Doll and M. Näbauer, "Experimental Proof of Magnetic Flux Quantization in a Superconducting Ring," *ibid.*, p. 51.

[2] P. A. M. Dirac, "The Physicist's Picture of Nature," *Scientific American*, Vol. 208, 1963, p. 45.

A physical vector field has a unique direction at every point where it does not vanish. Consequently, the rule (a) implies that the lines of a physical vector field do not intersect one another.

The lines in Fig. 61 describe the electric field produced jointly by two charges; they are *curved*. By contrast, a map of the field produced by either charge alone would consist of *straight* radial lines.

The field of a charged parallel-plate capacitor, pictured in Fig. 63(a), is the sum of all the fields produced individually by each "excess" proton located on one plate and each "excess" electron on the other plate.[3] The field lines in that figure describe the total electric field produced by all these charges taken together—we must stress the fact that the individual field lines do not represent the individual fields contributed to the total field by individual protons and electrons. The fringing of the field near the edges of a capacitor can be reduced by a "guard ring," as in Fig. 63(b). (The guard ring and the main capacitor can be charged and discharged by separate circuits.) To simplify our diagrams, we omit guard rings, and picture the electric field of a charged capacitor as in Fig. 63(c).

We write N for what is called "the net number of field lines crossing a signed surface S." To evaluate N, we count the number of points on S where the lines cross S from back to front, count the number of points on S where lines cross S from front to back, and subtract the second number from the first. To illustrate, consider the signed bowl-shaped surface and the three field lines shown

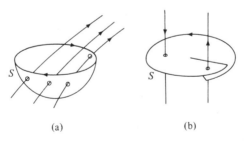

Fig. 64. Field lines crossing a bowl-shaped and a ramp-shaped surface.

in Fig. 64(a); in this case we have one back-to-front crossing and three front-to-back crossings, so that $N = 1 - 3 = -2$. Another example is shown in Fig. 64(b), where the signed surface S resembles a circular ramp. The field line at the left crosses S from front to back; the line at the right crosses S from back to front *twice*. Consequently, $N = -1 + 2 = 1$.

Let S be a small flat surface having the area da, located at a point P, and facing at the angle γ_n to the direction of the vector field $\mathbf{F}$ at P (see Fig. 65,

---

[3] More precisely, this field is the resultant of the fields of each proton and each electron located in or on the plates. But it is, in fact, determined by the "excess" charges, because the resultant of the fields of the other charges is zero.

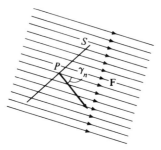

Fig. 65. Diagram leading to the formal equation $dN = (dN)_{max} \cos \gamma_n$ for the number of field lines crossing a small flat surface.

where **F** is parallel and S is perpendicular to the page). Let dN be the net number of lines of **F** crossing S. Depending on the orientation of S, this number will be positive, negative, or zero. It is perhaps obvious that dN has its maximum value, say $(dN)_{max}$, when S faces in the direction of **F**, while for other orientations we may write, at least formally,

$$dN = (dN)_{max} \cos \gamma_n. \tag{1}$$

The ratio $(dN)_{max}/da$ is, by definition, the average density of the field lines near P, and the limit of this ratio when $da \to 0$ is the line density at P. The rule (b) stated above can, therefore, be written as $(dN)_{max}/da = F$, or as

$$(dN)_{max} = F \, da, \tag{2}$$

with the understanding that $da \to 0$. In view of (1) we then have the formal relation $dN = F \cos \gamma_n \, da$, which holds for any orientation of S and can be written as

$$dN = F_n \, da. \tag{3}$$

An equation such as (1) is, of course, formal rather than rigorous. For example, let 11 field lines cross S when it faces in the direction of **F**, so that $(dN)_{max} = 11$. If S is then turned so as to face at 60 deg with **F**, Equation (1) will tell us that S will be crossed by $5\frac{1}{2}$ lines—an obvious impossibility. A way out of this complication is to cut in half the unit used for measuring F. The numerical value of F will then be doubled, the density of lines of **F** will be doubled, the value of $(dN)_{max}$ will become 22 instead of 11, and the value of $(dN)_{max} \cos 60°$ will become 11 instead of $5\frac{1}{2}$. In fact, if we express F in terms of smaller and smaller units of measurement, we can make (1), (2), and (3) as nearly exact as we please. But in practice it is more convenient to keep the standard units and to extend the terminology so as to provide for "fractions" of field lines.

Flux lines and lines of force. Some of the physical vector fields discussed in this book are the conduction current density **J**, the "electric" fields

E and **D**, and the "magnetic" fields **B** and **H**. The lines of **J**, **D**, and **B** are often called *flux lines*, while the lines of **E** and **H** are called *lines of force*. But this terminology is not altogether standard, and we will usually say *field lines* for any of these lines. Flux lines and lines of force satisfy in principle the rules (a) and (b) given above.

Direction lines. We will occasionally use lines called *direction lines*; they are imagined lines that satisfy rule (a) about directions, but do not necessarily satisfy rule (b) about density. According to (a), the direction of a direction line of a vector field at a point P is the same as the direction of a flux line or a line of force at P.

If direction lines start or stop (as in Fig. 61), then flux lines also start or stop. By contrast, if direction lines are continuous, then there are two possibilities: The flux lines may be continuous (as in Fig. 62), or they may start and stop (as in parts of Fig. 91[32]). The significance of these statements should become clearer as we study specific examples.

Equations of field lines. To illustrate the computation of field lines, we consider the flow of charge past a cylindrical hole. The direction of a line of **J** at a point P is, by definition, the direction of the field **J** at P. Consequently, if a field map should be superimposed upon the pointer map of Fig. 54[14], the slope dy/dx of the line at a point P would satisfy the equation

$$\frac{dy}{dx} = \frac{J_y}{J_x}, \tag{4}$$

where J_x and J_y are cartesian components of **J**. In the present example, the simplest way of solving the differential equation (4) is to go over to the cylindrical coordinates ρ and ϕ. Since $x = \rho \cos \phi$ and $y = \rho \sin \phi$, the left-hand side of (4) becomes

$$\frac{dy}{dx} = \frac{dy/d\phi}{dx/d\phi} = \frac{(d\rho/d\phi) \sin \phi + \rho \cos \phi}{(d\rho/d\phi) \cos \phi - \rho \sin \phi}. \tag{5}$$

Using (12[14]) on the right-hand side of (4), we then get the equation

$$\frac{(d\rho/d\phi) \sin \phi + \rho \cos \phi}{(d\rho/d\phi) \cos \phi - \rho \sin \phi} = -\frac{b^2 \sin 2\phi}{\rho^2 - b^2 \cos 2\phi}, \tag{6}$$

which reduces to

$$\frac{(\rho^2 + b^2)}{\rho(\rho^2 - b^2)} d\rho + \frac{\cos \phi}{\sin \phi} d\phi = 0, \tag{7}$$

and integrates to

$$\ln \frac{\rho^2 - b^2}{\rho} + \ln \sin \phi = \ln c, \tag{8}$$

where c is an arbitrary constant. This result can be written as

$$\left(\rho - \frac{b^2}{\rho}\right) \sin \phi = c. \tag{9}$$

The field lines in Fig. 62 were computed from (9) for the following values of the parameter c:

$$c = \pm \tfrac{1}{8}b, \pm \tfrac{3}{8}b, \pm \tfrac{5}{8}b, \ldots. \tag{10}$$

Note that, sufficiently far from the hole, (9) reduces to $\rho \sin \phi = c$, and hence to $y = c$.

The formulas for J_x and J_y, taken for granted in our proof of (9), will be derived in §69.

Field lines and units of measurement. In most physical problems, field lines are associated with units of measurement. According to the rule (b) for picturing field lines, we have, at least in principle,

$$\frac{\text{number of lines of } \mathbf{F}}{\text{meter}^2} = F, \tag{11}$$

where F is the magnitude of $\mathbf{F}$. Therefore, the unit of measurement associated with a line of $\mathbf{F}$ is

$$(\text{unit adopted for } F) \cdot (\text{meter}^2). \tag{12}$$

Thus, if our unit for the magnitude J of current density is the ampere/meter^2, the unit to be associated with lines of $\mathbf{J}$ is the ampere. This subject will come up in §25, concerned with graphical estimation of flux.

EXERCISE

1. A closed surface (S) lies in Fig. 62 in the copper. Use (2^{14}) and (3^{14}) to verify in detail that, since the state is steady, the integral of J_n, taken over (S), is zero.

18. UNIFORM VECTOR FIELDS

A vector field is said to be *uniform* in a region R if it has the same direction and the same magnitude at every point in R; if a uniform field does not change with time and this fact needs emphasis, we call the field *uniform* and *constant*,

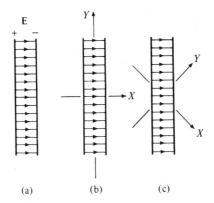

Fig. 66. A uniform vector field. In terms of the coordinate frame shown in part (b), the formula for this field is $\mathbf{E} = E_0 \mathbf{1}_x$.

or *uniform* and *steady*. Strictly uniform fields are abstractions, but fields that are nearly uniform and can be treated as uniform in limited regions arise in practice quite often. One example is the electric field inside a charged parallel-plate capacitor equipped with guard rings. If the capacitor is charged and oriented as in Fig. 66(a), the electric intensity **E** is uniform between the plates and directed from left to right. This **E** does not depend on the coordinate frame that we might choose for describing **E** in formulas; in other words, this **E** is an *invariant* vector field.

If we choose the frame XY shown in Fig. 66(b) and write E_0 for the constant magnitude of **E**, the formula for **E** at any point P between the plates becomes

$$\mathbf{E} = E_0 \mathbf{1}_x. \tag{1}$$

The value of E_0 depends on the density of charges on the plates.

The simplest mathematical field of the form (1) is

$$\mathbf{F} = \mathbf{1}_x. \tag{2}$$

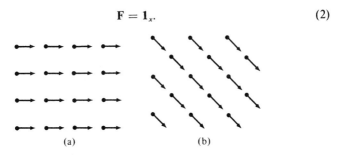

Fig. 67. Sets of base-vectors $\mathbf{1}_x$ belonging, respectively, to the coordinate frames shown in parts (b) and (c) of Fig. 66.

In this case $F_x = 1, F_y = 0, F_z = 0$, and $F = 1$. We call (2) "the field $\mathbf{1}_x$," or "the field $\mathbf{i}$." A pointer map of this field is given in Fig. 67 twice: in part (a) for the upright coordinate frame of Fig. 66(b) and in part (b) for the slanting frame of Fig. 66(c). It is apparent that the field $\mathbf{1}_x$ is *not* an invariant field, because it changes orientation when the coordinate frame is reoriented.

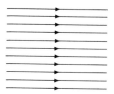

Fig. 68. A line map of the field $\mathbf{1}_x$ belonging to the coordinate frame shown in Fig. 66(b).

A line map of $\mathbf{1}_x$ is shown in Fig. 68 for the coordinate frame described in the footnote.[4] Similar layers of lines should be imagined above and below the page. In fact, since the magnitude of this field is everywhere unity, one field line must be imagined per every square meter of any plane perpendicular to the direction of this field. Except for the limits imposed by the size of the page, the lines of the field $\mathbf{1}_x$ have neither beginnings nor ends.

The most general formula for a uniform vector field referred to a cartesian coordinate frame is

$$\mathbf{F} = a\mathbf{1}_x + b\mathbf{1}_y + c\mathbf{1}_z, \qquad (3)$$

where a, b, and c are constants. The traditional way of writing (3) is

$$\mathbf{F} = a\mathbf{i} + b\mathbf{j} + c\mathbf{k}. \qquad (4)$$

EXERCISES

1. The pointers in Fig. 67 are spaced regularly. Is this necessary? Is this advisable? Explain. The lines in Fig. 68 are spaced regularly. Is this necessary? Is this advisable? Explain.

2. In terms of the coordinate frame shown in Fig. 66(b), the formula for the field **E** in the capacitor is (1). Express the same **E** in terms of the frame shown in Fig. 66(c).

3. Let $a = 1$, $b = 2$, and $c = 0$, and verify that, in our usual frame,[4] a map of the field (3) then has the form indicated in Fig. 69. Sketch line maps for the following cases: $a = 1, b = 1, c = 0$; $a = 1, b = -1, c = 0$; $a = -1, b = 2, c = 0$; and $a = 0, b = 1, c = 1$.

4. Show that (3) is the most general cartesian formula for a uniform vector field.

---

[4] The page is the xy plane or a plane parallel to it; the x axis (not shown) runs from left to right.

Figure 69

19. THE FIELD $x1_x$

To lead up to another type of vector field, we consider a positively charged *metal* plate far removed from other things. After the currents in the plate have subsided, the electric field inside the plate vanishes, the charges distribute

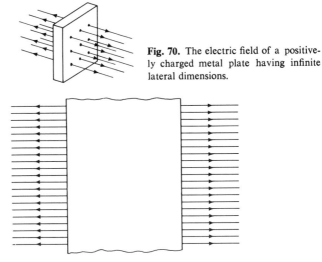

Fig. 70. The electric field of a positively charged metal plate having infinite lateral dimensions.

Fig. 71. Another view of the field shown in Fig. 70.

themselves evenly on the two faces of the plate (except near the edges), and the electric field produced jointly by all these charges is normal to the plate outside and near the plate (except near the edges). Line maps of this field are shown in Figs. 70 and 71 for the idealized case of a plate whose lateral dimensions are infinite. The lines begin on the faces of the plate because positive charges are located there. No lines are drawn inside the plate because the electric field vanishes there. Outside the plate the electric field is normal to the plate, and so are the field lines.

Next consider a *nonconducting* plate and suppose that positive charges are uniformly distributed *inside* it, as though the plate is made of compressed uniformly charged sawdust. The field lines, which begin on positive charges, will now begin inside the plate, as in Fig. 72, where the beginning points of the lines are distributed more or less uniformly. The electric field now vanishes on the midplane of the plate, but as we go from the midplane toward a face, the field strength increases, as implied in the figure by the increasing density of

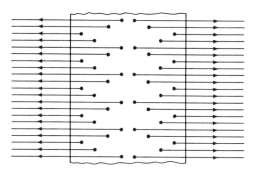

Fig. 72. The electric field of a nonconducting plate, charged uniformly in its interior. This field has the form $cx\mathbf{1}_x$ inside the plate and is uniform outside.

the lines. At the right of the plate the field is uniform; at the left it is also uniform but has the opposite direction. The particular straight lines beginning on particular charges are not intended to describe the individual fields produced by these particular charges; instead, the totality of lines describes the total field produced by all the charges taken together. The reader should imagine similar layers of lines above and below the page. We will see in Chapter 14 that the lines shown in Fig. 72 are called " lines of **D**," but for the present their name is not important.

If the midplane of the plate is the yz plane, the formula for the field pictured in Fig. 72 *inside the plate* proves to be

$$\mathbf{F} = x\mathbf{1}_x, \qquad (1)$$

except for a numerical coefficient that depends on the charge density in the plate. The cartesian components of (1) at a point $P(x, y, z)$ are $F_x = x$, $F_y = 0$, and $F_z = 0$. The magnitude of (1) at P is $|x|$ and is equal to the distance from P to the yz plane. The direction of (1) at P, which is determined by the sign of the x coordinate of P, is away from the yz plane, unless x is zero.

The middle part of Fig. 72 was drawn as follows. On the planes which are perpendicular to the page and on which $x = 0, 1, 2,$ and 3, the magnitude of the field $x\mathbf{1}_x$ is 0, 1, 2, and 3, respectively. Therefore, these planes should be crossed, respectively, by 0, 1, 2, and 3 field lines per square meter. Consequently, we can get the correct line density on these planes, and also the correct average line density at points between them, if we start the lines midway between these planes, where $x = \frac{1}{2}, \frac{3}{2}, \frac{5}{2}$, and so on. The first step in sketching the right-hand half of Fig. 72 was to draw several equally spaced horizontal lines, all starting at the same distance from the yz plane. A similar set was then started between these lines three times farther from the yz plane, another set five times farther, and so on. The other half of the map was drawn by symmetry.

In this book, we call the beginning point of a line of a vector field $\mathbf{F}$ a *source* of $\mathbf{F}$ and denote it by a small full circle; the end point of a line we call a *sink* of $\mathbf{F}$ and denote it by a small empty circle.

If a vector field has the form $x\mathbf{1}_x$ throughout a region R, one can picture this field by imagining that R contains only sources of this field and no sinks, and that the sources are distributed in R with the uniform density of one source per cubic meter.

EXERCISES

1. Sketch a pointer map of the field $x\mathbf{1}_x$.
2. Assume that the field inside the plate of Fig. 72 is $\mathbf{F} = cx\mathbf{1}_x$, where c is a constant; write b for the thickness of the plate, and verify that then the formula for the entire field shown in the figure is

$$\mathbf{F} = \begin{cases} -\frac{1}{2}bc\mathbf{1}_x & \text{if } x < -\frac{1}{2}b \\ cx\mathbf{1}_x & \text{if } -\frac{1}{2}b \leq x \leq \frac{1}{2}b \\ \frac{1}{2}bc\mathbf{1}_x & \text{if } x > \frac{1}{2}b. \end{cases} \quad (2)$$

3. Sketch pointer maps and line maps of the fields $y\mathbf{1}_y$ and $x^2\mathbf{1}_x$; use the coordinate frame of the footnote on page 102.

20. CIRCULAR VECTOR FIELDS

For our next example we turn to magnetostatics. The axis of a long straight copper rod runs along the z axis of a cartesian frame; the cross section of the rod is circular and has the radius ρ_0. The rod carries a steady current in the $+z$ direction, the current density is uniform throughout the rod, and the return portion of the circuit is an arc of such a large radius that its magnetic effects can be ignored. We will see in Chapter 20 that the magnetic field intensity $\mathbf{H}$ produced by the current flowing in the rod is a circular vector field whose sense is related by the right-hand rule to the direction of the current and whose magnitude H at a point P depends on the distance ρ from P to the z axis. More precisely, H is directly proportional to ρ if P lies inside the rod, and inversely proportional to ρ if P lies outside; that is,

$$H = \begin{cases} \dfrac{\rho}{\rho_0} H_0 & \text{if } \rho \leqslant \rho_0 \\ \dfrac{\rho_0}{\rho} H_0 & \text{if } \rho \geqslant \rho_0. \end{cases} \quad (1)$$

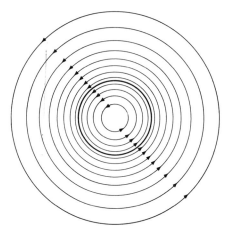

Fig. 73. A circular magnetic field whose magnitude is proportional to ρ inside and to $1/\rho$ outside the circle shown in the figure.

Here H_0 is the value of H at the surface of the rod. A line map of **H** is shown in Fig. 73, where the thick line is the outline of the cross section of the rod; in this figure the current flows in the rod toward the reader, who should imagine layers of similar field lines above and below the page. These lines have neither beginnings nor ends. The magnitude H of **H** is plotted in Fig. 74.

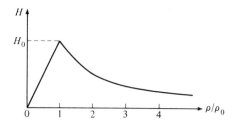

Fig. 74. The magnitude of the magnetic field of Fig. 73; ρ_0 is the radius of the circle shown in that figure.

The magnetic field inside the wire is one example of a two-dimensional circular vector field whose magnitude depends on the cylindrical coordinate ρ, but not on z or ϕ; the magnetic field outside the wire is another example.

Equation (1) specifies the dependence of H on ρ, but does not give the direction of **H**, which we described above in words. However, both the magnitude and direction of **H** can be described by the single formula (2^{35}).

21. THE FIELD $y\mathbf{1}_x$

If the cross section of the copper rod is oval rather than circular, the lines of **H** are also oval, both inside and outside the rod; their shapes inside the rod are suggested in Fig. 75. In each part of this figure the current density is uniform throughout the rod, as in Fig. 73, but the current flows *into* the page.

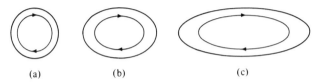

Fig. 75. Directions of magnetic fields *inside* metal rods having circular and oval cross sections and carrying current *into* the page.

Now suppose that the series of oval cross sections begun in Fig. 75 is continued, and the ovals keep the same width but get longer and longer In the limit, the rod then becomes a flat plate of finite thickness but infinite lateral dimensions. If we assume uniform current density throughout this plate, the line map of the field **H** inside the plate takes the form shown in some detail in

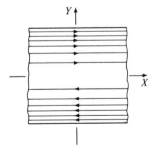

Fig. 76. A magnetic field of the form $cy\mathbf{1}_x$ inside a metal plate carrying a uniformly distributed current *into* the page. Similar sets of lines lie in planes parallel to the xy plane.

Fig. 76. In terms of the coordinates drawn in this figure, the field **H** inside the plate proves to be a numerical multiple of the field

$$\mathbf{F} = y\mathbf{1}_x. \tag{1}$$

More explicitly,

$$\mathbf{H} = ky\mathbf{1}_x, \tag{2}$$

where k is a positive constant. In words: The magnitude of **H** at a point P inside the plate is proportional to the distance from P to the midplane; furthermore, **H** points in the positive x direction if the y coordinate of P is positive, and reverses this direction for negative y's. The spacing of the field lines in Fig. 76 is discussed in §23.

We will see in §88 that, outside the (infinite) plate, the *direction* of **H** depends on y in the same way as inside the plate, but the *magnitude* of **H** is a constant, independent of y.

EXERCISE

1. Sketch a pointer map of the field $y\mathbf{1}_x$.

CHAPTER 7

The Flux Integral

In §22 we introduce an important surface integral, called the *flux integral*. In §23 this integral is interpreted pictorially in terms of field lines. Approximate evaluation of flux integrals by counting field lines is illustrated in §25.

22. FLUX ACROSS A SURFACE

The normal component of a vector field **F** relative to a signed surface S is

$$(19^{16}) \qquad F_n = F \cos \gamma_n. \qquad (1)$$

Consider a small element of S (area da) and compute F_n at a point on this element. The product

$$F_n \, da \qquad (2)$$

is called the *flux* of the field **F** across (through) the surface element, and the integral

$$\iint_S F_n \, da \qquad (3)$$

is called the *flux integral*, the *net flux*, or simply the *flux* of the field **F** across (through) the surface S. To summarize:

$$\text{net flux of } \mathbf{F} \text{ across } S = \iint_S F_n \, da, \qquad (4)$$

or, more briefly,

$$\text{flux} = \iint_S F_n \, da. \tag{5}$$

If the signature of S should be reversed, the numerical values of F_n everywhere on S will reverse their signs, and so will the numerical value of the flux.

The flux integral is an ordinary number (positive, negative, or zero), and hence flux is a *scalar*. Since flux pertains to a surface rather than a point, it is *not* a field quantity. If the flux of a vector field across some portions of S is positive and across the rest of S is negative, the term "net flux" may be preferable to "flux."

The current i flowing across a surface S is

$$(2^{15}) \qquad\qquad i = \iint_S J_n \, da. \tag{6}$$

Therefore, in view of (5), *the current across a surface is the flux of current density across this surface.*

The flux of the current density $\mathbf{J}$ is denoted by i or I; the flux of the magnetic induction $\mathbf{B}$ is denoted by ϕ or Φ, and sometimes by Φ_B. If the flux of a vector field has a simple name, the field itself is sometimes called the "density" of its flux. Thus the field $\mathbf{J}$ (whose flux is called "current") is called "current density," and the field $\mathbf{B}$ (whose flux is called "magnetic flux") is often called "magnetic flux density" rather than "magnetic induction," which was Faraday's term. Literally, the term "flux" means "flow," but its technical meaning is given by (4), whether or not anything is flowing.

The definition (4) applies to both closed and nonclosed surfaces, but if S is closed, the term "net outward flux" may be preferable to "net flux." Therefore, when dealing specifically with closed surfaces, we may write (4) as

$$\text{net outward flux of } \mathbf{F} \text{ across } (S) = \iint_{(S)} F_n \, da. \tag{7}$$

EXERCISES

1. Recall the elementary formula $\phi = BA$ for the magnetic flux pertaining to an area A and correlate it with (4). To begin with, define one of the two magnetic quantities, B or ϕ, without referring to the other. (Do not confuse the act of defining a physical quantity with the act of giving it a name.)

2. The unit vector $\mathbf{1}_n$ points in the direction of the positive normal to a small surface element whose area is da; show that (5) can be written as

$$\text{flux} = \iint_S \mathbf{F} \cdot \mathbf{1}_n \, da. \tag{8}$$

Let the vector $d\mathbf{a}$ point in the direction of $\mathbf{1}_n$ and have the magnitude da. Show that (5) can be written as

$$\text{flux} = \iint_S \mathbf{F} \cdot d\mathbf{a} \tag{9}$$

23. FLUX AND FLUX LINES

The formal expression for the number of lines of a vector field **F** that cross a small, flat, signed surface of area da is

$$(3^{17}) \qquad dN = F_n\, da. \tag{1}$$

Let us compute the net number of field lines, say N, that cross a signed surface S of any shape and size. To do this, we subdivide S into small elements, apply (1) to each element, and compute the totals of the two sides of (1). The total on the left is the sum of the dN's, namely N. The total on the right is the integral of F_n over the entire surface S. Consequently, the formal result is

$$N = \iint_S F_n\, da. \tag{2}$$

The qualifying remarks made in §17 about (1) apply equally well to (2): By using smaller and smaller units of measurement for F, the formula (2) can be made as nearly exact as we please, but in practice it is more convenient to keep the standard units and, if necessary, to speak of fractions of field lines. This recourse to tricks of terminology emphasizes the fact that the right-hand side of (2) is a well-defined mathematical quantity, while its left-hand side, namely N, has to do with pictures that are necessarily vague. In view of (2) and (4^{22}),

$$\left.\begin{array}{l}\text{net number of lines of }\mathbf{F}\\ \text{crossing a surface }S\end{array}\right\} = \text{net flux of }\mathbf{F}\text{ across }S. \tag{3}$$

For this reason the field lines of vector fields whose fluxes are of physical interest are often called *flux lines*. The formal equation (3) is useful both in making line maps and, once a map has been made, in evaluating flux integrals by counting field lines.

Example. We will now describe how to draw a line map of the field

$$(1^{21}) \qquad \mathbf{F} = y\mathbf{1}_x. \tag{4}$$

At a point $P(x, y, z)$, the density of the lines should be equal to the magnitude of the field at P, which is $|y|$. Our problem is to find the spacing of the lines that would give the correct line density, at least on the average.

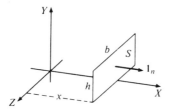

Figure 77

Let us compute the flux of (4) across the surface S (height h, width b) shown in Fig. 77 and facing toward the right. At any point on S we have $F_n = F_x$, so in this example $F_n = y$ and

$$\text{flux} = \iint_S F_n \, da = \int_{-\frac{1}{2}b}^{\frac{1}{2}b} \int_0^h y \, dy \, dz = \tfrac{1}{2} b h^2. \tag{5}$$

Therefore, according to (3), S should be crossed by $\tfrac{1}{2}bh^2$ lines. To determine their relative spacing, we set $b = 1$ meter. We then have $N = \tfrac{1}{2}h^2$, so 0, 1, 2, and 3 lines are required for h equal to $\sqrt{0}$, $\sqrt{2}$, $\sqrt{4}$, and $\sqrt{6}$ meters. These conditions can be met in the xy plane, on the average, by drawing the lines at distances from the zx plane proportional to $\sqrt{1}$, $\sqrt{3}$, $\sqrt{5}$, and $\sqrt{7}$. Figure 76[21] was scaled in this way.

EXERCISES

1. Sketch a line map of the field $y^2 \mathbf{1}_x$ after computing the relative spacing of the lines.

2. In the text we introduced the concept of field lines in statements (a) and (b) of §17 and then "derived" Equation (2) of the present section. Reverse this procedure, use (2) for the definition of N, and then "derive" statements (a) and (b) of §17.

24. LINES OF CURRENT DENSITY

The direction and the density of the lines of $\mathbf{J}$ at any point P give us, respectively, the direction and the magnitude of the *current density* at P. Also, in view of (6[22]) and (2[23]), the net number of lines of $\mathbf{J}$ crossing any surface S gives us the net *current* flowing across S. That is, if the net current flowing across S is i, then the net number of lines of $\mathbf{J}$ crossing S is given by the formal equation

$$N = i. \tag{1}$$

In the diagrams of this section we represent the current density of one ampere per square meter by the line density of one line of **J** per square meter. One can, therefore, say that in these diagrams each line of **J** represents one ampere, in conformity with (12^{17}). Some of the statements of this section will be made more precise in §40.

Examples. A straight copper rod carries a current of 3 amperes. This statement means that 3 amperes flow across every properly signed surface that forms a complete cross section of the rod, and consequently every such surface

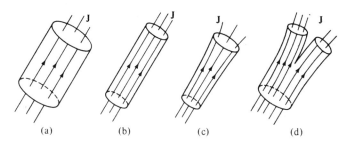

Fig. 78. Line maps of current density.

must be crossed by 3 lines of **J**. A line map of **J** for this case is shown in Fig. 78(a), where the lines have been spaced more or less evenly, to suggest that the magnitude J of **J** is constant throughout the rod.

Now suppose that an imagined flat surface, smaller than the cross section of the rod, is so placed inside and at right angles to the rod of Fig. 78(a) that none of the three lines of **J** crosses it. We then have a disagreement with (1)—no lines of **J** cross this surface, yet the current across it is not zero. This kind of discrepancy is characteristic of line maps and emphasizes the fact that they are merely pictures—suggestive pictures, yet only pictures. (The discrepancy can, of course, be reduced in this example by changing the unit for measuring current from the ampere to the microampere and imagining a diagram with three million lines instead of three.)

Figure 78(b) is a map of **J** when 3 amperes flow in a thinner rod; the magnitude of the current density is now larger than before, and the lines of **J** are more crowded. In Fig. 78(c), the spreading of the lines in the thicker part of the rod is consistent with the decrease of J there.

In the case of branching conductors the reasoning is much the same. Figure 78(d) describes a current of 5 amperes that divides into a 3-ampere current and a 2-ampere current. Figure 79 is a rough map of **J** for the case when a battery (dotted box) sends 1 ampere and 2 amperes, respectively, into two resistors connected across it in parallel.

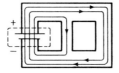

Fig. 79. Rough map of current density for a battery and two resistors connected across it in parallel.

Next, consider a capacitor that is being charged by a constant-current generator and take the charging current as 2 amperes. A sequence of rough line maps of **J** for this case is shown in Fig. 80. The first diagram pertains to the initial instant, when the generator (dotted box) is turned on but the capacitor is still uncharged. The other diagrams pertain to two later instants and display the fact that charges are accumulating on the plates.

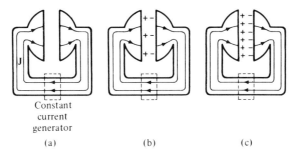

Fig. 80. A constant-current generator charging a parallel-plate capacitor.

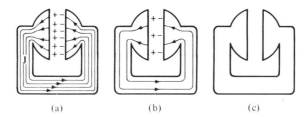

Fig. 81. A discharging parallel-plate capacitor.

Figure 81 describes in a similar way a capacitor discharging through a resistor when the self-inductance of the circuit can be ignored. As shown in part (a), the initial value of the discharge current is taken as 4 amperes. In part (b) the discharge is half completed; in part (c) it is practically completed.

Figures 80 and 81 suggest that the lines of **J** begin in regions where the net charge is decreasing with time and end in regions where the net charge is increasing. Figure 79 suggests that, if charge densities do not change with time, the lines of **J** have neither beginnings nor ends.

Figure 81 illustrates the fact that when a time-varying vector field is described by a sequence of line maps, each map must be drawn afresh. For instance, there is no way of *gradually* deforming the four lines of **J** in Fig. 81(a) into the two lines shown in Fig. 81(b).

Kirchhoff's current law. This law states that, in the steady d. c. case, the sum of the currents flowing toward any junction of conductors is equal to the sum of currents flowing away from this junction. Figure 79 and similar line maps of **J** suggest that this law can be expressed in pictorial terms as follows: In a line map describing a complete d. c. circuit in a steady state, the lines of **J** are closed.

EXERCISES

1. A 3-ampere current divides in a branching conductor into two equal currents. Sketch a line map of **J**.
2. Why are the diagrams of this section called line maps of current density and not line maps of current?
3. Why do the currents in Figs. 80(b) and 81(b) flow in opposite directions despite the identical charge distributions on the capacitor? How do the time rates of change of charge on the left-hand plate compare in the two cases?
4. A charged capacitor is shorted when $t = 0$. Assume that the discharge is oscillatory with a period T, ignore damping, and sketch line maps of **J** for the instants $t = 0, \frac{1}{4}T, \frac{1}{2}T$, and $\frac{3}{4}T$. Indicate the charges on the plates by plus and minus signs.

25. GRAPHICAL ESTIMATION OF FLUX

Line maps can be used for estimating flux integrals by counting field lines. To illustrate, we return to the flow of charge past a cylindrical hole. A line map of **J** for this case is shown in Figs. 62[17] and 82. We will use Fig. 82 to estimate the current flowing across the surface S_4, first shown in Fig. 56[15]. In §15 the depth and width of S_4 were denoted by h and l, but in Fig. 82 we have set $l = b$, where b is the radius of the hole. For the moment, we will call the field lines lying in the plane of the page the "visible" lines. To get a three-dimensional picture of **J**, we must imagine similar layers of "invisible" lines above and below the page. As before, we write J_0 for the magnitude of **J** far from the hole.

Let us first find how much current is represented in Fig. 82 by a single visible or invisible field line. According to (10[17]), the visible lines lie $\frac{1}{4}b$ meters apart far from the hole. The current density far from the hole is uniform, and hence the layers of lines should also lie $\frac{1}{4}b$ meters apart in depth. The line density far from the hole is then equal to $16/b^2$ lines per square meter. This line

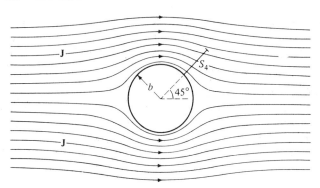

Fig. 82. A diagram for estimating the current flowing across the surface S_4 by counting lines of J that cross S_4.

density represents the current density of J_0 amperes per square meter, and hence each visible or invisible line represents $\frac{1}{16}b^2 J_0$ amperes.

Next consider a rectangular surface normal to the page and h meters deep. If n visible lines cross it, what is the total number of lines, say N, that cross it? Since the layers of lines lie $\frac{1}{4}b$ meters apart, the surface intercepts $4h/b$ layers. Each layer contributes n lines, and hence $N = 4nh/b$. And since each line represents $\frac{1}{16}b^2 J_0$ amperes, we conclude that if a rectangular surface of depth h is crossed by n visible lines, the estimated current across it is $\frac{1}{4}nbhJ_0$ amperes.

The surface S_4 in Fig. 82 is crossed by four visible field lines; therefore, $n = 4$ and the estimated current flowing across S_4 is about bhJ_0 amperes. The exact value, found by letting $l = b$ in (13^{15}), is about $1.06 bhJ_0$ amperes. (Since so few lines are involved, the good agreement is, of course, fortuitous.) Estimates of this kind can sometimes be improved by interpolating extra lines, freehand, between the lines of the available map; this process amounts to decreasing the unit used for measuring the strength of the given field.

In the computations described above, we made the following assumption: if lines of J are continuous lines computed from (9^{17}), and if their density far from the hole is adjusted to give the correct magnitude of J far from the hole, then their density near any point P will give the correct magnitude of J near P. By methods too advanced for this book, one can justify this assumption in the present two-dimensional example and in other similar problems. A more complicated case will come up in §32.

Line maps of a flow past a cylinder can be drawn as in Fig. 83, where one line runs along the x axis. This line corresponds to the middle row of pointers in Fig. 52^{14}; it tells us that at points on the x axis outside the hole, J is directed toward the right. As we approach the point a from the left along this line, the neighboring lines veer away and the density of the lines decreases; this implies that the magnitude J of J decreases. But a line map is too crude a device to enable us to infer that J is *exactly zero* at a, as required by the formulas of

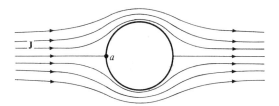

Fig. 83. An abbreviated map of J.

§14. (From the mathematical standpoint, the field line running along the x axis toward a includes all the points lying on this axis at the left of a but not the point a itself—in principle, this line has no beginning and no end.)

EXERCISES

1. Transfer the surface S_1 from Fig. 55^{15} to Fig. 82, make its width l equal to the radius of the hole, estimate the current across it by counting lines of **J**, and compare the result with (8^{15}) when $l = b$. Repeat for the case $l = \tfrac{1}{2}b$.
2. Consider the surface S_3 of Fig. 56^{15} in the manner of Exercise 1, but take $l = b$ and $l = 2b$. Use (11^{15}) to check.
3. Estimate by inspection the values of c used in (9^{17}) for computing Fig. 83.

26. FLUX ACROSS A CLOSED SURFACE

Let (S) be the closed surface of a region R. A field line crossing (S) from back to front (from the inner to the outer side) is said to be leaving R. A line crossing (S) from front to back is said to be entering R. If N_1 lines are leaving and N_2 lines are entering R, the quantity $N_1 - N_2$ is called the (net) number of lines emerging from R. In the case of closed surfaces the following statement replaces (3^{23}):

$$\left\{ \begin{array}{l} \text{net number of lines of } \mathbf{F} \text{ emerging} \\ \text{from the region enclosed by } (S) \end{array} \right\} = \left\{ \begin{array}{l} \text{net outward flux} \\ \text{of } \mathbf{F} \text{ across } (S). \end{array} \right. \tag{1}$$

The formula for computing fluxes across closed surfaces is

$$(7^{22}) \qquad \text{net outward flux of } \mathbf{F} \text{ across } (S) = \iint_{(S)} F_n \, da. \tag{2}$$

We will now evaluate the fluxes of three simple vector fields across the cubical surface (S) shown in Fig. 84 and centered on the point $P(x_0, y_0, z_0)$. The

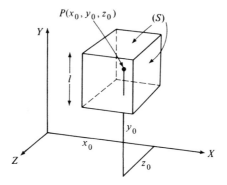

Figure 84

right-hand face of (S) looks in the $+x$ direction, and hence the normal component of any vector field **F** relative to it is simply F_x. The left-hand face looks in the opposite direction and, therefore, at points on that face, we have $F_n = -F_x$. All in all,

$$\iint_{(S) \text{ of cube}} F_n \, da = \iint_{\text{right face}} F_x \, dy \, dz + \iint_{\text{upper face}} F_y \, dz \, dx + \iint_{\text{near face}} F_z \, dx \, dy$$
$$- \iint_{\text{left face}} F_x \, dy \, dz - \iint_{\text{lower face}} F_y \, dz \, dx - \iint_{\text{far face}} F_z \, dx \, dy. \tag{3}$$

Example. Let $\mathbf{F} = \mathbf{1}_x$, so that $F_x = 1$, $F_y = 0$, and $F_z = 0$. Equation (3) then gives

$$\iint_{(S) \text{ of cube}} F_n \, da = \iint_{\text{right face}} dy \, dz - \iint_{\text{left face}} dy \, dz = l^2 - l^2 = 0. \tag{4}$$

In pictorial terms,

$$\left. \begin{array}{c} \text{net number of lines of } \mathbf{1}_x \\ \text{emerging from the cube} \end{array} \right\} = 0. \tag{5}$$

This statement is, of course, consistent with the idea that the lines of $\mathbf{1}_x$ have no beginnings and no ends.

Second example. Let $\mathbf{F} = x\mathbf{1}_x$, so that $F_x = x$, $F_y = 0$, and $F_z = 0$. In this case (3) becomes

$$\iint_{\substack{(S) \text{ of} \\ \text{cube}}} F_n \, da = \iint_{\substack{\text{right} \\ \text{face}}} x \, dy \, dz - \iint_{\substack{\text{left} \\ \text{face}}} x \, dy \, dz. \tag{6}$$

On the right-hand face $x = x_0 + \tfrac{1}{2}l$, on the left-hand face $x = x_0 - \tfrac{1}{2}l$, and, therefore,

$$\iint_{\substack{(S) \text{ of} \\ \text{cube}}} F_n \, da = (x_0 + \tfrac{1}{2}l)l^2 - (x_0 - \tfrac{1}{2}l)l^2 = l^3. \tag{7}$$

Thus, the flux is equal to the volume of the cube. In pictorial terms,

$$\left.\begin{array}{l}\text{net number of lines of } x\mathbf{1}_x \\ \text{emerging from the cube}\end{array}\right\} = l^3. \tag{8}$$

This result implies that the lines of the field $x\mathbf{1}_x$ have beginnings, as we already know from Fig. 72[19].

Third example. If $\mathbf{F} = x^3\mathbf{1}_x$, we find (Exercise 1) that

$$\iint_{\substack{(S) \text{ of} \\ \text{cube}}} F_n \, da = (3x_0^2 + \tfrac{1}{4}l^2)l^3. \tag{9}$$

In this case the flux depends not only on the size of the cube but also on its location. In terms of field lines,

$$\left.\begin{array}{l}\text{net number of lines of } x^3\mathbf{1}_x \\ \text{emerging from the cube}\end{array}\right\} = (3x_0^2 + \tfrac{1}{4}l^2)l^3. \tag{10}$$

Fourth example. Next, let us find the flux of the field $\mathbf{F} = z\mathbf{1}_z$ across the spherical surface of Fig. 85. According to (20^{16}), we have, at P,

$$F_n = (z\mathbf{1}_z) \cdot (\mathbf{1}_n) = z(\mathbf{1}_z \cdot \mathbf{1}_n) = z \cos \theta = r_0 \cos^2 \theta. \tag{11}$$

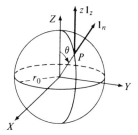

Figure 85

120   The Flux Integral   § 26

Using the formula for da implied in (18^Λ), we get

$$\iint\limits_{\substack{(S) \text{ of} \\ \text{sphere}}} F_n \, da = \iint\limits_{\substack{(S) \text{ of} \\ \text{sphere}}} r_0 \cos^2 \theta \, da = \int_0^{2\pi} \int_0^{\pi} r_0^3 \cos^2 \theta \sin \theta \, d\theta \, d\phi = \tfrac{4}{3} \pi r_0^3. \qquad (12)$$

This flux is equal to the volume of the sphere, a result analogous to (7).

EXERCISES

1. Verify (9).
2. Show that the flux of the field $x^2 \mathbf{1}_x$ across the cubical surface in Fig. 84 is $2x_0 l^3$, and hence has the same sign as x_0; verify that this fact is consistent with a result of Exercise 3[19].
3. Evaluate by detailed integration the flux of the field $\mathbf{1}_z$ across the spherical surface in Fig. 85.

CHAPTER 8

Source Density and the Divergence

In §§27 and 28 we continue with the description of vector fields by field lines and, on this pictorial basis, identify certain surface integrals with certain volume integrals. These integrals come into play in such laws as Kirchhoff's current law, the law of conservation of charge, and Coulomb's law of interaction between electric charges. In §29 we give the name "divergence" to the limit of a certain integral and note that the mathematical concept of divergence fits in with the pictorial concept of source density, discussed in §28. The computation of divergence by differentiation rather than integration is illustrated in §30.

27. SOURCES AND SINKS OF VECTOR FIELDS

As before, we call the points where the field lines of a vector field **F** begin and end the *sources* and the *sinks* of **F**, respectively, and indicate sources by full circles and sinks by empty circles, as illustrated for a rather artificial case in Fig. 86. When dealing with any particular field **F**, we write

$\sigma_1 =$ number of sources located in a region R,
$\sigma_2 =$ number of sinks located in R,
$\sigma = \sigma_1 - \sigma_2 =$ net number of sources located in R.

For example, in Fig. 86,

$$\sigma_1 = 5, \quad \sigma_2 = 3, \quad \sigma = 2. \tag{1}$$

Note that we treat a sink as a negative source.

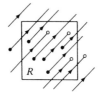

Figure 86

If a field line enters a region R at a point P on the surface (S) of R, we call P a *point of entry*; if a line leaves R at P, we call P a *point of exit*, and we write

N_1 = number of points of exit,

N_2 = number of points of entry,

$N = N_1 - N_2$ = net number of lines emerging from R.

In Fig. 86,

$$N_1 = 6, \qquad N_2 = 4, \qquad N = 2. \tag{2}$$

Field lines that begin or end *on* the surface of R are referred to in Exercise 2.

In Fig. 86 we have $\sigma = 2$ and $N = 2$. The relation $\sigma = N$ holds in general, as one can see as follows. In Fig. 87 a single field line begins and ends outside a region R, so $\sigma = 0$. As illustrated in this figure for three special cases, the

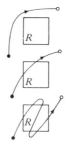

Figure 87

number of points of exit is then equal to the number of points of entry, so $N = 0$ and hence $\sigma = N$. Similarly, if a line begins outside R and ends inside, we have $\sigma = -1$ and $N = -1$; if it begins and ends inside, we have $\sigma = 0$ and $N = 0$; if it begins inside and ends outside, we have $\sigma = 1$ and $N = 1$. Consequently,

$$\sigma = N \tag{3}$$

for any single field line, and hence also for any set of field lines. In words: For any field **F**,

Sources and Sinks of Vector Fields   § 27

$$\begin{Bmatrix} \text{net number of sources} \\ \text{located in } R \end{Bmatrix} = \begin{Bmatrix} \text{net number of lines} \\ \text{emerging from } R. \end{Bmatrix} \quad (4)$$

Two equally good line maps of the same vector field need not agree in every detail. In particular, the numbers σ and N for a specific region may have the values σ' and N' according to one map and different values, say σ'' and N'', according to the other. But (3) will hold in either case; that is, $\sigma' = N'$ and $\sigma'' = N''$.

The right-hand side of (3) is evaluated by a "surface count"—by counting entry and exit points *on the surface* of a region. The left-hand side of (3) is evaluated by a "volume count"—by counting sources and sinks *inside* the region. The pictorial identity (3) will soon lead us to a mathematical identity between a surface integral and a volume integral. In the meantime we combine (3) with (7^{22}) and get the formula

$$\sigma = \iint_{(S)} F_n \, da, \quad (5)$$

where (S) is the surface of the region R.

In view of (4), Equations (5^{26}), (8^{26}), and (10^{26}) can be stated as follows: The net number of sources of the field $\mathbf{F}$ contained in the cube of Fig. 84^{26} is zero if $\mathbf{F} = \mathbf{1}_x$; it is l^3 if $\mathbf{F} = x\mathbf{1}_x$; it is $(3x_0^2 + \frac{1}{4}l^2)l^3$ if $\mathbf{F} = x^3\mathbf{1}_x$.

EXERCISES

1. An exhibition hall has counting turnstiles at its entrances and exits. By analogy with (3), what are two distinct ways of determining the number of people in this hall at some particular time?

2. A field line may begin on, end on, or be tangent to the surface of R. These special cases can be fitted in with (3) by suitable conventions. Suggest such a convention.

3. According to (3), the unit of measurement for σ must be the same as that for N, and hence must be given by (12^{17}). Assume that x is expressed in meters, and check the units of the σ's given in the last sentence of this section.

28. SOURCE DENSITY

Let σ be the net number of sources of a field $\mathbf{F}$ in a region R, whose volume is v. The ratio σ/v is then called the average density of sources of $\mathbf{F}$ in R. For example, since the volume of the cube in Fig. 84^{26} is l^3 cubic meters, we conclude from the last sentence of §27 that, in this cube, the average density of sources of the field $\mathbf{1}_x$ is zero, the average density of sources of the field $x\mathbf{1}_x$ is one source per cubic meter, and the average density of sources of the field $x^3\mathbf{1}_x$ is $3x_0^2 + \frac{1}{4}l^2$ sources per cubic meter.

Let a point P lie in a region R that is bounded by the surface (S) and has the volume v. Next imagine that (S) contracts upon P without affecting the distribution of the sources and sinks of $\mathbf{F}$. The limit of the ratio σ/v as $v \to 0$ is called the (net) *source density* of $\mathbf{F}$ at P. To illustrate, we consider the center P of the cube of Fig. 84, denote its coordinates by (x, y, z) instead of (x_0, y_0, z_0), and require that $l \to 0$. The average source densities listed in the preceding paragraph then lead to the following conclusions, pertaining to an arbitrarily chosen point whose coordinates are (x, y, z):

$$\text{if } \mathbf{F} = \mathbf{1}_x, \quad \text{source density} = 0, \tag{1}$$

$$\text{if } \mathbf{F} = x\mathbf{1}_x, \quad \text{source density} = 1, \tag{2}$$

$$\text{if } \mathbf{F} = x^3\mathbf{1}_x, \quad \text{source density} = 3x^2. \tag{3}$$

These densities are expressed in terms of (net) sources per cubic meter.

According to the definition stated above,

$$\text{source density} = \lim_{v \to 0} \left(\frac{\sigma}{v}\right). \tag{4}$$

To make $\mathbf{F}$ appear explicitly in (4), we use (5^{27}) and write

$$\text{source density of } \mathbf{F} \text{ at } P = \lim_{v \to 0} \left(\frac{1}{v} \iint_{(S)} F_n \, da\right); \tag{5}$$

here (S) is the surface and v the volume of a region that includes P and shrinks upon P under the mathematical conditions hinted at in the remarks on shrinking regions in §5.

Next, suppose that we are given the source density of a field $\mathbf{F}$ and wish to compute the net number σ of sources of $\mathbf{F}$ contained in a region R. The formal expression for the net number of sources contained in a small element of R having the volume dv is source density times dv, and, therefore,

$$\sigma = \iiint_R (\text{source density}) \, dv. \tag{6}$$

This formula is similar to Equation (4^3), which pertains to charge density. Because of (5^{27}), we may write (6) as

$$\iiint_R (\text{source density}) \, dv = \iint_{(S)} F_n \, da. \tag{7}$$

We introduced the concept of source density on the rather vague pictorial basis of field lines and the beginning points and end points of these lines. Further pictorial arguments have led us to (5). This fact permits us now to remove all vagueness by adopting (5) as a clear-cut mathematical *definition* of source density.

EXERCISES

1. Show that

$$\text{if } \mathbf{F} = x^n \mathbf{1}_x, \quad \text{source density} = nx^{n-1}. \tag{8}$$

2. Compute in detail the volume integral of the function $3x^2$ over the cubical region of Fig. 84[26]. Explain why one can expect the result to agree with the right-hand side of (9[26]).

29. DIVERGENCE

In this section we introduce the scalar field called the "divergence" of a vector field **F**. It is denoted by the symbol "div **F**" and is defined at a point P by the equation

$$\text{div } \mathbf{F} = \lim_{v \to 0} \frac{1}{v} \left(\iint_{(S)} F_n \, da \right). \tag{1}$$

Here (S) is the surface and v the volume of a region R that includes the point P and contracts upon it. The right-hand side of (1) is a scalar pertaining to the point P; if the field **F** is an invariant field, the right-hand side of (1) is independent of the choice of a coordinate frame. In short, *the divergence of an invariant vector field is itself an invariant scalar field.*

Comparison of (1) with (5[28]) shows that the divergence of a vector field **F** is identical with the (net) source density of **F**. In fact, what we have done just now is to introduce a new word (divergence) and a new symbol (div **F**), but not a new mathematical concept.

Equations (1), (2), (3), and (8) of §28 can now be abbreviated as follows:

$$\text{div } \mathbf{1}_x = 0, \tag{2}$$

$$\text{div } x\mathbf{1}_x = 1, \tag{3}$$

$$\text{div } x^3 \mathbf{1}_x = 3x^2, \tag{4}$$

$$\text{div } x^n \mathbf{1}_x = nx^{n-1}. \tag{5}$$

126 Source Density and the Divergence

The term "source density" is in some ways better than "divergence." For one thing, the statement that the divergence of a certain field is different from zero may give the beginner the impression that the lines of this field diverge in the sense of "making angles" with one another. This impression is wrong. For instance, the lines of the field $x^2 \mathbf{1}_x$ are all parallel, even though the divergence of this field is not identically zero.

The term "emergence," suggested to the author by Mr. T. W. Odell, would fit the situation better than "divergence," and one might wish that it were in general use.

Another way of writing (1) is given in (44^A). The "del" notation for divergence is described in Appendix D.

EXERCISE

1. Show that, if c is a numerical constant, and $\mathbf{F}$ and $\mathbf{G}$ are vector fields, then div $c\mathbf{F} = c$ div $\mathbf{F}$ and div $(\mathbf{F} + \mathbf{G}) =$ div $\mathbf{F} +$ div $\mathbf{G}$.

30. COORDINATE FORMULAS FOR DIVERGENCE

It is perhaps apparent from our pictorial discussion that the source density of a field $\mathbf{F}$, and hence also the divergence of $\mathbf{F}$, have nothing to do with any coordinate frame installed somewhere in space, unless, of course, $\mathbf{F}$ has been described in terms of such a frame to begin with. If $\mathbf{F}$ is invariant, so is the right-hand side of (1^{29}). We will now show, however, that the formula for the divergence can be greatly simplified in appearance with the help of a cartesian coordinate frame. The last step in our computation will be to let $v \to 0$, and, therefore, we will keep only the lowest-order terms throughout the work.

We write

$$\mathbf{F} = F_x \mathbf{1}_x + F_y \mathbf{1}_y + F_z \mathbf{1}_z, \tag{1}$$

fix our attention on an arbitrarily chosen point $P(x, y, z)$, and consider the brick-shaped "elementary cell," say R, centered on P and oriented as in

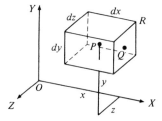

Figure 88

Fig. 88. We then compute the integral of F_n over the surface (S) of R, put the result into (1^{29}), and finally let each edge of R shrink to zero.

On the right-hand face of R we have $F_n = F_x$ and $da = dy\,dz$. If we let dx stay constant and let $dy \to 0$ and $dz \to 0$, this face will be shrinking toward its midpoint, marked Q, where F_x has the value denoted below by $F_x(Q)$; the integral of F_x over this face will approach the product $F_x(Q)\,dy\,dz$. That is, when $dy \to 0$ and $dz \to 0$, we have

$$\iint_{\substack{\text{right}\\\text{face}}} F_n\,da = \iint_{\substack{\text{right}\\\text{face}}} F_x\,dy\,dz \to \iint_{\substack{\text{right}\\\text{face}}} F_x(Q)\,dy\,dz \to F_x(Q)\,dy\,dz. \tag{2}$$

This result is not quite what we want, because we are interested primarily in the point $P(x, y, z)$, and not in the point $Q(x + \frac{1}{2}dx, y, z)$. Therefore, we recall (6^7), ignore higher-order terms, and write

$$F_x(Q) = F_x(P) + \frac{1}{2}\frac{\partial F_x}{\partial x}\,dx. \tag{3}$$

It then follows from (2) that, when $dx \to 0$, $dy \to 0$, and $dz \to 0$, we have

$$\iint_{\substack{\text{right}\\\text{face}}} F_n\,da \to \left[F_x(P) + \frac{1}{2}\frac{\partial F_x}{\partial x}\,dx\right]dy\,dz. \tag{4}$$

At the midpoint, say Q', of the left-hand face, we have $F_n = -F_x$. Therefore, when $dy \to 0$ and $dz \to 0$, we get

$$\iint_{\substack{\text{left}\\\text{face}}} F_n\,da = -\iint_{\substack{\text{left}\\\text{face}}} F_x\,dy\,dz \to -\iint_{\substack{\text{left}\\\text{face}}} F_x(Q')\,dy\,dz \to -F_x(Q')\,dy\,dz. \tag{5}$$

The coordinates of Q' are $(x - \frac{1}{2}dx, y, z)$, and consequently the relation corresponding to (4) is

$$\iint_{\substack{\text{left}\\\text{face}}} F_n\,da \to \left[-F_x(P) + \frac{1}{2}\frac{\partial F_x}{\partial x}\,dx\right]dy\,dz. \tag{6}$$

Accordingly, apart from higher-order terms,

$$\iint_{\substack{\text{right}\\\text{face}}} F_n\,da + \iint_{\substack{\text{left}\\\text{face}}} F_n\,da = \frac{\partial F_x}{\partial x}\,dx\,dy\,dz. \tag{7}$$

When the two remaining pairs of faces are included in the integration, we get

$$\iint_{(S)} F_n \, da = \left(\frac{\partial F_x}{\partial x} + \frac{\partial F_y}{\partial y} + \frac{\partial F_z}{\partial z}\right) dx \, dy \, dz, \tag{8}$$

apart from higher-order terms.

The volume of the brick-shaped region is $v = dx \, dy \, dz$, so, putting (8) into (1^{29}), we get

$$\text{div } \mathbf{F} = \lim_{v \to 0} \left(\frac{\partial F_x}{\partial x} + \frac{\partial F_y}{\partial y} + \frac{\partial F_z}{\partial z}\right); \tag{9}$$

here the condition $v \to 0$ stands for the three conditions $dx \to 0$, $dy \to 0$, and $dz \to 0$. Had we kept the higher-order terms, the quantities dx, dy, and dz would appear in the parentheses in (9), but they would vanish when $v \to 0$, and hence our final result is the following *exact* formula for computing divergence in terms of cartesian coordinates:

$$\text{div } \mathbf{F} = \frac{\partial F_x}{\partial x} + \frac{\partial F_y}{\partial y} + \frac{\partial F_z}{\partial z}. \tag{10}$$

The formulas that correspond to (10) in cylindrical and spherical coordinates are given in (57^A) and (65^A), in terms of symbols that are defined in §§34 and 36.

We deduced (10) from Equation (1^{29}), which defines the divergence. So far as appearances go, (10) is much the simpler; but (1^{29}) displays the pictorial meaning of the divergence—and therefore also its physical meaning—more clearly. In fact, (1^{29}) hides nothing: all we need to remember is that the flux integral gives the net number of field sources enclosed by (S); it is then apparent that div $\mathbf{F}$ is the density of these sources and can, therefore, be visualized with the help of field lines. One reason why the pictorial meaning of the right-hand sides of such equations as (10) and (57^A) is not equally plain is that they mix together the intrinsic properties of, say, an invariant vector field (which do not depend on the choice of a coordinate frame) and the special properties of a particular coordinate frame. We have here an illustration of the fact that the explicit use of coordinates may make a field equation simpler to manage in computations, but may at the same time obscure its pictorial and physical content.

Example. Let us use Equation (10) to compute the divergence of the field $\mathbf{F} = x^n \mathbf{1}_x$. We have

$$F_x = x^n, \quad F_y = 0, \quad F_z = 0, \tag{11}$$

so

$$\frac{\partial F_x}{\partial x} = nx^{n-1}, \quad \frac{\partial F_y}{\partial y} = 0, \quad \frac{\partial F_z}{\partial z} = 0. \tag{12}$$

Consequently, in this example,

$$\frac{\partial F_x}{\partial x} + \frac{\partial F_y}{\partial y} + \frac{\partial F_z}{\partial z} = nx^{n-1}, \tag{13}$$

and we conclude that

$$\text{div } x^n \mathbf{1}_x = nx^{n-1}, \tag{14}$$

in agreement with (5^{29}).

An example of a physical equation involving a divergence is (11^{14}), which can be written as

$$\text{div } \mathbf{J} = 0. \tag{15}$$

We will see in Chapter 10 that this equation is the field form of Kirchhoff's junction law for steady currents.

EXERCISE

1. Compute div $y\mathbf{1}_x$, and use Fig. 76^{21} as a pictorial check.

31. THE DIVERGENCE THEOREM

We now return to § 28 and the equation

$$\iiint_R (\text{source density}) \, dv = \iint_{(S)} F_n \, da, \tag{1}$$

to which we were led by pictorial considerations. Since the right-hand sides of (5^{28}) and (1^{29}) are the same, we may write (1) as

$$\iiint_R \text{div } \mathbf{F} \, dv = \iint_{(S)} F_n \, da. \tag{2}$$

This mathematical identity is called the *divergence theorem* or the *flux theorem*; it also shares with many other theorems the name *Gauss's theorem*. It is proved rigorously in more advanced books; here we will merely add a few remarks to the pictorial arguments we have used so far.

Suppose that the region R is so small that the values of div $\mathbf{F}$ are nearly the same at all points in it. We can then write, ignoring higher-order terms,

$$(\text{div } \mathbf{F}) \cdot (\text{volume of } R) = \iint_{(S)} F_n \, da. \tag{3}$$

If R has the shape of a brick, this result is the same as Equation (8^{30}), which we *did* prove for small brick-shaped regions.

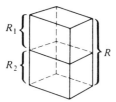

Figure 89

Next we will illustrate the fact that if (2) holds for each of two regions into which a region R can be subdivided, then it will automatically hold for the whole region. Suppose that R has the shape of the double brick shown in Fig. 89 and that (2) holds for each of its halves, R_1 and R_2, so that

$$\iiint_{R_1} \text{div } \mathbf{F} \, dv = \iint_{(S_1)} F_n \, da \tag{4}$$

and

$$\iiint_{R_2} \text{div } \mathbf{F} \, dv = \iint_{(S_2)} F_n \, da, \tag{5}$$

where (S_1) is the six-faced surface of R_1 and (S_2) is the six-faced surface of R_2. Let us now add (4) and (5). On the left, we get the integral of div $\mathbf{F}$ over the whole region R. On the right, the portion of the integral in (4) pertaining to the lower face of R_1 cancels the portion of the integral in (5) pertaining to the upper face of R_2, and what remains is just the integral of F_n over the six faces of R. Consequently, when we add (4) and (5), we get Equation (2) for the whole region R.

The arguments of the last two paragraphs, when properly refined, can be combined into a rigorous proof of the divergence theorem, provided the field $\mathbf{F}$ satisfies certain continuity conditions that are usually satisfied in physical problems.

32. FLUX THROUGH A CONTOUR

We have dealt so far with flux *across a surface*. In this section we show that if (and only if) the divergence of a vector field is identically zero, one can also give a useful meaning to the term "flux *through a contour*." To do this, we compare the fluxes across different surfaces spanning the same contour.

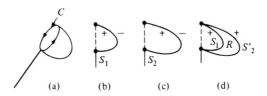

Figure 90

Figure 90 shows a signed contour C that resembles the rim of a butterfly net, and an open surface S_1 that spans C and resembles the net itself; S_2 is another surface spanning C and resembling a deeper net; in parts b, c, and d of the figure the contour C is shown edge-on. Let R be the region located between S_1 and S_2. The surface of R, say (S), is a closed surface that consists of the portion S_1 and the portion S_2'; the latter coincides with S_2 but has the opposite signature. Consequently, for any field **F**,

$$\iint_{S_2'} F_n \, da = -\iint_{S_2} F_n \, da. \tag{1}$$

The flux of **F** emerging from R is

$$\iint_{(S)} F_n \, da = \iint_{S_1} F_n \, da + \iint_{S_2'} F_n \, da. \tag{2}$$

Therefore, in view of (1) and the divergence theorem,

$$\iiint_R \operatorname{div} \mathbf{F} \, dv = \iint_{S_1} F_n \, da - \iint_{S_2} F_n \, da. \tag{3}$$

Now, if

$$\operatorname{div} \mathbf{F} = 0, \tag{4}$$

Equation (3) becomes

$$\iint_{S_1} F_n \, da = \iint_{S_2} F_n \, da. \tag{5}$$

Consequently, if $\operatorname{div} \mathbf{F} = 0$, then the flux of **F** across any surface spanning a contour C is the same as that across any other surface spanning C. Under these circumstances it is often convenient to speak of this flux as the flux *through* the contour C or the flux *linking* C. A field whose divergence is zero is called "solenoidal."

The physical field called "magnetic induction" or "magnetic flux density" and denoted by **B** is assumed in Maxwell's theory to satisfy the equation

$$\operatorname{div} \mathbf{B} = 0. \tag{6}$$

Accordingly, one may say "magnetic flux through a loop of thin wire" or "magnetic flux linking the wire loop" instead of saying "magnetic flux across any arbitrarily chosen surface that spans the wire loop." To find the numerical value of the flux of a given solenoidal field **F** through a given contour C, one must usually compute the flux of **F** across some particular surface spanning C; one would naturally choose a surface for which the computation is simplest.

Stop-and-go flux lines. The flux of a field **F** across a surface S can be computed in principle by counting the (net) number of lines of **F** that cross S. We have proved that if div $\mathbf{F} = 0$ and if the surfaces S_1, S_2, S_3, and so on all span the same contour C, then the flux of **F** across each of these surfaces is the same. This means pictorially that all these surfaces are pierced by the same (net) number of lines of **F**. If so, then, strictly speaking, the lines of **F** can have no beginnings or ends in any finite region: they must either be closed, as in Fig. 73[20], or come from infinity and go out to infinity again, as in Fig. 62[17].

Now, as we remarked in §29, the word "divergence" is but another name for (net) source density. Therefore, the equation div $\mathbf{F} = 0$ can be interpreted

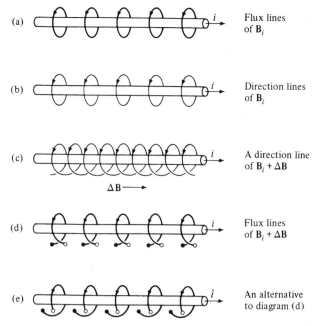

Fig. 91. Parts (a) and (b) of this diagram describe the field $\mathbf{B}_i$ caused by a current i flowing in a wire. Parts (c), (d), and (e) include the effect of an additional magnetic field, parallel to the wire and caused by permanent magnets located nearby.

pictorially to mean that as many lines of **F** begin in any region R as end there. This statement is strictly equivalent to that in the preceding paragraph, but it does not stress explicitly the continuity of the lines of **F**. To illustrate the usefulness of this change of emphasis, we consider an example described by Iona[1] and pertaining to the equation div **B** = 0.

The current i in Fig. 91 flows in a long straight wire and produces a certain magnetic flux density, say $\mathbf{B}_i$. The five *flux lines* of $\mathbf{B}_i$ shown in Fig. 91(a) are closed circular lines, as in Fig. 73[20]. The *direction lines* of $\mathbf{B}_i$, shown in Fig. 91(b) and defined in §17 for vector fields in general, are also circular.

Next, suppose that a small constant field $\Delta\mathbf{B}$, parallel to the wire, is superimposed upon $\mathbf{B}_i$ by permanent magnets, and write **B** for the resulting magnetic flux density. A direction line of **B** is then a helix, wound around the wire as in Fig. 91(c). What makes this example particularly interesting is that the number of turns of the helix per meter of length of the wire becomes *larger* if $\Delta\mathbf{B}$ is made *smaller*; to put this roughly, the density of the direction lines of **B** increases with decreasing $\Delta\mathbf{B}$. Accordingly, if we are to get a reasonable flux density, we must not allow a flux line to follow a direction line all the way, but must keep stopping it and restarting it at a different point. One suitable set of flux lines is shown in Fig. 91(d), where each line runs along a single turn of the direction line and then skips a turn. An alternative set is shown in Fig. 91(e). Note that the density of these flux lines does not change when we cross any plane containing the axis of the wire—and this is perhaps all that one can ask of a picture that is useful but not perfect.

We have shown that the equation div **F** = 0 does not necessarily mean pictorially that the *flux lines* of **F** have no beginnings or ends. Note, however, that even in the case of the magnetic fields pictured in Fig. 91, the *direction lines* have no beginnings or ends.

EXERCISE

1. Let div **F** = 0 and derive (5) for the case when S_1 and S_2 intersect each other, as in Fig. 92, where the circular contour C is shown edge-on. Does (5) follow from (4) if C is not flat?

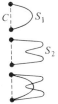

Figure 92

[1] Mario Iona, "Lines With Ends to Describe div **B** = 0," *American Journal of Physics*, Vol. 31, p. 398 (1963).

CHAPTER 9

Cylindrical and Spherical Coordinates

Several important electric and magnetic fields are expressed most conveniently in terms of cylindrical or spherical polar coordinate frames. In this chapter we introduce the base-vectors of these frames and describe a few fields in terms of these vectors as well as in terms of cylindrical and spherical components. The formulas for the divergence in terms of cylindrical and spherical components are (57^A) and (65^A). The derivations of these formulas are similar to that in the cartesian case.

The cylindrical and spherical components of an invariant vector field are scalar fields, but since they depend on the orientation of the coordinate frame, they are not invariant scalar fields—just as in the cartesian case.

33. THE VECTORS 1_ρ, 1_ϕ, AND 1_z

The "cylindrical" base-vectors pertaining to a point $P(\rho, \phi, z)$ are shown in Fig. 93. The vector 1_ρ points in the direction of the most rapid increase of the coordinate variable ρ at P. (It is perpendicular to the z axis and points away from it.) The vector 1_ϕ points in the direction of the most rapid increase of ϕ at P. (It points counterclockwise if viewed from a point on the z axis whose z coordinate is larger than that of P.) The familiar vector 1_z points in the direction of the most rapid increase of z. The directions of 1_ρ at different points are, in general, different, and so are the directions of 1_ϕ. At points *on* the z axis the directions of 1_ρ and 1_ϕ cannot be defined in general, although in special cases they may be specified by a limiting process.

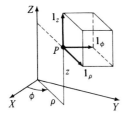 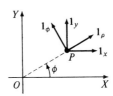

Fig. 93. The "cylindrical" base-vectors at a point P.

Figure 94

If we project the vectors of Fig. 93 onto the xy plane and include the vectors $\mathbf{1}_x$ and $\mathbf{1}_y$, we get Fig. 94. It is apparent that

$$\mathbf{1}_\rho = \mathbf{1}_x \cos\phi + \mathbf{1}_y \sin\phi, \qquad (1)$$

$$\mathbf{1}_\phi = -\mathbf{1}_x \sin\phi + \mathbf{1}_y \cos\phi. \qquad (2)$$

These equations can be written as

$$\mathbf{1}_\rho = \frac{x}{\rho}\mathbf{1}_x + \frac{y}{\rho}\mathbf{1}_y, \qquad (3)$$

$$\mathbf{1}_\phi = -\frac{y}{\rho}\mathbf{1}_x + \frac{x}{\rho}\mathbf{1}_y, \qquad (4)$$

where the coordinates x, y and ρ all pertain to the point P.

The base-vectors $\mathbf{1}_\rho$ and $\mathbf{1}_\phi$ pertaining to a point P are functions of the coordinates of P. To compute the partial derivatives of $\mathbf{1}_\rho$ and $\mathbf{1}_\phi$ with respect to ρ and ϕ, we first express these vectors in terms of the constant vectors $\mathbf{1}_x$ and $\mathbf{1}_y$. For example, in view of (1),

$$\frac{\partial}{\partial \phi}\mathbf{1}_\rho = \frac{\partial}{\partial \phi}(\mathbf{1}_x \cos\phi + \mathbf{1}_y \sin\phi) = -\mathbf{1}_x \sin\phi + \mathbf{1}_y \cos\phi. \qquad (5)$$

Therefore, in view of (2),

$$\frac{\partial}{\partial \phi}\mathbf{1}_\rho = \mathbf{1}_\phi. \qquad (6)$$

Other derivatives of this kind are listed in Appendix A, beginning with (28$^\text{A}$).

When a vector field $\mathbf{F}$ is resolved in terms of $\mathbf{1}_\rho$, $\mathbf{1}_\phi$, and $\mathbf{1}_z$, the scalar coefficients of these vectors are denoted by F_ρ, F_ϕ, and F_z, and the formula for $\mathbf{F}$ reads

$$\mathbf{F} = F_\rho \mathbf{1}_\rho + F_\phi \mathbf{1}_\phi + F_z \mathbf{1}_z. \qquad (7)$$

The three subscripted *F*'s are called *cylindrical components* of **F**; each of them may be a function of all the coordinate variables, as well as of the time *t*. The function F_ϕ is sometimes called the "circular" or "azimuthal" component and F_ρ the "radial" component of **F**, but the term "radial component" is also used in another sense, described in §36.

EXERCISES

1. Verify by dot multiplication that the right-hand sides of (3) and (4) are indeed *unit* vectors and that, at any point where they can be defined, they are mutually perpendicular. Show also that

$$F_\rho = \mathbf{F} \cdot \mathbf{1}_\rho, \quad F_\phi = \mathbf{F} \cdot \mathbf{1}_\phi \quad F_z = \mathbf{F} \cdot \mathbf{1}_z. \tag{8}$$

and

$$F = \sqrt{F_\rho^2 + F_\phi^2 + F_z^2}. \tag{9}$$

2. Verify Equations (32^A) and use the middle equation to compute the acceleration of a particle moving uniformly in a circle.
3. Derive the formulas for $\mathbf{1}_x$ and $\mathbf{1}_y$ in (19^A) and (20^A).
4. Show in detail that

$$F_x = F_\rho \cos\phi - F_\phi \sin\phi = \frac{x}{\rho}F_\rho - \frac{y}{\rho}F_\phi, \tag{10}$$

$$F_y = F_\rho \sin\phi + F_\phi \cos\phi = \frac{y}{\rho}F_\rho + \frac{x}{\rho}F_\phi, \tag{11}$$

and

$$F_\rho = F_x \cos\phi + F_y \sin\phi = \frac{x}{\rho}F_x + \frac{y}{\rho}F_y, \tag{12}$$

$$F_\phi = -F_x \sin\phi + F_y \cos\phi = -\frac{y}{\rho}F_x + \frac{x}{\rho}F_y. \tag{13}$$

5. Show that

$$F_\rho = (\mathbf{1}_x \cdot \mathbf{1}_\rho)F_x + (\mathbf{1}_y \cdot \mathbf{1}_\rho)F_y + (\mathbf{1}_z \cdot \mathbf{1}_\rho)F_z. \tag{14}$$

Then write $\mathbf{1}_\rho$ for **G** in (25^16) and explain, without explicit evaluation of the coefficients, why the coefficients of F_x and F_y in (12) and (13) are the same as the respective coefficients of $\mathbf{1}_x$ and $\mathbf{1}_y$ in (3) and (4). Note the other similarities between the equations in Exercise 4 and Equations (19^A) and (20^A).

34. TWO-DIMENSIONAL RADIAL VECTOR FIELDS

Imagine a long straight metal rod that has a circular cross section and is positively charged, and assume that the ends of the rod are so far away that their effects can be ignored. In the static case the lines that begin on positive charges will all begin on the surface of the rod. As shown in Fig. 95, they will be

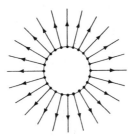

Fig. 95. The electric field outside a uniformly charged rod; it is proportional to $\frac{1}{\rho}\mathbf{1}_\rho$.

straight and perpendicular to the rod—the only shape and direction consistent with the symmetry and uniformity of the charge distribution. Figure 95 applies equally well to any plane perpendicular to the rod—the field is two-dimensional. As we go away from the rod, the density of the lines decreases, implying that the electric field gets weaker. In this example, the field strength at any point P outside the rod is, in fact, proportional to $1/\rho$, where ρ is the distance from P to the axis of the rod, which we take to be the z axis. This field is pictured in perspective in Fig. 96.

Fig. 96. Another view of the field shown in Fig. 95.

If the rod is an insulator and is charged uniformly throughout its interior, the field lines begin inside it, as they do in the plate of Fig. 72[19]. We will see in §63 that the field strength outside the rod still varies as $1/\rho$, but inside the rod it is directly proportional to ρ.

The "first-power" two-dimensional radial vector field. Radial fields of the kind illustrated above can be conveniently described with the help of the unit vector $\mathbf{1}_\rho$. For example, the formula

$$\mathbf{F} = \rho \mathbf{1}_\rho \tag{1}$$

describes a vector field whose direction at any point P not lying on the z axis is perpendicular to and away from this axis, and whose magnitude at P is equal to the distance ρ from P to this axis. A pointer map of (1) is shown in Fig. 97, where the length of the pointer pertaining to a point P (having its tail at P) is proportional to the distance from P to the axis of symmetry.

Fig. 97. A pointer map of the field $\rho \mathbf{1}_\rho$.

A line map of the field (1) is shown in Fig. 98. To get a basis for drawing such a map, we compute the flux of the field (1) across an imagined closed surface (S) that has the shape of a pill box, as in Fig. 99. On the top and bottom of (S) we have $F_n = 0$. On the curved part of (S) we have $\rho \mathbf{1}_\rho = \rho_0 \mathbf{1}_\rho$, so $F_n = \rho_0$. The area of the curved part is $2\pi\rho_0 h$, and hence the net flux emerging from the pill box is $2\pi\rho_0^2 h$, which is, therefore, also the net number of sources of the field $\rho \mathbf{1}_\rho$ contained in the box. The volume of the box is $\pi\rho_0^2 h$, and consequently the average source density in the box is $(2\pi\rho_0^2 h)/(\pi\rho_0^2 h) = 2$. Since this result does not depend on the size of the box, we may omit the word

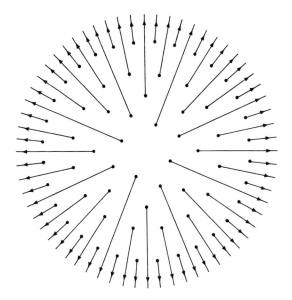

Fig. 98. A line map of the field $\rho \mathbf{1}_\rho$.

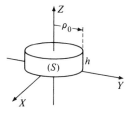

Figure 99

"average" and write

$$\left.\begin{array}{c}\text{source density of}\\ \text{the field } \rho \mathbf{1}_\rho\end{array}\right\} = 2\,\frac{\text{sources}}{\text{cubic meter}}. \tag{2}$$

Consequently, the sources of the field $\rho \mathbf{1}_\rho$ are distributed in space uniformly. Equation (2), to which we were led above by pictorial arguments, may be written as

$$\operatorname{div} \rho \mathbf{1}_\rho = 2. \tag{3}$$

The cylindrical components of the field $\rho \mathbf{1}_\rho$ are $(\rho, 0, 0)$. In view of (3^{33}),

$$\rho \mathbf{1}_\rho = x \mathbf{1}_x + y \mathbf{1}_y, \tag{4}$$

and consequently the cartesian components of $\rho \mathbf{1}_\rho$ are $(x, y, 0)$.

Remark on pointer maps. Suppose that, in an elementary problem on planar mechanics, a force is applied to a body at a point $P(x, y)$. The lever arm of this force about the point O (the origin of the coordinate frame) is then the vector whose x component is x and whose y component is y; that is, the lever arm is just the vector $\rho \mathbf{1}_\rho$, given by (4). Now, in mechanics, the lever arm about O of a force applied at P is usually represented by a pointer whose tail is at O and tip at P. But in field theory, for consistency with other pointer maps, the vector $\rho \mathbf{1}_\rho$ pertaining to a point P is best pictured by a pointer that has its tail at P and points away from the axis of symmetry, as in Fig. 97.

The "inverse first-power" two-dimensional radial vector field.
The field

$$\mathbf{F} = \frac{1}{\rho} \mathbf{1}_\rho \tag{5}$$

is directed away from the z axis; its magnitude at any point P is equal to $1/\rho$, as pictured in the pointer map of Fig. 100. To get a basis for drawing a line

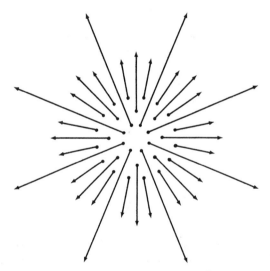

Fig. 100. A pointer map of the field $\frac{1}{\rho} \mathbf{1}_\rho$.

map, we compute the flux of (5) across the surface (S) of the pill box of Fig. 99. The values of F_n are zero on the top and bottom of (S), and $1/\rho_0$ on the curved part. Accordingly, the flux in question is $(1/\rho_0)(2\pi\rho_0 h)$, and, therefore,

$$\left. \begin{array}{l} \text{net number of sources of the field} \\ \frac{1}{\rho} \mathbf{1}_\rho \text{ located in the box} \end{array} \right\} = 2\pi h. \tag{6}$$

Thus, the number of sources contained in the box does not depend on the radius of the box. It follows that all the sources lie on the z axis, and hence the line map of the field $\rho^{-1}\mathbf{1}_\rho$, shown in Fig. 101, is especially simple. It also follows that the source density of this field at points not lying on the z axis is zero;

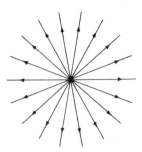

Fig. 101. A line map of the field $\dfrac{1}{\rho}\mathbf{1}_\rho$.

that is,

$$\left.\begin{array}{l}\text{source density of}\\[2pt]\text{the field }\dfrac{1}{\rho}\mathbf{1}_\rho\end{array}\right\} = 0 \quad \text{for } \rho > 0. \tag{7}$$

Accordingly,

$$\operatorname{div}\frac{1}{\rho}\mathbf{1}_\rho = 0 \quad \text{for } \rho > 0. \tag{8}$$

The average source density in the box is $(2\pi h)/(\pi\rho_0^2 h) = 2/\rho_0^2$, and becomes infinite when $\rho_0 \to 0$. Therefore, we postpone the case $\rho = 0$ until we develop the symbols required for describing this case (note Exercise 4^{64}).

Fields of the form $f(\rho)\mathbf{1}_\rho$. The general expression for two-dimensional radial vector fields is $f(\rho, \phi)\mathbf{1}_\rho$, but we will need only a simpler form, namely,

$$\mathbf{F} = f(\rho)\mathbf{1}_\rho. \tag{9}$$

The fields $\rho\mathbf{1}_\rho$ and $\rho^{-1}\mathbf{1}_\rho$ are special cases of (9). The cartesian components of (9) are

$$F_x = \frac{xf(\rho)}{\rho}, \qquad F_y = \frac{yf(\rho)}{\rho}, \qquad F_z = 0, \tag{10}$$

or, written a little differently,

$$F_x = f(\rho)\cos\phi, \qquad F_y = f(\rho)\sin\phi, \qquad F_z = 0. \tag{11}$$

142 Cylindrical and Spherical Coordinates

EXERCISES

1. Sketch separate line maps of the three fields $\rho\mathbf{1}_\rho$, $x\mathbf{1}_x$, and $y\mathbf{1}_y$ involved in (4), and verify that their source densities are consistent.

2. Derive (57^A) after the pattern of §30, but using the "cylindrical" elementary cell of Fig. 102.

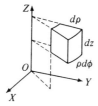

Fig. 102. An elementary cell pertaining to a cylindrical coordinate frame.

3. Verify (3) twice: using cartesian coordinates and (10^{30}), and using (57^A). Verify the divergence theorem for the case when $\mathbf{F} = \rho\mathbf{1}_\rho$ and R is the drum-shaped region of Fig. 99.

4. Let

$$\mathbf{F} = f(\rho)\mathbf{1}_\rho, \tag{12}$$

write f for $f(\rho)$ and f' for $df(\rho)/d\rho$, and show that, for $\rho > 0$,

$$\frac{\partial F_x}{\partial x} = \frac{1}{\rho^3}(y^2 f + x^2 \rho f'), \qquad \frac{\partial F_x}{\partial y} = -\frac{xy}{\rho^3}(f - \rho f'), \tag{13}$$

$$\frac{\partial F_y}{\partial x} = -\frac{xy}{\rho^3}(f - \rho f'), \qquad \frac{\partial F_y}{\partial y} = \frac{1}{\rho^3}(x^2 f + y^2 \rho f'), \tag{14}$$

$$\frac{\partial F_x}{\partial z} = \frac{\partial F_y}{\partial z} = \frac{\partial F_z}{\partial x} = \frac{\partial F_z}{\partial y} = \frac{\partial F_z}{\partial z} = 0, \tag{15}$$

$$\operatorname{div} f(\rho)\mathbf{1}_\rho = \frac{1}{\rho}(f + \rho f'), \tag{16}$$

and

$$\frac{\partial F_z}{\partial y} - \frac{\partial F_y}{\partial z} = 0, \qquad \frac{\partial F_x}{\partial z} - \frac{\partial F_z}{\partial x} = 0, \qquad \frac{\partial F_y}{\partial x} - \frac{\partial F_x}{\partial y} = 0. \tag{17}$$

Which of these equations may hold even if $\rho = 0$?

35. TWO-DIMENSIONAL CIRCULAR VECTOR FIELDS

Let us return to the magnetic intensity **H** produced by a uniformly dense current flowing in a long straight rod having a circular cross section. This field is pictured in Fig. 73[20]. If the radius of the rod is ρ_0 and the radial coordinate of a point P is ρ, then the magnitude of H at P is

$$(1^{20}) \qquad H = \begin{cases} \dfrac{\rho}{\rho_0} H_0 & \text{if } \rho \leq \rho_0 \\ \dfrac{\rho_0}{\rho} H_0 & \text{if } \rho \geq \rho_0, \end{cases} \qquad (1)$$

where H_0 is the value of H at the surface of the rod.

Now, if the axis of the rod is taken to be the z axis, then the direction of **H** at P is the same as the direction of the vector $\mathbf{1}_\phi$ erected at P, and the complete formula for **H** can be written as

$$\mathbf{H} = \begin{cases} \dfrac{\rho}{\rho_0} H_0 \mathbf{1}_\phi & \text{if } \rho \leq \rho_0 \\ \dfrac{\rho_0}{\rho} H_0 \mathbf{1}_\phi & \text{if } \rho \geq \rho_0. \end{cases} \qquad (2)$$

Thus, **H** belongs to the family of fields included in the formula

$$\mathbf{F} = f(\rho)\mathbf{1}_\phi \qquad (3)$$

and called "circular" vector fields. The lines of these fields are closed, and therefore

$$\operatorname{div} f(\rho)\mathbf{1}_\phi = 0. \qquad (4)$$

EXERCISES

1. Verify the following formulas, where S is the flat rectangular surface whose back is shown in Fig. 103:

$$\text{flux of } \rho \mathbf{1}_\phi \text{ across } S = \tfrac{1}{2}h(\rho_2^2 - \rho_1^2), \qquad (5)$$

$$\text{flux of } \frac{1}{\rho} \mathbf{1}_\phi \text{ across } S = h \ln \frac{\rho_2}{\rho_1}. \qquad (6)$$

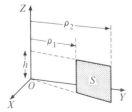

Figure 103

2. Show that in a line map of the field $\rho \mathbf{1}_\phi$ the radii of the consecutive field lines should be proportional to $\sqrt{1}, \sqrt{3}, \sqrt{5}, \cdots$.

3. Let
$$\mathbf{F} = f(\rho)\mathbf{1}_\phi, \tag{7}$$

write f for $f(\rho)$ and f' for $df(\rho)/d\rho$, and show that, for $\rho > 0$,

$$\frac{\partial F_x}{\partial x} = \frac{xy}{\rho^3}(f - \rho f'), \qquad \frac{\partial F_x}{\partial y} = -\frac{1}{\rho^3}(x^2 f + y^2 \rho f'), \tag{8}$$

$$\frac{\partial F_y}{\partial x} = \frac{1}{\rho^3}(y^2 f + x^2 \rho f'), \qquad \frac{\partial F_y}{\partial y} = -\frac{xy}{\rho^3}(f - \rho f'), \tag{9}$$

$$\frac{\partial F_x}{\partial z} = \frac{\partial F_y}{\partial z} = \frac{\partial F_z}{\partial x} = \frac{\partial F_z}{\partial y} = \frac{\partial F_z}{\partial z} = 0, \tag{10}$$

$$\operatorname{div} f(\rho)\mathbf{1}_\phi = 0, \tag{11}$$

and

$$\frac{\partial F_z}{\partial y} - \frac{\partial F_y}{\partial z} = 0, \qquad \frac{\partial F_x}{\partial z} - \frac{\partial F_z}{\partial x} = 0, \qquad \frac{\partial F_y}{\partial x} - \frac{\partial F_x}{\partial y} = \frac{1}{\rho}(f + \rho f'). \tag{12}$$

4. Recall Exercise 8[14] and show that, when electric charge flows past a cylindrical hole, the cylindrical components of the current density are

$$J_\rho = J_0\left(1 - \frac{b^2}{\rho^2}\right)\cos\phi, \qquad J_\phi = -J_0\left(1 + \frac{b^2}{\rho^2}\right)\sin\phi, \qquad J_z = 0. \tag{13}$$

Then use (57[A]) to verify that div $\mathbf{J} = 0$.

36. THE VECTORS $\mathbf{1}_r$, $\mathbf{1}_\theta$, AND $\mathbf{1}_\phi$

A triplet of base-vectors pertaining to a spherical coordinate frame and to a point P is shown in Fig. 104. The unit vectors $\mathbf{1}_r$, $\mathbf{1}_\theta$, and $\mathbf{1}_\phi$ point, respectively,

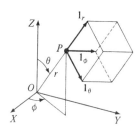

Fig. 104. The "spherical" base-vectors at a point P.

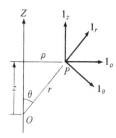

Figure 105

in the directions of the most rapid increase, at P, of the coordinate variables r, θ, and ϕ.

Figure 105, whose plane contains the point P and the z axis, shows the vectors 1_ρ, 1_z, 1_r, and 1_θ pertaining to P. It is apparent that

$$1_r = 1_\rho \sin\theta + 1_z \cos\theta, \tag{1}$$

$$1_\theta = 1_\rho \cos\theta - 1_z \sin\theta. \tag{2}$$

Another way of writing (1) and (2) is

$$1_r = \frac{\rho}{r} 1_\rho + \frac{z}{r} 1_z, \tag{3}$$

$$1_\theta = \frac{z}{r} 1_\rho - \frac{\rho}{r} 1_z. \tag{4}$$

It now follows from (3^{33}) and (4^{33}) that

$$1_r = \frac{x}{r} 1_x + \frac{y}{r} 1_y + \frac{z}{r} 1_z, \tag{5}$$

$$1_\theta = \frac{xz}{\rho r} 1_x + \frac{yz}{\rho r} 1_y - \frac{\rho}{r} 1_z, \tag{6}$$

$$1_\phi = -\frac{y}{\rho} 1_x + \frac{x}{\rho} 1_y, \tag{7}$$

where x, y, z, ρ, and r are the various coordinates of P.

Except for possible indeterminacies at points on the polar axis, any vector field $\mathbf{F}$ can be expressed at any point in the form

$$\mathbf{F} = F_r 1_r + F_\theta 1_\theta + F_\phi 1_\phi, \tag{8}$$

where F_r, F_θ, and F_ϕ are scalar fields, called the *spherical components* of $\mathbf{F}$. In particular, F_ϕ is called the "circular" component of $\mathbf{F}$; when confusion with the cylindrical component F_ρ is not likely, F_r is called the "radial" component of $\mathbf{F}$.

Arguments similar to those of Exercises 4^{33} and 5^{33} show that the formulas expressing the spherical components of a vector field $\mathbf{F}$ in terms of its cartesian components can be obtained by writing F's for 1's in (5), (6), and (7). That is,

$$F_r = \frac{x}{r} F_x + \frac{y}{r} F_y + \frac{z}{r} F_z, \tag{9}$$

$$F_\theta = \frac{xz}{\rho r} F_x + \frac{yz}{\rho r} F_y - \frac{\rho}{r} F_z, \tag{10}$$

$$F_\phi = -\frac{y}{\rho} F_x + \frac{x}{\rho} F_y \tag{11}$$

To illustrate the use of these transformations, we return for a moment to the fourth example in §26. The problem was to find the normal component of the field

$$\mathbf{F} = z\mathbf{1}_z \tag{12}$$

relative to a spherical surface having the radius r_0 and centered on the origin, as in Fig. 85[26]. This component is simply the component F_r of (12). According to (12), we have $F_x = 0$, $F_y = 0$, and $F_z = z$. Therefore, according to (9),

$$F_r = \frac{z^2}{r} = \frac{(r\cos\theta)^2}{r} = r\cos^2\theta. \tag{13}$$

When we set $r = r_0$, we get (11^{26}).

EXERCISES

1. Investigate the right-hand sides of (5), (6), and (7) along the lines of Exercise 1^{33}. Also, show that

$$F_r = \mathbf{F}\cdot\mathbf{1}_r, \quad F_\theta = \mathbf{F}\cdot\mathbf{1}_\theta, \quad F_\phi = \mathbf{F}\cdot\mathbf{1}_\phi, \tag{14}$$

and

$$F = \sqrt{F_r^2 + F_\theta^2 + F_\phi^2}. \tag{15}$$

2. Check (25^A) and (26^A).
3. Check the expressions for $\mathbf{1}_x$, $\mathbf{1}_y$, and $\mathbf{1}_z$ given in (22^A), (23^A), and (24^A).
4. Check the nine derivatives in (34^A), (35^A), and (36^A).
5. The cartesian components of the current density $\mathbf{J}$ for the case of a spherical cavity in a conductor are given in Exercise 2^{16}. Show that in this case

$$J_r = J_0\left(1 - \frac{r_0^3}{r^3}\right)\cos\theta, \quad J_\theta = -J_0\left(1 + \frac{r_0^3}{2r^3}\right)\sin\theta, \quad J_\phi = 0. \tag{16}$$

Then verify that (a) at points on the surface of the cavity the flow is tangential to the surface; (b) at points in the xy plane the flow is normal to this plane; and (c) at points on the equator of the cavity $J = 1.5J_0$.

6.′ Use (16) and (65^A) to show that div $\mathbf{J} = 0$.
7. Derive (65^A) after the pattern of §30 but using the "spherical" elementary cell, shown in Fig. 262^A.

37. SPHERICALLY SYMMETRIC VECTOR FIELDS

Fields of the form

$$\mathbf{F} = f(r)\mathbf{1}_r, \tag{1}$$

where $f(r)$ is a scalar function of the coordinate variable r, are called *spherically symmetric* or, when confusion with fields of the form $f(\rho)\mathbf{1}_\rho$ is not likely, simply *radial* vector fields. We will consider two examples.

The spherical "first-power" vector field. If $f(r) = r$ the field (1) is denoted by $\mathbf{r}$; that is, the definition of the symbol $\mathbf{r}$ reads

$$\mathbf{r} = r\mathbf{1}_r. \tag{2}$$

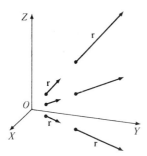

Figure 106

The magnitude r of the field $\mathbf{r}$ at a point P is thus equal to the distance of P from the origin. Pointers representing this field at six points in space are drawn in Fig. 106. According to (5^{36}), another way of writing (2) is

$$\mathbf{r} = x\mathbf{1}_x + y\mathbf{1}_y + z\mathbf{1}_z. \tag{3}$$

The remarks on pointer maps that follow (4^{34}) apply also to the field $\mathbf{r}$. This field satisfies the equation

$$\text{div } \mathbf{r} = 3, \tag{4}$$

which implies that the source density of the field $\mathbf{r}$ is uniform and equal to 3 sources per cubic meter. The field $\mathbf{r}$ is of interest in dealing, for example, with the electric field inside a nonconducting sphere, uniformly charged throughout its interior.

The spherical "inverse-square" vector field. Another important spherically symmetric field is

$$\mathbf{F} = \frac{1}{r^2} \mathbf{1}_r. \qquad (5)$$

Its normal component relative to the spherical surface of radius r_0 shown in Fig. 107 is $1/r_0^2$, and its net outward flux across this surface is $(1/r_0^2)(4\pi r_0^2)$

Figure 107

$= 4\pi$. Therefore, this surface encloses 4π sources of the field. This number does not depend on the radius of the surface, so all the sources must lie at the origin O. Consideration of source density then leads to the equation

$$\operatorname{div} \frac{1}{r^2} \mathbf{1}_r = 0 \quad \text{if } r > 0. \qquad (6)$$

A formula that includes the origin is given in (19^{64}).

An "ideal" line map of the field (5) would consist of "4π straight lines," which all start at the origin O, run to infinity, and pierce any spherical surface centered on O at points uniformly distributed over this surface.

EXERCISES

1. Show in three ways that div $\mathbf{r} = 3$. First, use (2) and (65^A); next, use (3) and (49^A); finally, find the flux of $\mathbf{r}$ across the surface of some convenient region, find the average source density, and so on.
2. Use (65^A) to verify (6).

CHAPTER 10

Conservation of Charge

In this chapter we express in field-theoretic terms the law of conservation of electric charge. In circuit theory, this law leads to Kirchhoff's first law—the "current" or "junction" law. Since Kirchhoff's current law is already familiar to the reader, we begin by translating it into field-theoretic language. The result is Equation (9^{38}), which is a special form of the law of conservation of charge, restricted to situations in which all the relevant physical quantities stay constant. The complete law of conservation of charge is formulated in §40.

38. KIRCHHOFF'S CURRENT LAW

Let us say for the moment that a current is positive if it flows outward from a junction of wires and negative if it flows toward the junction. The outward directions can be indicated by arrows, as in Fig. 108. These arrows do *not*

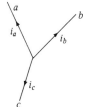

Figure 108

indicate the directions of the currents i_a, i_b, and i_c that flow in the branches labeled a, b, and c. The current i_a, for example, may flow either with or against the arrow marked on branch a; however, if it flows with the arrow, we take the number i_a to be positive, while in the second case we take i_a to be negative.

Kirchhoff's current law then requires that, if the currents in Fig. 108 remain constant, we have

$$i_a + i_b + i_c = 0. \tag{1}$$

This equation implies that, unless all three currents are zero, one or two of them must be negative and flow toward the junction.

Figure 108 is magnified in Fig. 109, which also shows an imagined surface that completely encloses the junction. The three portions of this surface located inside the wires are marked S_a, S_b, and S_c; the rest of it, located in air, is marked

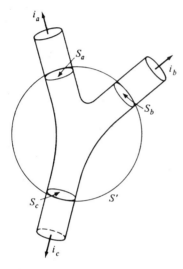

Figure 109

S'. In view of (2^{15}), the expressions for the currents in terms of current density are

$$i_a = \iint_{S_a} J_n \, da, \quad i_b = \iint_{S_b} J_n \, da, \quad i_c = \iint_{S_c} J_n \, da. \tag{2}$$

Therefore, (1) can be written as

$$\iint_{S_a} J_n \, da + \iint_{S_b} J_n \, da + \iint_{S_c} J_n \, da = 0. \tag{3}$$

At all points on S' we have $J = 0$, so that

$$\iint_{S'} J_n \, da = 0. \tag{4}$$

Combining (3) and (4), and remembering that S_a, S_b, S_c, and S' form together a closed surface, say (S), we conclude that

$$\iint_{(S)} J_n \, da = 0. \tag{5}$$

Equation (5) means that the net outward current flowing across a closed surface is zero. We have proved (5) only for the special case shown in Fig. 109. However, if the relevant currents and current densities are constant, one can modify the details of our arguments and derive (5) for *every* imagined closed surface (S), regardless of the configuration of the circuit and the shape and location of (S), including closed surfaces that lie either partly or completely inside a conductor, or completely outside it.

When applied to **J**, the divergence theorem (2^{31}) reads

$$\iiint_R \text{div } \mathbf{J} \, dv = \iint_{(S)} J_n \, da, \tag{6}$$

where R is the region bounded by (S). It follows that (5) can be written as

$$\iiint_R \text{div } \mathbf{J} \, dv = 0. \tag{7}$$

Furthermore, having claimed (without detailed proof) that (5) holds for *every* closed surface (S), we can replace (7) by

$$\iiint_{\substack{\text{every}\\\text{region}}} \text{div } \mathbf{J} \, dv = 0. \tag{8}$$

Recalling (45^5) and (46^5), and assuming that all the relevant quantities (current densities, densities of surface charges, and so on) remain constant with time, we finally conclude that

$$\text{div } \mathbf{J} = 0 \tag{9}$$

at every point in a conductor. Equation (9) states Kirchhoff's current law in its field form. It means pictorially that, under the conditions stated above, the direction lines of **J** have no beginnings or ends.

We have anticipated (9) for special cases in (11^{14}) and in Exercise 6^{36}.

39. LEAKAGE CURRENT IN A CABLE

To illustrate the use of the equation

$$\text{div } \mathbf{J} = 0, \tag{1}$$

we return to the leakage current in a cable under the simplifying assumptions listed in footnote 2[13]. We consider a portion of the cable l meters long (Fig. 110), restrict our attention to the insulating sheath, introduce a cylindrical coordinate frame whose z axis is the axis of the core, and write

$$\mathbf{J} = J_\rho \mathbf{1}_\rho + J_\phi \mathbf{1}_\phi + J_z \mathbf{1}_z. \tag{2}$$

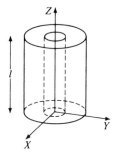

Fig. 110. The insulating sheath of an armored cable.

Since we are ignoring magnetic fields, we are also ignoring any tendency of the leakage charges to move in circles; accordingly,

$$J_\phi = 0. \tag{3}$$

Furthermore, since we are ignoring any potential drop along the core, we are ignoring any tendency of the leakage charges to move parallel to the cable; accordingly,

$$J_z = 0. \tag{4}$$

Consequently, (2) reduces to

$$\mathbf{J} = J_\rho \mathbf{1}_\rho. \tag{5}$$

Consider now a point $P(\rho, \phi, z)$ in the insulation. The current density at P does not depend on ϕ because the cable is axially symmetric. Nor does it depend on z, because we are ignoring any potential drop along the core, so our problem is two-dimensional. Therefore, the factor J_ρ in (5) can depend only on ρ, and (5) becomes

$$\mathbf{J} = f(\rho)\mathbf{1}_\rho. \tag{6}$$

Our next step is to identify the function $f(\rho)$ in (6) for which div $\mathbf{J} = 0$. Using (57$^\text{A}$) and remembering (3), (4), and (6), we get

$$\frac{1}{\rho}\frac{d}{d\rho}[\rho f(\rho)] = 0. \tag{7}$$

The factor $1/\rho$ is finite in the insulation; therefore, we may cancel it out and conclude that the product $\rho f(\rho)$ is a constant, say k. Consequently, $f(\rho) = k/\rho$ and

$$\mathbf{J} = \frac{k}{\rho}\,\mathbf{1}_\rho. \tag{8}$$

A line map of this inverse first-power field is shown in Fig. 111.

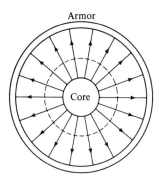

Fig. 111. A line map of the density of the leakage current in the insulation of a cable.

It remains to express k in terms of the leakage current. Let i be the current that leaks from the core to the armor in a portion of the cable l meters long. Let S be an imagined open-ended cylindrical surface, coaxial with the cable, as indicated by the dotted circle in Fig. 111; the "length" of S (at right angles to the page) is l; its radius is ρ_0, say. Take the side of S facing the armor as positive, so that, on S, $J_n = J_\rho = k/\rho_0$. The current i is the flux of $\mathbf{J}$ across S, so

$$i = \iint_S J_n\,da = \int_0^l \int_0^{2\pi} \frac{k}{\rho_0}(\rho_0\,d\phi\,dz) = 2\pi k l. \tag{9}$$

Consequently, $k = i/2\pi l$, and (8) becomes

$$\mathbf{J} = \frac{i}{2\pi l}\frac{1}{\rho}\,\mathbf{1}_\rho. \tag{10}$$

The ratio i/l is the leakage current per unit of length of the cable (amperes/meter).

Suppose that, as in §13, a battery is connected between the copper core and the steel armor at one end of the cable, but the other end of the core is not connected electrically to anything. The leakage current i pertaining to any portion of the cable flows toward this portion in the core and flows back to the battery in the armor. Therefore—except in the idealized case of a core and an armor that have infinite conductivities—we conclude from Ohm's law that the existence of a leakage current implies the presence of an electric field in the core; in particular, the z component of **E** does not vanish there, and does not permit ignoring the potential drop along the core.

In §70 we will discuss interfaces between different mediums and will find that the component of **E** tangential to the interface of two mediums is continuous. This means in the present case that $E_z \neq 0$ not only in the core and armor, but also in the insulation. Thus, as long as we are allowing for leakage currents, Equation (4)—and with it the formulas (8) and (10)—cannot be true in principle, even though they are good approximations in practice. In the exercises which follow, the reader should assume that the conductivities of the core and the armor are high enough, and that the conductivity of the insulation is low enough, to justify using (4) and (8).

EXERCISES

1. Contrast our reason for writing $J_\phi = 0$ with our reason for saying that J_ρ is independent of ϕ.
2. The insulating sheath of an armored cable consists of several axially symmetric layers; each is homogeneous but has a different conductivity. Show that (10) still holds at every point in the insulation.
3. Derive (10) "without calculus," using the simplest arguments that you can think of.

40. CONSERVATION OF CHARGE

The law of conservation of charge states that electric charge cannot be created or destroyed, but can only be moved from place to place. In different words, it states that electric charge cannot appear in or disappear from a region without crossing the boundary of this region. The following wording is most convenient for our purposes: *The time rate of decrease of the net charge contained in a region R is equal to the time rate at which, in effect, positive charge is leaving R across the boundary (S) of R.* We will now write this law in its field form, restricting ourselves to stationary conductors and insulators.

Let the region R have a fixed shape and remain stationary relative to the observer. Let $q(t)$, or simply q, be the net charge contained in R at the instant t. Let $i(t)$, or simply i, be the net current flowing at the instant t across the surface (S) of R; that is, let i be the time rate at which positive charge is, in effect, crossing (S) from back to front at the instant t. The statement made above in italics can then be written as

$$i = -\frac{dq}{dt}. \qquad (1)$$

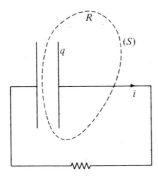

Fig. 112. A circuit illustrating the equation $i = -dq/dt$.

To illustrate, we consider a capacitor discharging through a resistor. The dotted line in Fig. 112 indicates the closed surface (S) of a region R that includes the right-hand plate but not the other plate. We write q for the charge on the right-hand plate and regard the current i as positive if it crosses (S) from back to front. Equation (1) states in this example that the current (amperes) flowing out of the right-hand plate is equal to the time rate of decrease (coulombs per second) of the charge located on this plate.

Equation (1) holds, of course, whether the charge q in Fig. 112 is positive or negative. If q is positive, the numerical value of q will be decreasing toward zero, the numerical value of dq/dt will be negative, and (1) will tell us that the current i is positive and hence flows in the direction of the arrow. Similarly, if q is negative, it will be increasing toward zero, dq/dt will be positive, and (1) will tell us that the current i is negative and hence flows oppositely to the arrow.[1]

We will now rewrite (1) in a different form. Suppose that the interior of a body is charged and denote the charge density by ρ (not to be confused with a cylindrical coordinate). In general, this density varies from place to place and from one instant to another; for example, if cartesian coordinates are used, we have, in general,

$$\rho = \rho(x, y, z, t). \tag{2}$$

The net charge contained in any region R at any instant t is

$$q = \iiint_R \rho \, dv. \tag{3}$$

---

[1] In circuit theory the letter q is used not for the charge located on one of the plates (as we are using it), but for what is called the "charge stored in the capacitor." In the case of a series RC circuit, for example, the signs are then adjusted as follows: One of the two senses along the circuit is arbitrarily chosen as positive; the current is regarded as positive if it has this sense, and the charge stored in the capacitor is regarded as positive if the capacitor has been charged by a positive current. The equation that then replaces (1) is $i = dq/dt$.

While the charges originally placed in the body rearrange themselves, driven by their mutual attractions and repulsions, they constitute currents. If the body is connected to batteries, additional currents may flow in it. We will write **J** for the total current density, whatever its causes may be. The net current *leaving* a fixed region R at the instant t is then

$$i = \iint_{(S)} J_n \, da, \tag{4}$$

where (S) is the boundary of R.

In view of (3) and (4), we may write (1) as

$$\iint_{(S)} J_n \, da = -\frac{d}{dt} \iiint_R \rho \, dv, \tag{5}$$

which is the "integral form" of the law of conservation of charge. Note that (5) holds at every instant t for any fixed region R, whatever its location relative to conducting bodies or insulators; for example, R may lie completely or only partly inside a conductor, and it may include both conducting materials and insulators. Since R is fixed, the formula (43^5) permits us to rewrite (5) as

$$\iint_{(S)} J_n \, da = -\iiint_R \frac{\partial \rho}{\partial t} \, dv. \tag{6}$$

The steps that lead from (6) to a field equation are similar to those that we took in §38. Using the divergence theorem, we write (6) as

$$\iiint_R \left(\text{div } \mathbf{J} + \frac{\partial \rho}{\partial t} \right) dv = 0. \tag{7}$$

This equation holds for every fixed region R, and hence we conclude that the integrand is identically zero and that, at all points in space and at all times,

$$\text{div } \mathbf{J} = -\frac{\partial \rho}{\partial t}. \tag{8}$$

Equation (8) is the field form of the law of conservation of charge.

To bring out the pictorial significance of (8), suppose that the charge density ρ is decreasing at all points in a region R. The net charge contained in R will then be decreasing and, according to (8), the divergence of **J** at all points in R will be positive. The "natural" pictorial interpretation of the preceding sentence is that the lines of **J** begin in R and end in regions where the net charge is increasing. Strictly speaking, however, all that (8) implies pictorially is that, if the net charge in a region R is decreasing, then more lines of **J** begin in R than end there. But since field lines can at best give only an imperfect description of a vector field, the more precise statement does not ordinarily add much to clarity.

If the relevant current and charge densities do not vary with time, then (8) reduces to the field form of Kirchhoff's current law, namely,

$$(9^{38}) \qquad \qquad \text{div } \mathbf{J} = 0. \qquad \qquad (9)$$

EXERCISE

1. Verify (1) for the case of a capacitor that is being charged.

CHAPTER 11

Electric Intensity and the Laws of Coulomb, Ohm, and Joule

So far in this book we have barely mentioned the *causes* of conduction currents. As remarked in the chart in the preface, these currents are driven by electric fields. In this chapter we turn from electric currents to electric fields, but without as yet any detailed discussion of *their* causes.

This book is restricted to two types of electric fields—electrostatic fields and electric fields induced magnetically. We describe them briefly in §41. The definition of the term "electric intensity" follows in §42. We then return to conduction currents in isotropic conductors and derive the field form of Ohm's law in §44. The chapter ends with a derivation of the field form of Joule's law for the time rate at which heat is generated per unit volume at any point in a current-carrying conductor.

In defining such quantities as electric intensity at a point and electromotive force in a path, we will speak of hypothetical "test charges" and will imagine placing them at particular points, or carrying them along actual or imagined paths. As long as we restrict ourselves to electric fields in vacuum, this procedure is straightforward. But when we come to ponderable bodies—conductors and insulators—the reader may well have misgivings about this procedure. The fact is that, while detailed analysis of electric fields in solids is quite complicated, the conclusions obtained rather simply with the help of imagined test charges are adequate for many purposes—and, in particular, for the purposes of this book. One may well say that macroscopic electric fields within material bodies are theoretical devices and not directly measurable quantities.[1]

---

[1] Report of the Coulomb's Law Committee of the A.A.P.T., *American Journal of Physics*, Volume 18 (1950), p. 17.

41. ELECTRIC FIELDS

Suppose that a small uncharged body is held at a point P and that, when an electric charge is placed on it, an extra force is found to act upon it, an extra push or pull in one direction or another. We then say that P is located in an *electric field*. For example, if an uncharged pith ball is hung on a thread in the air midway between the (vertical) plates of a charged capacitor, the thread will be vertical; but if the ball is positively charged, it will deflect as in Fig. 113. Therefore, we conclude that the pith ball is located in an electric field.

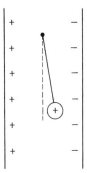

Figure 113

We reserve the general term "electric field" (and also "magnetic field") for qualitative statements. Quantitative terms, such as "electric intensity" and "electric displacement density," will be defined presently. In this book we restrict ourselves to electromagnetic phenomena that take place in conductors and insulators that are at rest relative to the observer, and we also confine ourselves to electric fields of only two types:

- Electric fields associated with static charges; they are derivable from Coulomb's inverse-square law and are called *electrostatic* fields.
- Electric fields associated with time-varying magnetic fields; they are derivable from Faraday's law of induction, and therefore we call them *Faraday* fields.

In particular, we will take it for granted that chemical cells have electromotive forces defined in §1 and used in such equations as (5¹), but we will not be concerned with regions in which these emf's are produced by intricate atomic interactions.

Consider the space *outside* an electric generator whose two terminals protrude from its boxlike metal shell as in Fig. 114, and assume that this shell shields the space outside it from all electromagnetic effects taking place inside. An ordinary rotating-armature generator may approximate such an ideally shielded generator if its stray magnetic field is sufficiently small. It is then immaterial whether the box contains an electromagnetic generator of the rotating

Figure 114

kind, a chemical cell, an electrostatic machine, or some other device. In any case, the electric field *outside* the box is produced directly by the charges located on the terminals and is, therefore, a field of the *electrostatic* type. If these charges should change with time (as they would if the box contains an a.c. generator), this field would actually be a *quasi-static* field, provided one can ignore the radiative effects that we will study later and that become important at high frequencies.

If a shielded generator is of the constant-voltage kind and if a coil is connected across its terminals, the electric field outside the box will be produced directly by (a) the charges located on the terminals and (b) the charges that develop on the surface of the coil. This field is, therefore, also electrostatic. (In addition, the current flowing in the coil produces a magnetic field outside the box.)

By contrast, consider a region containing a short-circuited secondary winding of a transformer. The alternating magnetic field produces there a periodic electric field of the Faraday type. The surface charges that develop periodically on this winding and elsewhere produce a periodic electric field of the electrostatic type. Consequently, in addition to a magnetic field, this region is pervaded by a superposition of two types of electric fields.

42. ELECTRIC INTENSITY

Electric fields exert forces on electric charges. The *force-exerting property* of these fields leads to the concept of a certain vector field, denoted by **E**, defined below by Equation (5), and called "electric field strength," "electric field intensity," or simply *electric intensity*.

We will have to speak of "test bodies" carrying "test charges." A test body is a small body, usually hypothetical; it carries a known *positive* charge, say q^* coulombs, and can be placed (in principle) at any point in the electric field whose force-exerting property is being explored. We write $\mathbf{F}_e$ for the force (newtons) exerted by the electric field on the test charge, and hence also on the test body itself. To begin with, we consider electric fields in free space.

Suppose that the three point charges (q_1, q_2, and q_3 coulombs) of Fig. 115 are kept in fixed positions in a vacuum. If a test charge (q^* coulombs) is put at a point P near these charges (or near any collection of immovable charges), experiment shows that

Fig. 115. The force exerted by positive static charges on a test chage q^* placed at a point P.

(a) The *direction* of the force $\mathbf{F}_e$ acting on the test charge is independent of the (positive) numerical value of q^*.
(b) The *magnitude* of $\mathbf{F}_e$ is directly proportional to the numerical value of q^*.

In brief, experiment shows that both the direction and the magnitude of the vector

$$\frac{\mathbf{F}_e}{q^*} \tag{1}$$

are independent of the numerical value of q^*. Consequently, this vector has to do with the electric field at P and not with the test charge. This vector is the *electric intensity* at P and is denoted by $\mathbf{E}(P)$ or simply $\mathbf{E}$. To summarize: The electric intensity $\mathbf{E}$ is a vector field, defined at any point P by the equation

$$\mathbf{E} = \frac{\mathbf{F}_e}{q^*}, \tag{2}$$

where $\mathbf{F}_e$ is the force that would be exerted by the electric field on a test charge of q^* coulombs placed at P. Another way of writing (2) is

$$\mathbf{F}_e = q^* \mathbf{E}. \tag{3}$$

The field $\mathbf{E}$ can be pictured by pointer maps and line maps. Lines of $\mathbf{E}$ are often called "lines of (electric) force." The magnitude of $\mathbf{E}$ is denoted by E.

Note that $\mathbf{E}$ is an *electromechanical* rather than a purely electrical quantity. The MKS-Giorgi unit for measuring it that follows directly from (2) is the newton per coulomb. In view of the equivalence

$$1 \frac{\text{newton}}{\text{coulomb}} = 1 \frac{\text{volt}}{\text{meter}}, \tag{4}$$

the volt per meter is an alternative unit, which we will later find equally suggestive.

Experiment shows that the force exerted by a charge q_1 on a charge q is not affected by the presence of other charges acting on q. Accordingly, the total force exerted on a charge q jointly by several other charges can be found by the parallelogram rule from the forces that would be exerted on q by these charges acting one at a time. It then follows from (2) that electric intensities obey the parallelogram rule; we anticipated this fact when we called **E** a *vector* field.

We have assumed so far that the test charge q^* does not affect the positions of the charges responsible for the electric field that is being explored, but this condition does not always hold. For example, when a test charge is put between the plates of a capacitor, the charges located on the plates rearrange themselves, and the electric field they produce is no longer what it was in the absence of the test charge. To minimize this effect, the test charge must be small. In fact, a definition of **E** that is in principle better than (2) is

$$\mathbf{E} = \lim_{q^* \to 0} \frac{\mathbf{F}_e}{q^*}. \tag{5}$$

We will, however, continue to use (2) and (3), and will ask the reader to keep in mind the condition $q^* \to 0$.

Figure 115 pertains to an electric field produced directly by electric charges. But the experimental results (a) and (b) listed above hold also for electric fields produced by time-varying magnetic fields, and so does the parallelogram rule. Therefore, the definition (5) of **E** applies to electric fields of either type, and so does its shortened form (2). Practical measurements of **E** are usually based not directly on the definition (5), but on equations derived from it.

When we wish to stress the fact that an electric field is electrostatic, we write its intensity as $\mathbf{E}^{es}$ or $\mathbf{E}'$; also, for emphasis, we sometimes write $\mathbf{E}^f$ or $\mathbf{E}''$ in the case of Faraday fields.

Coulomb's law in free space. Suppose that a charge q is located at the origin O of a coordinate frame, and assume that, except for this charge, all space is empty. To compute from Coulomb's law the electric intensity at a point P, lying r meters from O, we proceed as follows.

Imagine placing at P a (positive) test charge q^*. Coulomb's law states that the force of interaction between two point charges is directly proportional to the product of the charges and inversely proportional to the square of the distance between them; that is, the force $\mathbf{F}_e$ acting on q^* is proportional to qq^*/r^2. If the charges are located in free space and if q and q^* are expressed in coulombs and r in meters, the experimental value of the constant of proportionality proves to be 9×10^9 meters/farad, to within a fraction of one per cent. If q is positive, the charge q^* will be repelled by q, so that the force acting on q^* will have the direction of the vector $\mathbf{1}_r$ erected at P; consequently,

$$\mathbf{F}_e = 9 \times 10^9 \frac{qq^*}{r^2} \mathbf{1}_r; \tag{6}$$

in particular, $\mathbf{F}_e$ points away from O. If q is negative, the right-hand side of (6) reverses direction, so that (6) is correct for either sign of q. It now follows from (2) that, in terms of the MKS-Giorgi units of measurement, the formula for the electric intensity produced *in free space* by a point charge q, located at the origin O, is, very nearly,

$$\mathbf{E} = 9 \times 10^9 \frac{q}{r^2} \mathbf{1}_r. \tag{7}$$

The factor of proportionality is so large because minute fractions of a coulomb produce appreciable electric fields at considerable distances away (Exercise 1).

Dielectrics. If a point P lies in a vacuum or in a rarefied gas, placing a test charge at P is conceptually simple. If P lies in a solid dielectric, say glass, then a small cavity must be imagined to be made around P before a test charge q^* can be placed at P; however, because of "polarization," the force acting on q^* then proves to depend on the *shape* of the cavity, and only a properly oriented needle-shaped cavity will do for determining $\mathbf{E}$ at P (note Fig. 166[62]). In liquids the situation is more complicated.

EXERCISES

1. Show that one centimeter away from a point charge of 1 picacoulomb (10^{-12} coulomb) the electric intensity is nearly 90 volts per meter.

2. Let m be the mass of the pith ball in Fig. 113, q the charge on it, and α the angle that the (massless) thread makes with the vertical. Derive the formula $E = (mg/q) \tan \alpha$ for the magnitude of the electric intensity at the point of equilibrium of the center of the ball. When is this formula a good approximation to the value of E at this point in the absence of the pith ball?

3. Verify the equivalence (4). Verify that, in our system of units, the factor 9×10^9 in (7) can be expressed in meters per farad.

4. Two equal and opposite point charges, q and $-q$, held at a fixed distance l apart, form together an *electric dipole*; for definiteness, we take q to be positive. Let $\mathbf{1}$ be the unit vector pointing from $-q$ toward q. The vector $ql\mathbf{1}$ is then called the *electric moment* of the dipole; we denote it by $\mathbf{p}_e$:

$$\mathbf{p}_e = ql\mathbf{1}. \tag{8}$$

An electric dipole is held in a uniform electric field of intensity $\mathbf{E}$. Show that this field will apply to it a torque of magnitude $qlE \sin \gamma$, where γ is the angle between $\mathbf{E}$ and $\mathbf{p}_e$.

5. Show that the definition of $\mathbf{E}$ is not affected by the choice of sign of the test charge q^*. It is convenient, however, always to use positive test charges.

43. CHARGE-DRIVING FORCES

If we tie a knot in an insulated wire and connect its ends across a battery, how are the electrons guided through the knot? Even if the wire is straight, how do they know which way to go? When electrons approach the junction of several wires, how do they know which branch to enter? The answers to these questions hinge on the fact that currents in wires are guided by the *forces* acting on the moving charges that constitute the currents.[2] The remark "An electron sees a place of higher potential, and so it goes there" is both vague and misleading; for example, the electrons circulating in the shorted secondary winding of a transformer certainly cannot *all* be moving from places of lower potential to places where the potential is higher. If one can say that an electron is "aware" of anything at an instant t, it is aware only of the *forces* that act *on it* at the place where it is located at this instant.

Experiment shows that to produce a current in a stationary wire an *electric field* must be set up inside the wire by some external means. We will call this field for a moment the "applied" field, to distinguish it from the atomic fields ever present in conductors and nonconductors alike. The applied electric fields considered in this book are restricted to the two types described in §41, and they include the fields of the electrostatic type produced by the surface charges that usually develop on current-carrying conductors.

To put things roughly, the applied electric field exerts forces on the conduction electrons and accelerates them. But, as remarked in §1, their speeds do not increase for long, because they keep colliding with the atoms comprising the conductor and, therefore, move in a stop-and-go fashion. The kinetic energy they gain while accelerating between collisions is transferred to the atomic lattice during the collisions and shows up as heat. This picture implies that the current density J at a point P in a stationary conductor is a function of the intensity E of the applied electric field at P. But this picture is too rough to specify the exact form of this function because it ignores the details of the interaction of the conduction electrons with the atomic lattice. We will see in §44 that in the simplest case ("ohmic" conductors) we have $\mathbf{J} = g\mathbf{E}$, where g is the conductivity at P.

Charge-driving forces. If $\mathbf{E}$ is the electric intensity at a point P, then the force of electric origin acting on a test charge q^* placed at P is

$$(3^{42}) \qquad \mathbf{F}_e = q^*\mathbf{E}. \qquad (1)$$

Now a question: If we should write -1.60×10^{-19} coulombs (the electronic charge) for q^*, will (1) give correctly the force (newtons) acting on an *electron*

---

[2] Similarly, electromagnetic waves in metallic channels are guided indirectly by the forces acting on the charges moving in the walls of the channels.

located at P? The answer is not obvious. Indeed, the charge q^* in (1) is an idealized test charge devoid of all physical properties except a Coulomb interaction with other charges; but the electron, besides carrying a charge, has several very special properties of its own. These properties, however, can be expected to show up only in situations involving the details of atomic structure, and hence one might surmise that (1) will hold for an electron that is not interacting with atoms to any appreciable extent. Experiment shows that this surmise is correct: when interactions with atoms can be ignored altogether, and even when dealing with conduction electrons in isotropic ohmic conductors pervaded by electrostatic fields and electric fields of the Faraday type, we may use the equation

$$\mathbf{F}_e = q\mathbf{E} \qquad (2)$$

to find the force acting on an *actual* electron. It is the simplicity of this situation that permitted the field theory to make progress in the problem of conduction long before the discovery of the electron, and that permits us now to revert to the conventional assumption that positive charges do the moving.

In regions pervaded by electric fields of types other than the two defined in §41, the electric intensity $\mathbf{E}$ is still defined by (1), which refers to a hypothetical test charge q^*. But the force acting on an *actual* charge q is then not necessarily equal to $q\mathbf{E}$. For example, the forces acting on the actual charges at the electrodes of a chemical cell depend on the finer details of the structure of the ions and the solids and liquids that are involved, such as the relative positions of filled and unfilled electronic energy levels. An extraneous test charge would not fit into this scheme of things, and hence the field that drives the actual charges across the interfaces of a cell cannot be adequately explored with the help of such a test charge.

Magnetic fields. A *constant* magnetic field exerts a mechanical force only on a moving electric charge (§84). This force is perpendicular to the velocity of the charge; therefore, we call it a "deflecting" rather than a "driving" force.

A *time-varying* magnetic field exerts forces on charges both "directly" and "indirectly." First, if the charge is moving, there is a "charge-deflecting" force of the kind mentioned just above. Second, whether or not the charge is moving, there may also be a "charge-driving" force, given by (2), where $\mathbf{E}$ is now the intensity of the electric field of the Faraday type, induced by the changing magnetic field. A striking example of a combination of the two kinds of forces is provided by the *betatron*. In a betatron an electron is accelerated along a circular orbit by a "driving" force exerted by an electric field of the Faraday type, produced by the time-varying magnetic field; at the same time, the electron is prevented from flying off on a tangent by a "deflecting" force exerted on it by the magnetic field directly.

44. OHM'S LAW

We will now derive the field form of Ohm's law for isotropic conductors. Our reasoning will be based on the circuit form of this law and on the assumptions that the force equation (2^{43}) applies to conduction electrons and that at any point in an isotropic conductor the direction of **J** is the same as that of **E**. Ohm's law is not a universal law of nature, and cannot be derived without special assumptions. We ignore all magnetic fields in the conductor.

Figure 116 shows an isotropic but not necessarily homogeneous conductor carrying a steady current. We consider a small cylindrical portion of this conductor that includes the point P and is so oriented that the curved part of its

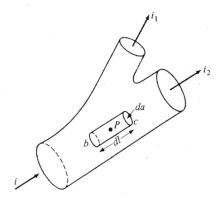

Fig. 116. Diagram leading to the field form of Ohm's law, namely $\mathbf{J} = g\mathbf{E}$.

surface is tangent to the lines of **J** and its flat ends are perpendicular to these lines. We take its length dl and its cross-sectional area da to be so small that the cylinder is essentially straight and that the conductivity g, the electric intensity **E**, and the current density **J** are each essentially constant throughout it. Our assumption that **J** has the same direction as **E** implies that **E** is parallel to the axis of the cylinder.

The little cylinder is so oriented that no charge is crossing the curved part of its surface. We can, therefore, imagine without loss of generality that this part of its surface is covered with a thin insulating film and only its two flat ends are connected electrically to the rest of the conductor. In other words, we may regard the cylinder as a separate circuit element (a rod of conductivity g, length dl, and cross-sectional area da) carrying a uniformly dense current; we may apply to this rod Ohm's law in its circuit form, namely,

$$(1^1) \qquad i_{bc} = \frac{V_{bc}}{R_{(bc)}}, \qquad (1)$$

where b and c are the end points of the rod. The resistance of the little rod is

(3²) $$R_{(bc)} = \frac{1}{g}\frac{dl}{da},\qquad(2)$$

and hence

$$i_{bc} = g\frac{da}{dl}V_{bc}.\qquad(3)$$

We will now translate (3) into field-theoretic terms. The flat ends of the cylinder are normal to **J**; hence, the current i_{bc} is equal to Jda, and (3) can be written as

$$J = g\frac{V_{bc}}{dl}.\qquad(4)$$

Now, as we remarked in §1 and as we will elaborate in the next chapter, the potential difference (volts) between two points is the work-per-unit-charge (joules per coulomb) involved in transferring electric charge from one of these points to the other. According to (2⁴³), the force acting on a positive charge q located in the cylinder is parallel to the axis of the cylinder and has the magnitude qE. Hence, the work (force times distance) required to move the charge q, inside the cylinder, from one end to the other is $qE\,dl$. Accordingly, the work-per-unit-charge required for the transfer is $E\,dl$. Hence $V_{bc} = E\,dl$, and (4) becomes

$$J = gE.\qquad(5)$$

Since the vectors **J** and **E** have the same direction and g is positive, we have

$$\mathbf{J} = g\mathbf{E}.\qquad(6)$$

Finally, we let $dl \to 0$ and $da \to 0$, and make the little cylinder contract upon P. Since this process does not affect (6), we conclude that (6) is in fact an exact field-theoretic counterpart of the circuit form of Ohm's law.

If magnetic fields in a conductor cannot be ignored, Ohm's law becomes $\mathbf{J} = g\,(\mathbf{E} + \mathbf{v}_d \times \mathbf{B})$. The extra term is due to the "Lorentz force," to be discussed later. A striking example of the effect of this (usually weak) force is mentioned on page 374.

Line maps of electric fields. We already have had several examples of line maps of electric fields, such as Fig. 61[17] for an electric doublet, Fig. 66[17] for a parallel-plate capacitor, and Fig. 95[34] for a charged metal rod. We will now use Ohm's law to add another example—the field **E** outside a cylindrical hole in a current-carrying block of copper. The current density for this case is given by (2¹⁴), (3¹⁴), and (4¹⁴). Therefore, if we write g for the conductivity of the copper block, the magnitude of **E** far from the hole is

$$E_0 = \frac{J_0}{g} \tag{7}$$

and the formula for **E** at any point outside the hole is

$$\mathbf{E} = E_0\left(1 - b^2 \frac{x^2 - y^2}{(x^2 + y^2)^2}\right)\mathbf{1}_x - 2E_0 b^2 \frac{xy}{(x^2 + y^2)^2}\mathbf{1}_y. \tag{8}$$

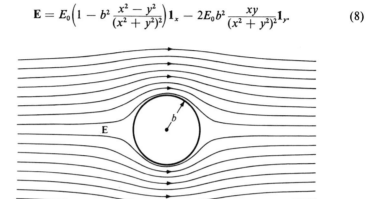

Fig. 117. A line map of the field **E** near a cylindrical hole in a current-carrying conductor.

A line map of (8) is shown in Fig. 117, which is the same as Fig. 62¹⁷, except for labeling. (We have set g equal to unity.) In cylindrical coordinates, according to (13³⁵),

$$E_\rho = E_0\left(1 - \frac{b^2}{\rho^2}\right)\cos\phi, \quad E_\phi = -E_0\left(1 + \frac{b^2}{\rho^2}\right)\sin\phi, \quad E_z = 0. \tag{9}$$

The field **E** *inside* the hole is pictured in Fig. 148⁵⁶. It cannot be found directly from Ohm's law, because there both **J** and g are zero, so that (6) leaves **E** indeterminate.

Time-varying electric fields. Suppose that the electric intensity **E** in a conductor changes with time. It is then perhaps obvious that, except at frequencies comparable with the natural atomic frequencies, (6) will remain valid even if **E** depends on t. However, the electric field that drives the charges in the conductor may then have two distinct parts: a part produced electrostatically and a part induced by a time-varying magnetic field. To stress this fact, we might rewrite (6) as

$$\mathbf{J} = g\mathbf{E} = g(\mathbf{E}' + \mathbf{E}''), \tag{10}$$

where the fields $\mathbf{E}'$ and $\mathbf{E}''$, respectively, are of the electrostatic and the Faraday type.

The equation $\mathbf{J} = g\mathbf{E}$ implies that if $\mathbf{E}$ is a periodic function of t, then $\mathbf{J}$ is a similar periodic function and is in phase with $\mathbf{E}$. As an example, consider a coil of wire connected across the terminals of a shielded a.c. generator. The intensity $\mathbf{E}$ of the total electric field that drives the charges at a point P in the coil then has the form

$$\mathbf{E} = \mathbf{E}'_{\text{generator}} + \mathbf{E}'_{\substack{\text{surface}\\\text{charges}\\\text{on coil}}} + \mathbf{E}''. \tag{11}$$

The first term on the right-hand side is caused by the generator. The second term is contributed by the surface charges that periodically develop on the coil; it has to do with the *distributed capacitance* of the coil. The last term is contributed by the time-varying magnetic field caused by the alternating current flowing in the coil; it has to do with the *self-inductance* of the coil. The equation $\mathbf{J} = g\mathbf{E}$ then implies that the current density $\mathbf{J}$ at P is in phase with the *total* electric intensity $\mathbf{E}$ that drives the charges at P. Accordingly, $\mathbf{J}$ need not be in phase with any one of the three terms on the right-hand side of (11). Since the generator is well shielded, the first term on that side can be ascribed to surface charges on the terminals of the generator.

Convection currents. Currents that flow in rarefied gases (say in electron tubes) or consist of electric charges dragged by insulators (say the belts of a Van de Graaff machine) are called *convection currents*. The equation $\mathbf{J} = g\mathbf{E}$ pertains to conduction currents. For convection currents, $\mathbf{J}$ and $\mathbf{E}$ need not be related in any simple way; in particular, if $\mathbf{E}$ is periodic, $\mathbf{J}$ need not be in phase with $\mathbf{E}$. For example, when the ionosphere is subjected to the periodic field of a radio wave, the ion currents are not in phase with $\mathbf{E}$.

EXERCISES

1. A silver wire, 2 mm^2 in cross section, carries 5 amperes. Find the value of E in the wire.

2. Starting with the equation $J = gE$, derive the formula $R = l/gA$ for the resistance of a homogeneous rod.

3. The wire shown twice in Fig. 118 carries a steady and uniformly dense current. The respective conductivities of the left and right portions of this wire are g_0 and $\frac{1}{2}g_0$. As suggested in the figure, E is then twice as strong at the right of the interface as it is at the left.

    Assume that, when the switch is first closed, the entire wire is suddenly pervaded by a uniform electric field of intensity $\mathbf{E}_0$, directed from left to right.

170   Electric Intensity

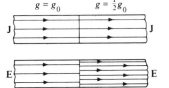

Figure 118

Then compute the initial currents in the two parts of the wire and describe the mechanism that eventually equalizes them.

4. A straight wire has a uniform cross section and carries a current of uniform density. It consists of three consecutive portions whose conductivities are g_0, $2g_0$, and g_0. Draw rough line maps of **J** and **E**, and describe the mechanism that causes the changes of **E**.

5. The coil of insulated wire shown in Fig. 119 has negligible self-inductance. Assume that, when it is first connected across a battery, it suddenly becomes pervaded by a uniform electric field of intensity E_0, directed from left to

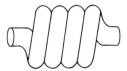

Figure 119

right. The total electric intensity at a point P in the coil at any later instant of time then has the form $\mathbf{E} = \mathbf{E}_0 + \mathbf{E}_s$, where $\mathbf{E}_s$ is contributed by surface charges. Describe the mechanism that eventually causes **E** at P to be directed *along* the wire; illustrate by diagrams.

45. JOULE'S LAW

A current of i amperes flows in an isotropic conductor whose resistance is R ohms. Let h be the time rate (joules per second, or simply watts) at which heat is generated in the entire conductor. The law formulated in 1840 by James Prescott Joule (1818–1889) then states that

$$h = i^2 R. \tag{1}$$

We will now translate this law into its field form and will compute the rate, say h_v, at which heat is generated, *per unit volume*, at any point in the conductor.

We return to the small cylinder surrounding the point P in Fig. 116, consider it, as before, to be a conducting rod connected electrically to the rest of the conductor only at its flat ends, and write dh for the time rate of heat generation (watts) in this cylinder as a whole. The resistance of the cylinder is $dl/(g\,da)$, the current in it is $J\,da$, and hence, according to (1),

$$dh = (J\,da)^2 \frac{1}{g}\frac{dl}{da} = \frac{1}{g}J^2\,dv, \tag{2}$$

where dv is the volume of the cylinder. Dividing (2) by dv, we find that the time rate at which heat is generated per unit volume in the neighborhood of P is

$$h_v = \frac{1}{g}J^2. \tag{3}$$

This result does not involve dl or dh, and hence does not change when $dl \to 0$ and $da \to 0$. Therefore, we conclude that (3), which is Joule's law in its field form, gives us the time rate of heat generation per unit volume (watts per cubic meter) *at* the point P.

Combining (3) with Ohm's law (6^{44}), we get the alternative forms

$$h_v = gE^2 = JE. \tag{4}$$

Since the vectors **J** and **E** pertaining to the same point in an isotropic conductor have the same direction, the equation $h_v = JE$ can be written as

$$h_v = \mathbf{J} \cdot \mathbf{E}. \tag{5}$$

When the rate of heat generation per unit volume (watts per cubic meter) is denoted by h_v, the formula for the rate of heat generation (watts) in a region R is

$$h = \iiint_R h_v\,dv. \tag{6}$$

More explicit forms of (6) are

$$h = \iiint_R \mathbf{J} \cdot \mathbf{E}\,dv = \iiint_R JE\,dv = \iiint_R gE^2\,dv = \iiint_R \frac{1}{g}J^2\,dv. \tag{7}$$

The R in these equations denotes a region, and should not be confused with the R in the formula $h = i^2 R$.

EXERCISES

1. Use some version of (7) to turn back to circuit theory and to show that $h = i^2 R$ for a homogeneous rod of length l, cross-sectional area A, and resistance R.

2. The conductivity of a certain rod (length l, cross-sectional area A, ends marked a and b) varies linearly from the value g_a at a to the value g_b at b; that is,

$$g(x) = g_a + \frac{(g_b - g_a)x}{l}, \tag{8}$$

where x is measured from a; this rod carries the current i. Compute the rate of heat generation in this rod. Check by circuit theory, using (1) and (9^2).

CHAPTER 12

Electromotive Force

In §1 we considered a chemical cell connected as in Fig. 1[1] and defined its electromotive force as the work that the cell does per coulomb of positive charge that it moves from its negative to its positive plate. The term "electromotive force" is not restricted to chemical cells, and is used for the work done per unit charge whenever charge is moved from one place to another. In this chapter we introduce the integral that represents the work-per-unit-charge. As we do this, we will speak of carrying hypothetical test charges along various mathematical paths, and the discussion might seem abstract. We will show in §53, however, that the integral introduced in this chapter is just the quantity measured in the laboratory by a voltmeter.

As before, we restrict ourselves to electrostatic fields and to electric fields of the Faraday type, so that the equation $F_e = q^*E$ will hold not only for hypothetical test charges but also for actual charges, such as conduction electrons. This restriction implies, of course, that, although we have referred above to electromotive forces produced by chemical action, we will exclude from consideration the regions in which such action is taking place; indeed, this action depends on certain very special properties of electrons which the idealized test charges do not have. We assume once again that the test charges are so small that we may ignore any effects that they may have on the electric field which they are used to explore.

We will now proceed to derive the formulas for the work that we must do to carry a test charge from a point a in an electric field to a point b along a prescribed path and to illustrate the fact that for some fields this work depends on the choice of the path leading from a to b, whereas for other fields the work is independent of the path. As before, we restrict ourselves to paths that are at rest relative to the observer.

46. WORK

To introduce the standard sign convention used in mechanics for *work*, we will first speak of a man walking at a constant speed on level ground with a dog on a leash and will ignore the work that he would have to do by muscular exertion if he were walking alone. If the dog tends to stay behind the man (and the man has to pull him forward), the force applied by the man to the dog has a positive forward component; the work being done *by the man* is then taken to be *positive*. But if the dog tends to run ahead of the man (and the man has to restrain the dog's forward motion), the force applied by the man to the dog has a negative forward component; the work being done *by the man* is then *negative*. Again, the dog may pull the leash at right angles to the man's path; the forward component of the force applied by the man to the dog is then zero, and the work being done by the man is zero. In general, while the man walks from a point a to a point b, the dog may be attracted by some objects and repelled by others; therefore, the total work done by the man during the journey from a to b may be positive, negative, or zero. In particular, if the man should walk with the dog around a level city block and return to his starting point, the total work that he will have done during the complete journey may be positive, negative, or zero, depending on the forces of attraction or repulsion exerted on the dog by various objects—other dogs, stray cats, and so on.

Next we turn to the hypothetical process of carrying a test charge q^* at a constant speed from a point a in an electric field of intensity $\mathbf{E}$ to a point b, along a prescribed path. The field will exert on the charge the force $q^*\mathbf{E}$, which may help us on some parts of the path and oppose us on others. Accordingly, the total work, say w, that we will have to do in this process may be positive, negative, or zero.

To derive a formula for work, we consider first two neighboring points P and Q whose separation dl is so small that the field $\mathbf{E}$ is essentially constant at all points on the straight line PQ, and so is the force $q^*\mathbf{E}$ exerted on the test charge by the electric field. As in elementary mechanics, the work required to

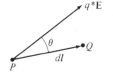

Figure 120

move the test body from P to Q at constant speed is then $\pm q^*\mathbf{E}\cos\theta\, dl$, where θ is the angle defined in Fig. 120; the sign depends on whether we mean the work done by us or the work done by the electric field. We choose to speak of the work that *we* do, denote it by dw, and write

$$dw = -q^* E \cos \theta \, dl. \tag{1}$$

The reader should verify that the numerical value of this dw has the correct sign whether the electric field helps us (when $\cos \theta$ is positive) or opposes us (when $\cos \theta$ is negative). The tangential component of $\mathbf{E}$ along the path PQ is

$$(17^{16}) \qquad E_t = E \cos \theta, \tag{2}$$

and, therefore,

$$dw = -q^* E_t \, dl. \tag{3}$$

Still another expression for dw is

$$dw = -q^* \mathbf{E} \cdot d\mathbf{l}, \tag{4}$$

where $d\mathbf{l}$ is the vector step from P to Q. Our units for measuring q^*, E, and dl are the coulomb, the newton per coulomb, and the meter; consequently, equations (1), (3), and (4) give dw in joules.

Now imagine taking the charge q^* from the point a in Fig. 121, along the path γ to the point b, which does not necessarily lie near a. To compute the

Figure 121

work, say $w_{a\gamma b}$, that we must do in this process, we divide the path γ into small steps, compute dw for each step, and find the limit of the sum of the dw's as the length of each step approaches zero. Since $dw = -q^* E_t \, dl$ and q^* is a constant, we then have

$$w_{a\gamma b} = -q^* \int_{a\gamma b} E_t \, dl. \tag{5}$$

The integral in (5) is called the *line integral* of E_t along the path $a\gamma b$.[1] If we carry the charge from a to b in Fig. 121 along the path δ, the formula for the work that we do is

$$w_{a\delta b} = -q^* \int_{a\delta b} E_t \, dl. \tag{6}$$

---

[1] An "ordinary" integral is a special case of a line integral, which arises, for example, when the path of integration runs along the x axis.

The examples that follow illustrate the fact that, depending on the type of electric field, the work required to carry a charge from a to b in an electric field may or may not depend on the choice of the path leading from a to b.

47. UNIFORM ELECTRIC FIELDS (WORK INDEPENDENT OF PATH)

Let

$$\mathbf{E} = E_0 \mathbf{1}_x, \tag{1}$$

where E_0 is a positive constant. This field, pictured in Fig. 122, can be produced inside a capacitor whose plates are parallel to the yz plane. The (x, y) coordinates of the points a and b in the figure are $(\rho_0, 0)$ and $(0, \rho_0)$.

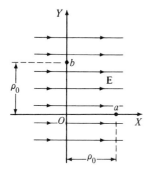

Fig. 122. A uniform electric field of the form $E_0 \mathbf{1}_x$.

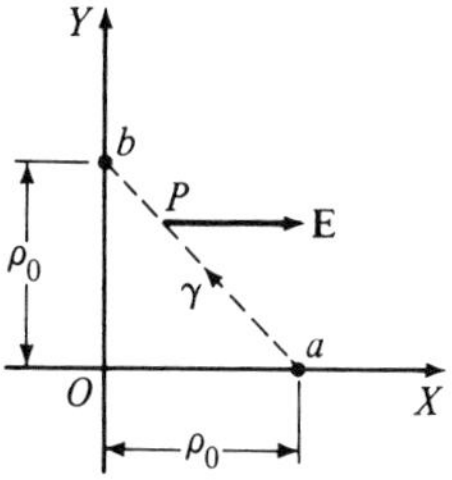

Figure 123

Example. Let us first compute the work w_{ayb} that we must do in carrying a charge q^* from a to b along the straight path γ shown in Fig. 123. The forward component of $\mathbf{E}$ at any point P on the path is then

$$E_t = -\frac{1}{\sqrt{2}} E_0, \tag{2}$$

and (5^{46}) becomes

$$w_{ayb} = \frac{q^* E_0}{\sqrt{2}} \int_{ayb} dl. \tag{3}$$

The integral in (3) is the sum of the steps dl for the path $a\gamma b$, which is $\sqrt{2}\, \rho_0$. Consequently,

$$w_{a\gamma b} = q^* E_0 \rho_0. \tag{4}$$

The integral can be evaluated also by changing from the "line variable" l, measured along the path, to the coordinate variable x. As we go along γ from a to b, the value of x decreases from ρ_0 to zero, and the increment of x corresponding to a forward step dl along this path is

$$dx = -\frac{1}{\sqrt{2}} dl. \tag{5}$$

Consequently,

$$\int_{a\gamma b} dl = \int_{\rho_0}^{0} (-\sqrt{2})\, dx = \sqrt{2}\,\rho_0, \tag{6}$$

as before.

Second example. Next let us go from a to b along the path δ, shown in Fig. 124 (a quarter-circle in the xy plane, radius ρ_0, center at O). In this case

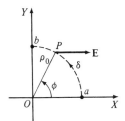

Figure 124

$$E_t = -E_0 \sin \phi, \tag{7}$$

and hence

$$w_{a\delta b} = q^* E_0 \int_{a\delta b} \sin \phi \, dl. \tag{8}$$

As we go along δ from a to b, the value of ϕ increases from zero to $\tfrac{1}{2}\pi$. The increment of ϕ corresponding to a forward step dl along this path is

$$d\phi = dl/\rho_0. \tag{9}$$

Therefore,

$$w_{a\delta b} = q^* E_0 \rho_0 \int_0^{(1/2)\pi} \sin \phi \, d\phi, \tag{10}$$

so that

$$w_{a\delta b} = q^* E_0 \rho_0. \tag{11}$$

Third example. Our next path, ϵ, is shown in Fig. 125 (three-quarters of a circle in the xy plane, radius ρ_0, center at O). Now the electric field hinders

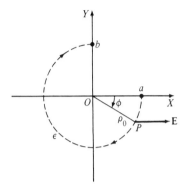

Figure 125

us during two-thirds of the journey, but helps us the rest of the way. Since the path ϵ runs clockwise, the formula for E_t is not (7) but

$$E_t = E_0 \sin \phi. \tag{12}$$

Accordingly,

$$w_{a\epsilon b} = -q^* E_0 \int_{a\epsilon b} \sin \phi \, dl. \tag{13}$$

As we go from a to b along the path ϵ, the angle ϕ changes from 0 to $-\tfrac{3}{2}\pi$, and the increment of ϕ corresponding to a forward step dl along ϵ is $d\phi = -dl/\rho_0$. Consequently,

$$w_{a\epsilon b} = q^* E_0 \rho_0 \int_0^{-(3/2)\pi} \sin \phi \, d\phi; \tag{14}$$

that is,

$$w_{a\epsilon b} = q^* E_0 \rho_0. \tag{15}$$

Remarks. According to (4), (11), and (15), we have

$$w_{ayb} = w_{a\delta b} = w_{a\epsilon b}. \tag{16}$$

These equalities are not accidental; we will see in §49 that the work required to carry a small charge from a point a to a point b in a field produced by *static charges* does not depend on the choice of the path. Consequently, if $\mathbf{E} = E_0 \mathbf{1}_x$, we can infer the equalities (16) directly from the fact that the electric field in question can be produced by static charges located on the plates of a capacitor, and we need not go into any of the details we considered in deriving (16). In particular, in this case, we need not specify the path taken from a to b and can write (4), (11), and (15) simply as

$$w_{ab} = q^*E_0\rho_0. \tag{17}$$

EXERCISES

1. The (x, y, z) coordinates of the points a, b, and c are $(\rho_0, 0, 0)$, $(0, \rho_0, 0)$ and $(0, 0, z_0)$. The path ζ leads from a straight to c and then straight to b. Let $\mathbf{E} = E_0\mathbf{1}_x$, and show in detail that $w_{a\zeta b} = q^*E_0\rho_0$.
2. Let $\mathbf{E} = cx\mathbf{1}_x$, where c is a constant. Evaluate the work integral separately for each of the paths pictured in Figs. 123, 124, and 125, and show that $w_{ayb} = w_{a\delta b} = w_{aeb} = \frac{1}{2}q^*c\rho_0^2$. Show also that $w_{a\zeta b} = \frac{1}{2}q^*c\rho_0^2$, where ζ is the path of Exercise 1.

48. THE INVERSE-SQUARE FIELD (WORK INDEPENDENT OF PATH)

If a single point charge of q coulombs is located in free space at the origin of a coordinate frame, we have

$$(7^{42}) \qquad \mathbf{E} = 9 \times 10^9 \frac{q}{r^2}\mathbf{1}_r. \tag{1}$$

We begin with two neighboring points, namely

$$P(r, \theta, \phi), \qquad Q(r + dr, \theta + d\theta, \phi + d\phi),$$

and compute the work that we must do to carry a test charge of q^* coulombs from P straight to Q. The formula for the vector step from P to Q is

$$(14^\text{A}) \qquad d\mathbf{l} = \mathbf{1}_r\, dr + \mathbf{1}_\theta\, r\, d\theta + \mathbf{1}_\phi\, r\sin\theta\, d\phi, \tag{2}$$

and, therefore,

$$\mathbf{E} \cdot d\mathbf{l} = 9 \times 10^9 \frac{q}{r^2}\, dr. \tag{3}$$

Consequently, according to (4^{46}),

$$dw = -9 \times 10^9 \frac{qq^*}{r^2}\, dr. \tag{4}$$

Electromotive Force

Next consider two points a and b that have the coordinates (r_a, θ_a, ϕ_a) and (r_b, θ_b, ϕ_b) and do not necessarily lie near each other. To compute the work needed to carry a charge q^* in the field (1) from a to b along some particular path, we divide this path into small steps and sum the dw's for these steps. Now, the dw in (4) depends only on r and dr, and not on θ, $d\theta$, ϕ, and $d\phi$. Therefore, to evaluate the sum of the dw's, we need only to integrate (4) with respect to r between the limits r_a and r_b. The result, namely,

$$w_{ab} = 9 \times 10^9 qq^* \left(\frac{1}{r_b} - \frac{1}{r_a} \right), \tag{5}$$

does not depend on the choice of the path leading from a to b. A more graphic basis for (5) is suggested in Exercise 1.

If the charge q does not lie at the origin of the given coordinate frame, we can change to a frame whose origin does lie at q and can then proceed as before.

Conservative fields. For *some* electric fields, the work needed to carry a test charge from an arbitrarily chosen point a to any other arbitrarily chosen point b depends only on the positions of a and b but not on the path taken from a to b. Such fields are called *conservative*, and we say for short that, in a conservative field, "Work does not depend on the path." We have found above from Coulomb's law that the electric field of a single static point charge located in free space is conservative.

EXERCISES

1. In Fig. 126 the point charge q is located at the origin, and the points a and b, as well as the path $a\alpha b$, lie in the xy plane. Replace the smooth path α by a sequence of radial segments interconnected by circular arcs centered on q,

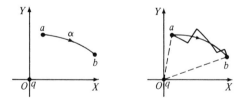

Figure 126

let the number of arcs increase without limit and their lengths all tend to zero, and show that for the field (1) the value of $w_{a\alpha b}$ depends only on the radial coordinates of a and b. Finally, derive (5) after removing the restrictions that q and the points a and b lie in the xy plane and that the path $a\alpha b$ lies in a plane.

2. Show that fields of the forms $\mathbf{E} = f(\rho)\mathbf{1}_\rho$ and $\mathbf{E} = f(r)\mathbf{1}_r$ are conservative.

49. ELECTROSTATIC FIELDS IN GENERAL (WORK INDEPENDENT OF PATH)

We will now outline a procedure for proving that *every electrostatic field is conservative*.

We have already proved this theorem for a single point charge in empty space. Next consider two charges, q_1 and q_2, held in fixed positions in free space, and write the intensity of the electric field produced by them jointly as

$$\mathbf{E} = \mathbf{E}_1 + \mathbf{E}_2, \tag{1}$$

so that

(5^{46})
$$w_{ayb} = -q^* \int_{ayb} (E_1)_t \, dl - q^* \int_{ayb} (E_2)_t \, dl. \tag{2}$$

We can evaluate the integrals in (2) separately, using two different coordinate frames. When computing the first integral, we put the origin at the charge q_1. The field E_1 then takes the form (1^{48}) with q^* replacing q, and hence the value of the first integral in (2) depends only on the positions of the points a and b relative to the charge q_1 but not on the choice of the path leading from a to b. Next we place the origin at q_2 and find in a similar way that the value of the second integral in (2) is also independent of the path taken from a to b. Consequently, the value of w_{ayb}, whatever it may be, does not depend on the choice of path. Since the points a and b can be chosen arbitrarily, it follows that the electric field produced by *two* fixed point charges in empty space is conservative.

This proof is readily extended to three point charges, four point charges, and so on. By careful mathematical reasoning, it can be extended also to *continuous* distributions of stationary charges.

In experiments not involving frequencies in the optical range, the electric fields produced by stationary charged or uncharged conductors or insulators can usually be treated as electrostatic fields (§1). Note that uncharged bodies may contribute to an electric field. For example, if a positively charged pith ball is placed near an uncharged conductor, electrons will congregate on the portion of the conductor nearest to the pith ball and will alter the electric field outside the conductor. Similarly, an uncharged insulator placed near electric charges will become "polarized" and will contribute to the electric field. Because of these effects, any test charges used to explore electric fields in the vicinity of conductors or insulators should be very small.

EXERCISE

1. Recall the second paragraph of §19 and explain why the equalities $w_{ayb} = w_{a\delta b} = w_{a\epsilon b} = w_{a\zeta b}$ of Exercise 2^{47} are not unexpected.

50. AN ELECTRIC FIELD PRODUCED MAGNETICALLY (WORK DEPENDS ON PATH)

Our next example involves an electric field of the Faraday type, produced in a copper washer placed near an a. c. electromagnet, as in Figs. 127 and 214[96]. The alternating magnetic field induces in the magnet and its vicinity an alternating *electric* field. The electric field induced inside the washer drives an alternating current in the washer, which can be regarded as a shorted single-turn secondary winding of a transformer.

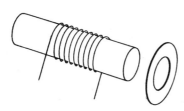 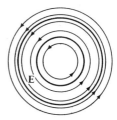

Fig. 127. A copper washer in the axially symmetric field of an a.c. electromagnet.

Fig. 128. The electric field of the Faraday type produced in and near the copper washer of Fig. 127. This figure pertains to some particular instant of time.

Suppose that the magnetic field is axially symmetric, that the washer is properly aligned with it, and that, at the instant in question, the current induced in the washer flows counterclockwise, if the washer is viewed in Fig. 127 from the right. According to Ohm's law (6[44]), the field E in the washer is then circular and has the counterclockwise sense. In the air near the washer, E is also circular, as in Fig. 128; this follows from the axial symmetry of the setup and the fact that the observed phenomena remain essentially the same when the washer is replaced by a larger or a smaller washer.

Now consider the work that we would have to do to carry a test charge q^* from a point a in the washer to a point b along the alternative paths γ, δ, and

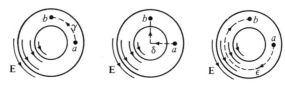

Figure 129

ϵ shown in Fig. 129, where only portions of the lines of E are drawn. The field E will periodically reverse its direction, but to avoid details that might obscure the main point, we assume that the test charge is carried from a to b so quickly that E does not change appreciably during the transfer.

If we take the path γ, the electric field will help us all the way, and we would do negative work; that is,

$$w_{a\gamma b} < 0. \tag{1}$$

If we take the path δ, we would move the test charge only at right angles to E and, therefore, do no work; accordingly,

$$w_{a\delta b} = 0. \tag{2}$$

Finally, if we take the path ϵ, the electric field will oppose us all the way, and we would do positive work; that is,

$$w_{a\epsilon b} > 0. \tag{3}$$

It follows that

$$w_{a\gamma b} \neq w_{a\delta b} \neq w_{a\epsilon b}. \tag{4}$$

This example illustrates the fact that *electric fields produced by time-varying magnetic fields are not conservative.*

EXERCISE

1. In Fig. 129, find paths ζ, η, and θ (each leading from a to b) such that $w_{a\zeta b} = \frac{1}{2} w_{a\gamma b}$, $w_{a\eta b} = 2w_{a\gamma b}$, and $w_{a\theta b} = -w_{a\gamma b}$. Convince yourself that in this case one can find a path leading from a to b for which w has *any* preassigned value.

51. CLOSED PATHS

Suppose that we take a test charge q^* from a point a in a static electric field to a point b along a certain curve, say η, and then take it back to a along the same curve. As we pass a point P on our way back, we must apply to the test charge a force of the same magnitude as was needed at P on the way from a to b, but the forward component of this force now has the opposite sign, because the direction of the motion is reversed. This argument applies at every point on the curve η and, consequently, if the back-and-forth journey is so rapid that the electric field (even if it is not strictly static) does not change appreciably in the meantime, we have the relation

$$w_{b\eta a} = -w_{a\eta b}. \tag{1}$$

Figure 130

The signed contour $a\gamma b\delta a$, or simply C, shown in Fig. 130, lies in an electric field of intensity **E**. Let w_C be the work that we must do to carry a charge q^* once around this contour from a back to a, in the positive sense of C. We then have

$$w_C = w_{a\gamma b} + w_{b\delta a}, \tag{2}$$

and, consequently,

$$w_C = w_{a\gamma b} - w_{a\delta b}. \tag{3}$$

Next consider the special case of an electrostatic field. Since such a field is conservative, we have $w_{a\gamma b} = w_{a\delta b}$, and (3) reduces to

$$w_C = 0. \tag{4}$$

In words: *No work is needed to carry a charge around any closed path in a purely electrostatic field*, provided this charge is so small that it does not perturb the field in question. Therefore, (4) can be written in this case as

(5[46])
$$-q^* \oint_C E_t^{es}\, dl = 0, \tag{5}$$

or simply as

$$\oint E_t^{es}\, dl = 0. \tag{6}$$

As illustrated in §50, an electric field produced by a time-varying magnetic field need not be conservative, and hence the formula

$$w_C = -q^* \oint_C E_t^f\, dl \tag{7}$$

cannot be simplified in general. But if the electric field is a combination of fields of the two types, we have

$$w_C = -q^* \oint_C (E_t^{es} + E_t^f)\, dl = -q^* \oint_C E_t^f\, dl. \tag{8}$$

The integral of the tangential component of a vector field around a closed path is often called the *circulation* of the field around the path, a term that comes from hydrodynamics.

EXERCISES

1. Suppose that the intensity of an electric field at a fixed instant of time is $\mathbf{E} = ky\mathbf{1}_x$, where k is a positive constant; the lines of $\mathbf{E}$ then resemble the field lines in Fig. 76[21].

   Let C be a rectangular contour in the xy plane, centered on an arbitrary point (x_0, y_0); its straight portions are parallel to the coordinate axes; it has the counterclockwise sense, if viewed from above the xy plane. Show that, for the field just described,

$$w_C = kq^*A, \qquad (9)$$

   where A is the area bounded by C. Since $w_C \neq 0$, we conclude that the field in question is not conservative, and cannot be produced electrostatically.

2. The contours $abca$ and $adba$ in Fig. 131 have a common portion. Show that

$$w_{abca} + w_{adba} = w_{adbca}. \qquad (10)$$

Figure 131

3. Use (10) and some limiting process to show that, if $\mathbf{E} = ky\mathbf{1}_x$, then (9) holds for every simple contour lying in the xy plane.

4. A contour C, which is not necessarily flat, lies in such an electric field that $w_C \neq 0$. The points a and b lie in the same field, but not on C. Show that the work required to carry a charge from a to b will depend on the choice of path. Consider also the case when a and b lie on C.

52. ELECTROMOTIVE FORCE

Suppose that we move a test charge q^* at a constant speed from a point a in an electric field to a point b along a stationary path γ, so that

(5[46]) $$\text{the work that } we \text{ do} = -q^* \int_{a\gamma b} E_t \, dl. \qquad (1)$$

The force exerted on the charge by the electric field at any point on the path is equal and opposite to that exerted by us; therefore,

$$\text{the work that } \textit{the field} \text{ does} = q^* \int_{a\gamma b} E_t \, dl. \tag{2}$$

This work depends on both the electric field and the size (coulombs) of the charge q^*. We are interested mainly in electric fields rather than test charges; therefore, we divide both sides of (2) by q^* and write

$$\text{the work that } \textit{the field} \text{ does } \textit{per unit charge} = \int_{a\gamma b} E_t \, dl. \tag{3}$$

Our unit for measuring work-per-unit-charge is the joule per coulomb, or simply the *volt*.

The integral on the right-hand side of (3) is usually called the *electromotive force* in the path γ leading from a to b; we denote it by the symbol $\text{emf}_{a\gamma b}$, which is defined by the equation

$$\text{emf}_{a\gamma b} = \int_{a\gamma b} E_t \, dl. \tag{4}$$

In words: The emf in a given path is the integral of the tangential component of **E** along this path. The physical interpretation of (4) follows from (3):

$$\text{emf}_{a\gamma b} = \begin{cases} \text{work done by the electric field, per coulomb} \\ \text{of transferred charge, when a small charge} \\ \text{is moved from } a \text{ to } b \text{ along the path } \gamma. \end{cases} \tag{5}$$

For another path, say δ, leading from a to b, we have

$$\text{emf}_{a\delta b} = \int_{a\delta b} E_t \, dl. \tag{6}$$

The product $E_t \, dl$ in (4) and (6) can be written also as $\mathbf{E} \cdot d\mathbf{l}$, where $d\mathbf{l}$ is a differential vector step along the path.

Electromotive forces are also called "electromotances" or "voltages." In the special case of electrostatic fields, still another term—*difference of potential*, or simply *potential difference*—comes into play; we take it up in §56. Note that "electromotive *force*" is an electromechanical rather than a purely mechanical quantity, which we express in joules per coulomb (our unit for measuring force is the newton).

In general, the values of the emf are different for different paths leading from one fixed point a to another fixed point b. For example, if the electric field is circular, as in Fig. 128[50], we conclude from (1^{50}), (2^{50}) and (3^{50}) that $\text{emf}_{a\gamma b} > 0$, $\text{emf}_{a\delta b} = 0$, and $\text{emf}_{a\epsilon b} < 0$, so that

$$\text{emf}_{a\gamma b} \neq \text{emf}_{a\delta b} \neq \text{emf}_{a\epsilon b}. \tag{7}$$

But in the special case of conservative electric fields, only the endpoints of the path matter. For example, if $\mathbf{E} = E_0 \mathbf{1}_x$ and if a and b are the points shown in Fig. 122[47], we infer from (16[47]) that

$$\text{emf}_{a\gamma b} = \text{emf}_{a\delta b} = \text{emf}_{a\epsilon b}, \tag{8}$$

so that each of the three emf's in (8) may be denoted simply by emf_{ab}.

If the path is closed and forms a signed contour C, we write emf_C for the electromotive force in the contour, and the definition (4) reads

$$\text{emf}_C = \oint_C E_t \, dl. \tag{9}$$

If the path of integration in (4) or (9) is moving relative to the observer or is changing its shape, and if a magnetic field is present in addition to the electric field, the force acting on a charge q becomes the Lorentz force (3[84]) rather than simply $q\mathbf{E}$, and the emf must be redefined accordingly.

EXERCISE

1. Restate the exercises of §51, and the answers, in terms of emf's rather than w's.

53. D. C. VOLTMETERS

When a voltmeter is connected to a circuit, it gives information about the electric field pervading this circuit and the space around it. We will now determine just what this information is when all the relevant fields are steady; but our procedure can be easily extended to time-varying fields. The conclusion is stated in Equation (9).

Our meter is an ordinary shuntless, high-resistance d. c. voltmeter, equipped with a permanent magnet and a moving-coil detector—but it has the zero mark in the middle of its scale, so that its readings may be negative as well as positive, depending on the direction of the current flowing in it. We label A and B the leads attached to the binding posts of the meter, marked, respectively, plus and minus; and we call the end of a lead connected to a circuit, rather than to the meter, the "far end" of the lead. Also, we assume for definiteness that when the meter is connected to the points a and b of a circuit, lead A is connected at a, as in Fig. 132.

We regard the current i in the meter as positive if it flows in the meter from lead A to lead B. The scale of the meter is presumably calibrated by the manufacturer in the usual way, so that

$$\text{reading of voltmeter} = iR, \tag{1}$$

where
$$R = \text{resistance of meter.} \tag{2}$$

Note that the reading of our meter is negative if i is negative.

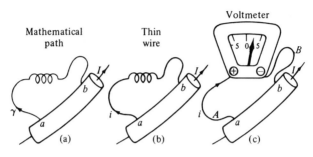

Figure 132

To begin with, we assume that the leads and the coil of the current detector are made of the same homogeneous and uniform wire, a restriction removed in Exercise 3. We write g for the conductivity and A for the cross-sectional area of this wire. If the length of this wire inside the meter between the binding posts is l' and the total length of the two leads is l'', then the resistance R of the meter is l'/gA, the resistance of the leads is l''/gA, and the total resistance between the far ends of the leads is $(l' + l'')/gA$. At this point we ignore the resistance of the leads compared to the resistance of the meter and replace the formula for the R in (1) and (2), namely $R = l'/gA$, by

$$R = \frac{l}{gA}. \tag{3}$$

where l is the total length of the wire comprising the coil and the leads, equal to $l' + l''$.

Theory. Figure 132(a) shows a conductor that forms a part of a d. c. circuit and carries a current I. We write **E** for the intensity of the electric field pervading the circuit and the space around it; the symbol **E** includes the contribution to this field made by the surface charges located on the various portions of the circuit. The line γ in the figure is a simple mathematical path of length l, leading from a point a on the circuit to another point b. The electromotive force in this path is

$$(4^{52}) \qquad \text{emf}_{a\gamma b} = \int_{a\gamma b} E_t \, dl. \tag{4}$$

The forward component E_t of **E** has, in general, different numerical values at different points on a path; in Fig. 133, for example, E_t is positive at c_1 and c_3, but negative at c_2.

Next suppose that, as in Fig. 132(b), the points a and b are connected by a thin uniform wire lying along the path γ, and that this wire forms the leads and coil of the voltmeter shown in Fig. 132(c). As soon as the connections

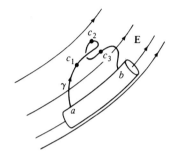

Figure 133

are made, extra surface charges develop on various parts of the main circuit and on the wire itself—charges that divert a current i from the main circuit into the wire and guide it along the wire. When we write $\mathbf{E}'$ for the intensity of the electric field produced by all these *extra* surface charges, we get the equation

$$\text{emf in the leads and coil} = \int_{a\gamma b} (E_t + E'_t)\, dl, \tag{5}$$

where E_t and E'_t are the forward components of $\mathbf{E}$ and $\mathbf{E}'$ at points *in the wire*. Note that the emf in the leads and coil is, in general, different from the emf in the mathematical path γ when the wire is not there.

To evaluate the integral in (5), we let $\mathbf{J}$ be the current density in the wire and assume that i is positive. Since the wire is uniform and thin, the field $\mathbf{J}$ is tangent to the wire throughout the wire and has a constant magnitude, say J_0, given by the equation $J_0 = i/A$. The field that drives the current in the wire is $\mathbf{E} + \mathbf{E}'$ and, therefore, by Ohm's law, $\mathbf{E} + \mathbf{E}' = \mathbf{J}/g$. Since $\mathbf{J}$ is tangent to the wire and has the constant magnitude J_0, it follows that, *in the wire*, the field $\mathbf{E} + \mathbf{E}'$ is tangent to the wire and has the constant magnitude J_0/g. Consequently, *in the wire*,

$$E_t + E'_t = \frac{J_0}{g} = \text{constant}, \tag{6}$$

even though E_t is not constant. It is a striking fact that, when a thin homogeneous wire is connected across a circuit, the extra surface charges arrange themselves automatically in such a way as to cause the total electric intensity *in the wire* to be tangent to the wire and to have the same magnitude throughout the wire. In view of (6),

$$\int_{a\gamma b} (E_t + E'_t)\, dl = \int_{a\gamma b} \frac{J_0}{g}\, dl = \frac{lJ_0}{g} = iR. \tag{7}$$

Consequently, (5) reduces to

$$\text{emf in the leads and coil} = iR. \tag{8}$$

If i is negative, a rewording of the argument takes us again from (5) to (8).

Our final result, which summarizes the rôle of a voltmeter, follows from (8) and (1):

$$\text{reading of voltmeter} = \text{emf in the leads and coil}. \tag{9}$$

More explicitly,

$$\text{reading of voltmeter} = \int_{a\gamma b} (E_t + E'_t)\, dl. \tag{10}$$

The path γ in (10) starts at a, and runs *inside the wire* along the lead A, along the turns of the coil inside the meter, and then along the other lead of the meter, ending at b. Note that this path lies *outside the circuit* to which the meter is connected.

EXERCISES

1. Derive (8) in detail for the case when the numerical value of i is negative, so that one cannot say that the magnitude of **J** is i/A.

2. Suppose that lead A in Fig. 132(c) is slightly removed from the point a, enough to break the electric contact but not enough to affect appreciably the integral of E_t along the leads and coil. Explain why the voltmeter will, nevertheless, read zero.

3. The coiled part of the wire comprising a shuntless voltmeter has a different cross section and a different conductivity from the leads. Derive (9) for this case.

4. In Fig. 134, the portion acb of a closed path lies in the leads and coil of a voltmeter, and the portion bda lies inside the main circuit. Under the conditions stated in the text, the field **E**′, produced by the extra surface charges, is electrostatic, and the integral of E'_t around the path $acbda$ is zero.

    Assume that the resistance of the voltmeter is so high that the presence of the meter has a negligible effect on the currents in the main circuit, and show that (10) can then be replaced by the equation

$$\text{reading of voltmeter} = \int_{a\gamma b} E_t \, dl. \tag{11}$$

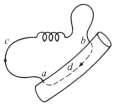

Figure 134

5. The wire comprising the leads and the coil of the voltmeter was described in the text as uniform and thin. One criterion of thinness in this case is that the radius of curvature of the wire is everywhere large compared to the radius of its cross section. Where was a condition of this sort implied in this section?

CHAPTER 13

Conservative Electric Fields

This chapter deals almost exclusively with conservative electric fields and with some of the special theoretical methods that have been developed for investigating them. We begin by describing two ways of testing whether or not a given electric field is conservative: an integral test and a differential test. The second of these tests involves the concept of the "curl" of a vector field, a concept that will be discussed more thoroughly in Chapter 19.

Next we show that if an electric field is conservative, then one can associate with it not only the *vector* field **E**—the electric intensity—but also a *scalar* field, say V, called "electrostatic potential." When **E** is given, V can be found by integration, except for an arbitrary additive constant. The reverse process—the computation of **E** when V is given—involves the concept of the "gradient" of a scalar field, defined in §57.

54. AN INTEGRAL TEST FOR CONSERVATIVE FIELDS

Let w be the work required to transfer a test charge from an arbitrarily chosen point a in an electric field to another arbitrarily chosen point b. According to the definition given in §48, the electric field is conservative if and only if the numerical value of w does not depend on the path taken from a to b. For a particular path, say γ, the work done by the field per unit charge is

$$(4^{52}) \qquad \text{emf}_{ayb} = \int_{ayb} E_t \, dl. \qquad (1)$$

Consequently, the following test is available to us when the formula for **E** is given: *To find whether or not an electric field is conservative, compute the* *integral in* (1) *for arbitrarily chosen points a and b, and see whether or not the value of this integral depends on the choice of the path leading from a to b.*

We have already applied this test in §48 to the most important special case —the inverse-square field of a single static point charge. We found that that field is conservative, and we then extended this result to fields produced by any distributions of static charges. Thus, the integral test described above has led us to a very general and important conclusion.

Suppose, however, that we are given a formula for **E** as a function of (x, y, z), but are *not* told that this field is produced by static charges. How can we then tell whether or not this field is conservative? The integral test is still available, of course, but if the formula for **E** is complicated, this test may be difficult or tedious to apply. Fortunately, there is an alternative test that involves differentiation rather then integration, and is usually very straightforward and simple. We will now introduce this test without pretense to rigor and will reintroduce it more carefully in Chapter 19.

55. A DIFFERENTIAL TEST FOR CONSERVATIVE FIELDS

A clue leading to a "differential" test comes from §50, which dealt with an electric field induced in a copper washer by a time-varying magnetic field. In that section we compared the amounts of work required to carry a test charge from a point a to a point b along the different paths shown in Fig. 129[50], and we concluded that this electric field is not conservative.

We could have come to the same conclusion in a superficially different way by focusing attention not on a hypothetical test charge, but on the conduction electrons in the washer. Figure 135 is a line map of the current density in the

Fig. 135. The current density in the copper washer of Fig. 128.

washer at an instant when **E** is directed as in Fig. 128[50]. The conduction electrons in the washer form an axially symmetric cloud, which at the instant in question is turning clockwise. The kinetic energy of these electrons is being converted into heat, and, therefore, we conclude that the field is exerting a *torque* on the cloud of conduction electrons and is doing work. Since the average paths of these electrons are closed, we can also conclude, as in §50, that the electric field shown in Fig. 128 is *not conservative*. Similar torques act on clouds of charges not in the copper washer in Fig. 127[50]. For example, *eddy currents* keep circulating in the iron core of the electromagnet; and if the core is laminated, they circulate in every conducting portion of it, however small this portion may be. The term "eddy currents" is usually reserved for unwanted currents, such as the currents in the iron core of a transformer; but one may also regard as an eddy current the useful current in the secondary winding and the load. When we speak below of the *torque-exerting property* of electric fields, we restrict ourselves to torques exerted on *symmetric* charge distributions; in particular, we exclude torques exerted on such nonsymmetric distributions as the electric dipole of Exercise 4[42].

The torque-exerting property of an electric field can be explored, in principle, with the aid of a hypothetical device that we will call an *electric pinwheel*. Such a pinwheel, pictured in Fig. 136, is a thin nonconducting circular disc, rigidly fastened to a central pin and carrying a positive charge q, which is spread uniformly along the rim.

Suppose that we hold the pinwheel in a two-dimensional radial electric field, as in Fig. 137; the field will then push the pinwheel radially outward, but, as suggested by symmetry, will not tend to spin it. Suppose, however, that we hold the pinwheel in the field pictured in Fig. 138 and described by the

Fig. 136. An electric pinwheel.

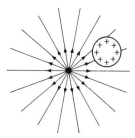

Fig. 137. A radial electric field exerts no torque on an electric pinwheel.

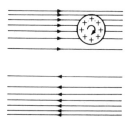

Fig. 138. An electric field of the form $cy\mathbf{1}_x$ exerts a torque on an electric pinwheel.

formula $\mathbf{E} = cy\mathbf{1}_x$, where c is a positive constant. The rightward push on the upper half of the disc will then be stronger than that on the lower half, and the field will exert a clockwise torque on the disc. Next suppose that we turn the disc in Fig. 138 counterclockwise through 360 deg, so that each element of charge on the rim will describe a circle and return to the point where it started. Because of the clockwise torque exerted by the field, we will do positive work in this process, and since each element of charge will have been carried around a *closed* path, we conclude that the electric field in question is not conservative. The argument based on the 360 deg turn applies to any electric field that exerts a torque on the pinwheel; therefore, we conclude that, if an electric field *can* exert a torque on an electric pinwheel, then it is not conservative. Conversely, *a conservative electric field cannot exert a torque on an electric pinwheel*. This observation leads to a "differential" test for conservative fields.

Torques on small pinwheels. To get compact formulas, we consider a small pinwheel and write q for the charge, r for the radius, and $a = \pi r^2$ for the area of the disc. We will assume that r is small enough to permit ignoring the higher-order terms in (2) and (3) below, and will wait until Chapter 19 for a more careful discussion.

We begin with two-dimensional fields that involve only x and y, and imagine holding the pin so that it points in the $+z$ direction. In Fig. 139, P is the center of the pinwheel and P' a point on the rim. The respective electric intensities at P and P' are $\mathbf{E}$ and $\mathbf{E}'$, say. We will need the component of $\mathbf{E}'$ tangential to the rim; according to Fig. 139(b), it is related to the cartesian components as follows:

$$E'_t = -E'_x \sin \alpha + E'_y \cos \alpha, \qquad (1)$$

where α is the angle defined in the figure. Now, the cartesian coordinates of P'

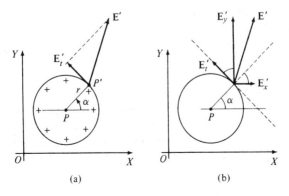

Fig. 139. Computation of the torque acting on an electric pinwheel.

relative to P are $r\cos\alpha$ and $r\sin\alpha$, and, therefore, apart from terms of higher order,

$$E'_x = E_x + \frac{\partial E_x}{\partial x} r \cos\alpha + \frac{\partial E_x}{\partial y} r \sin\alpha, \tag{2}$$

and

$$E'_y = E_y + \frac{\partial E_y}{\partial x} r \cos\alpha + \frac{\partial E_y}{\partial y} r \sin\alpha. \tag{3}$$

Consequently, (1) can be written as

$$E'_t = -\left(E_x \sin\alpha + \frac{\partial E_x}{\partial x} r \cos\alpha \sin\alpha + \frac{\partial E_x}{\partial y} r \sin^2\alpha\right)$$
$$+ \left(E_y \cos\alpha + \frac{\partial E_y}{\partial x} r \cos^2\alpha + \frac{\partial E_y}{\partial y} r \sin\alpha \cos\alpha\right). \tag{4}$$

Let a small element of the rim include the point P' and subtend at P the angle $d\alpha$. The charge on this element is $q\, d\alpha/2\pi$, the tangential component of the force exerted on it by the field is $(q/2\pi)E'_t\, d\alpha$; the lever arm of this component is r, and hence this force produces about the pin the torque

$$\frac{qr}{2\pi} E'_t\, d\alpha. \tag{5}$$

To emphasize the fact that the pin points in the $+z$ direction, we write T_z for the total torque exerted about the pin. To evaluate T_z, we integrate (5) with respect to α:

$$T_z = \frac{qr}{2\pi} \int_0^{2\pi} E'_t\, d\alpha. \tag{6}$$

The formula (4) for E'_t has six terms, each of which involves α through a trigonometric factor. The factors $\sin\alpha$, $\cos\alpha$, and $\sin\alpha\cos\alpha$ integrate to zero in (6); the factors $\sin^2\alpha$ and $\cos^2\alpha$ each integrate to π, and, therefore,

$$T_z = \frac{1}{2} qr^2 \left(\frac{\partial}{\partial x} E_y - \frac{\partial}{\partial y} E_x\right). \tag{7}$$

Note that the letters x, y, and z first appear in (7) in their cyclic order.

If the electric field in Fig. 139 should have a z component, this component might tend to tilt the pinwheel out of the xy plane, but would not tend to turn it about the pin. Therefore, (7) is actually not restricted to two-dimensional fields.

If the pin is held parallel to the x axis or the y axis, the formulas for the torques about the pin differ from (7) only by a cyclic permutation of the letters x, y, and z. Using the notation

$$k = \frac{2}{qr^2}, \tag{8}$$

we now write the full set of formulas as follows:

$$kT_x = \frac{\partial E_z}{\partial y} - \frac{\partial E_y}{\partial z}, \tag{9}$$

$$kT_y = \frac{\partial E_x}{\partial z} - \frac{\partial E_z}{\partial x}, \tag{10}$$

$$kT_z = \frac{\partial E_y}{\partial x} - \frac{\partial E_x}{\partial y}, \tag{11}$$

where (11) is the same equation as (7). In a more careful derivation of these torque formulas, one would let $r \to 0$ and $q \to \infty$ in such a way that k remains constant.

The right-hand sides of (9), (10), and (11) prove to be the cartesian components of a vector field called "the curl of $\mathbf{E}$," written as "curl $\mathbf{E}$," and defined in a coordinate-free way in §78. We adopt this terminology and notation without waiting for the proper definition and write

$$\text{curl}_x \mathbf{E} = \frac{\partial E_z}{\partial y} - \frac{\partial E_y}{\partial z}, \tag{12}$$

$$\text{curl}_y \mathbf{E} = \frac{\partial E_x}{\partial z} - \frac{\partial E_z}{\partial x}, \tag{13}$$

$$\text{curl}_z \mathbf{E} = \frac{\partial E_y}{\partial x} - \frac{\partial E_x}{\partial y}; \tag{14}$$

here the symbol $\text{curl}_x \mathbf{E}$ stands for the x component of curl $\mathbf{E}$, and so on. Another symbol for curl $\mathbf{E}$ is $\nabla \times \mathbf{E}$, described in Appendix D.

Earlier in this section we saw that a conservative field cannot exert a torque on a pinwheel; consequently, the right-hand sides of (9), (10), and (11) must all vanish for conservative fields. In Chapter 19 we will see that the converse of this statement is also true, and in anticipation of that result we now state the following theorem: *An electric field is conservative if and only if the curl of* $\mathbf{E}$ *is identically zero.*

Example. The formula for the intensity of the electric field in Fig. 138 is

$$\mathbf{E} = cy\mathbf{1}_x \quad (c = \text{constant} > 0). \tag{15}$$

Is this electric field conservative or not? To answer this question, we compute the cartesian components of the curl of **E**. Since

$$E_x = cy, \quad E_y = 0, \quad E_z = 0, \tag{16}$$

Equations (12), (13), and (14) give

$$\text{curl}_x \mathbf{E} = 0, \quad \text{curl}_y \mathbf{E} = 0, \tag{17}$$

and

$$\text{curl}_z \mathbf{E} = -c \neq 0. \tag{18}$$

Since the components of curl **E** do not all vanish, the field (15) is *not conservative*, as we already know from Fig. 138.

To express these results in terms of torques acting on small pinwheels, we turn to (9), (10), and (11), and get

$$T_x = 0, \quad T_y = 0, \quad T_z = -c/k = \text{constant}. \tag{19}$$

That is, no torque will act on the pinwheel if the pin is parallel to the x or the y axis, but a torque will act if the pin is parallel to the z axis, and this torque will not depend on the (x, y) coordinates of the pin. The reason why T_z is *negative* has to do with a right-hand rule for torques (§77).

Circular electric fields. Suppose that the z axis does not pass through the disc of the pinwheel, that $\mathbf{E} = f(\rho)\mathbf{1}_\phi$, and that $f(\rho)$ is positive, so that the lines of **E** run counterclockwise, as in Fig. 140. Parts (*a*) and (*b*) of this figure suggest, respectively, that if $f(\rho)$ *increases* with ρ fast enough, a counterclockwise torque will act on the pinwheel; but if $f(\rho)$ *decreases* with ρ fast enough, the sense of the torque will reverse. This implies that there is a special form

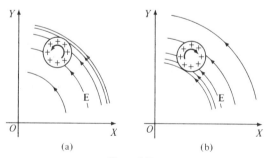

Figure 140

of $f(\rho)$ for which the torque T_z is zero. We will see in Exercise 3^{78} that this torque vanishes if $f(\rho) = c/\rho$, where c is a constant.

Suppose, however, that the pin is aligned with the z axis, as in Fig. 141, and the radius of the pinwheel is so chosen that the function $f(\rho)$ is not zero

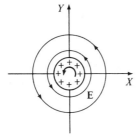

Fig. 141. An electric pinwheel aligned with the axis of a circular electric field.

at the rim of the disc. It is then apparent that a torque will act on the disc, and we conclude that *circular fields are not conservative*. This statement includes the field $(c/\rho)\mathbf{1}_\phi$, even though a torque then acts on the pinwheel only if the z axis passes through the disc.

EXERCISES

1. Let $\mathbf{E} = cx\mathbf{1}_y$ and show that $\operatorname{curl}_x \mathbf{E} = \operatorname{curl}_y \mathbf{E} = 0$, and $\operatorname{curl}_z \mathbf{E} = c$. Explain by line maps why the value of $\operatorname{curl}_z \mathbf{E}$ differs in sign from (18).

2. The field $\mathbf{E}$ outside a cylindrical hole in a current-carrying conductor is given by (8^{44}). Recall Equations (10^{14}) and show that this field is conservative.

56. ELECTROSTATIC POTENTIAL

So far we have relied on the reader's previous acquaintance with the concept of a potential difference. Now we will define it in field-theoretic terms and will also show that every conservative electric field has associated with it not only the *vector* field $\mathbf{E}$ but also a *scalar* field, denoted in this book by V. This field is called the "electric potential field," the "electric scalar potential," the "electrostatic potential," or, at the risk of an ambiguity, simply the "potential." Our unit for measuring V is the volt. The notation pertaining to this potential is not standard in the literature; therefore, in anticipation of the definitions given below, we will first list some terms and symbols used in this book.

If the numerical values of V, say V_a and V_b, at the points a and b satisfy the inequality $V_a > V_b$, one says that a is at a higher potential than b. In §1 we let the subscript ab mean "from a to b" and wrote i_{ab} for the current flowing from a to b in the resistor R in Fig. 1^1. The formula (1^1) for Ohm's law then reads $i_{ab} = V_{ab}/R$. Now, a current in a resistor flows from places of higher

200   Conservative Electric Fields   § 56

potential to places of lower potential, and consequently the symbol V_{ab} in this formula stands for the potential *drop* from a to b, as it does throughout this book. Accordingly, if $V_a > V_b$, the potential drop from a to b (which we denote by V_{ab}) is positive, and the potential rise from a to b (which we denote by $-V_{ab}$) is negative.

The emf in a path γ leading from a point a to a point b is

$$(4^{52}) \qquad \mathrm{emf}_{a\gamma b} = \int_{a\gamma b} E_t\, dl. \qquad (1)$$

If the electric field is conservative, the integral in (1) does not depend on the choice of the path leading from a to b, and hence we may abbreviate (1) to read

$$\mathrm{emf}_{ab} = \int_{ab} E_t\, dl. \qquad (2)$$

The *potential drop* from a to b, which we denote by V_{ab}, and which may be positive, negative, or zero, is *defined* by the equation

$$V_{ab} = \int_{ab} E_t\, dl. \qquad (3)$$

Thus, the terms "electromotive force" and "potential drop" are interchangeable in the case of conservative fields; but the term "potential drop" is then more suggestive, because it emphasizes the fact that the field under discussion is conservative. (If a field is *not* conservative, the quantity denoted above by emf_{ab} and V_{ab} does not have any clear-cut meaning, because then the value of the integral in (2) and (3) depends on the choice of path, and hence is not unique.)

In our notation, $-V_{ab}$ stands for the potential rise from a to b, and V_{ba} for the potential drop from b to a, so $V_{ba} = -V_{ab}$. More explicitly,

$$(\text{potential rise from } a \text{ to } b) = -\int_{ab} E_t\, dl = -V_{ab}. \qquad (4)$$

Comparison of (4) and (1^{52}) shows that the potential rise from a to b is equal to the work that we must do per unit charge in transferring a small positive charge from a to b. In (1^{52}) the path leading from a to b is indicated explicitly, because that equation holds in general. By contrast, (4) is restricted to conservative fields, and hence the integral in (4) does not depend on the choice of path.

Let us suppose that the points a, b, and c in Fig. 142 lie in an electric field of intensity E, and let us associate with these points the respective numbers V_a, V_b, and V_c, chosen as follows: The value of V_a is fixed arbitrarily, and the values of V_b and V_c are then computed from the formulas

Electrostatic Potential § 56

$$V_b = V_a - V_{ab} \tag{5}$$

and

$$V_c = V_a - V_{ac}, \tag{6}$$

where V_{ab} is given by (3), and

$$V_{ac} = \int_{ac} E_t \, dl. \tag{7}$$

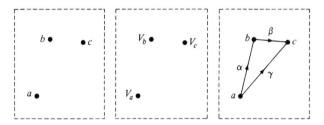

Fig. 142. Three points in an electric field, the potentials at these points, and paths connecting the points.

This process can be extended to every point in a conservative field; it amounts to associating with every point $P(x, y, z)$ a number $V(x, y, z)$, and thus defines a scalar field $V(x, y, z)$, which we usually denote simply by V. This field has an arbitrary feature, namely, the value chosen for V_a; but since

$$V_a - V_b = V_{ab} \tag{8}$$

and

$$V_a - V_c = V_{ac}, \tag{9}$$

and since the values of V_{ab} and V_{ac} are independent of the choice of V_a, we conclude that the difference between the values of V at any two arbitrarily chosen points—called the *difference of potential* or the *potential difference* between these points—is free from any arbitrariness. The reader should verify that this difference is also independent of the choice of the point (a, b, c, or any other point) at which we assign to V an arbitrary value (Exercise 1). Consequently the field V is arbitrary only to the extent of an additive constant, which cancels in the left-hand sides of such equations as (8) and (9). In any specific problem, the constant is usually chosen so as to simplify the appearance of the formula for V.

Surfaces on which V is constant are called *equipotential surfaces* or *equipotentials*.

The following formula, which follows from the definitions given above, is convenient to have on record: If $V(P)$ and $V(Q)$ are the respective potentials at the points P and Q, then

$$V(Q) = -\int_{PQ} E_t\, dl + V(P), \tag{10}$$

where the integral is evaluated over a path leading from P to Q.

Suppose that P and Q are neighboring points, and write dl for the length of the straight path leading from P to Q, and dV for the potential *rise* from P to Q. According to (10), we then have

$$dV = -E_t\, dl. \tag{11}$$

This equation can also be written as

$$dV = -\mathbf{E} \cdot d\mathbf{l}, \tag{12}$$

where $d\mathbf{l}$ is the vector step from P to Q.

Uniform electric fields. A simple electrostatic—and therefore conservative—field is the uniform field between the plates of a parallel-plate capacitor. We write

$$(1^{47}) \qquad \mathbf{E} = E_c \mathbf{1}_x \qquad (E_c = \text{constant}) \tag{13}$$

and

$$V(0, 0, 0) = V_0, \tag{14}$$

where V_0 is an arbitrary constant, and we proceed to compute the value of V at the point $P(x_0, y_0, z_0)$ in Fig. 143. A simple path of integration is $OABP$. We have $E_t = E_x = E_c$ on OA, $E_t = E_y = 0$ on AB, and $E_t = E_z = 0$ on BP, so (10) reduces to

$$V(x_0, y_0, z_0) = -\int_{OA} E_c\, dl + V_0 = -\int_0^{x_0} E_c\, dx + V_0 = -E_c x_0 + V_0, \tag{15}$$

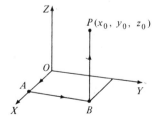

Figure 143

when we replace P by O and Q by A. To simplify appearances, we now set $V_0 = 0$, write the coordinates of P simply as (x, y, z), and get the formula

$$V(x, y, z) = -E_c x. \tag{16}$$

The equipotentials are planes parallel to the yz plane. They are pictured edge-on in Fig. 144 for equal steps of V.

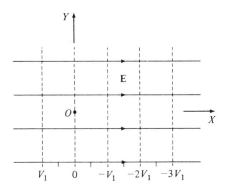

Fig. 144. Lines of the field $E_c \mathbf{1}_x$ and equipotential planes for this field, shown edge-on by the dashed lines.

Coulomb potential in free space. The intensity of the electric field produced in free space by a point charge of q coulombs located at the origin is

$$(7^{42}) \qquad \mathbf{E} = 9 \times 10^9 \frac{q}{r^2} \mathbf{1}_r. \tag{17}$$

The work that we must do per unit charge in carrying a small charge q^* from a point a to a point b in the field (17) is

$$(5^{48}) \qquad 9 \times 10^9 q \left(\frac{1}{r_b} - \frac{1}{r_a} \right), \tag{18}$$

where r_a and r_b are the respective radial coordinates of a and b. Consequently,

$$V(r_b) - V(r_a) = 9 \times 10^9 q \left(\frac{1}{r_b} - \frac{1}{r_a} \right). \tag{19}$$

To reduce (19) to its simplest form, we let V be zero at infinity. In fact, we let $r_a \to \infty$, let $V(r_a) \to 0$, write r for r_b, and get

$$V(r) = 9 \times 10^9 \frac{q}{r}. \tag{20}$$

The potential function (20) is associated with the Coulomb field (17) and is called the *Coulomb potential*. The equipotentials in this case are spherical surfaces centered on the charge q.

Cylindrical hole in a current-carrying conductor. For our next example of the computation of the potential field V, we return to the field

(8⁴⁴)
$$\mathbf{E} = E_0\left(1 - b^2 \frac{x^2 - y^2}{(x^2 + y^2)^2}\right)\mathbf{1}_x - 2E_0 b^2 \frac{xy}{(x^2 + y^2)^2}\mathbf{1}_y. \tag{21}$$

This field is conservative, so "work does not depend on path," and the existence of the field V is assured (recall Exercise 2⁵⁵). We will first compute the difference between the values of V at an arbitrarily chosen point $P(x_0, y_0)$ in Fig. 145,

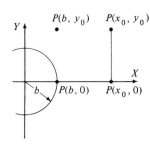

Figure 145

and at the point $P(b, 0)$, which lies on the surface of the hole and on the positive half of the x axis. To begin with, we assign to the potential at $P(b, 0)$ an arbitrarily chosen value V_0:

$$V(b, 0) = V_0. \tag{22}$$

To compute the potential drop from $P(b, 0)$ to $P(x_0, y_0)$, we use (10) and the path that leads from $P(b, 0)$ straight to $P(x_0, 0)$ and then straight to $P(x_0, y_0)$.

On the first portion of this path we have $E_t = E_x$ and $y = 0$, so, according to (21), $E_t = E_0[1 - (b^2/x^2)]$; since $dl = dx$, the integral in (10) is

$$-\int_b^{x_0} E_0\left(1 - \frac{b^2}{x^2}\right) dx = -E_0\left(x_0 - 2b + \frac{b^2}{x_0}\right). \tag{23}$$

On the remainder of the path we have

$$-\int_0^{y_0} E_y \, dy = 2E_0 b^2 \int_0^{y_0} \frac{x_0 y}{(x_0^2 + y^2)^2} \, dy = -E_0\left(\frac{x_0 b^2}{x_0^2 + y_0^2} - \frac{b^2}{x_0}\right). \tag{24}$$

The potential at $P(x_0, y_0)$ is the sum of the right-hand sides of (22), (23), and (24), so

$$V(x_0, y_0) = -x_0\left(1 + \frac{b^2}{x_0^2 + y_0^2}\right)E_0 + 2bE_0 + V_0. \qquad (25)$$

To simplify appearances, we now set the arbitrary constant V_0 equal to $-2bE_0$ and omit the subscripts in the symbols x_0 and y_0, getting the formula

$$V(x, y) = -x\left(1 + \frac{b^2}{x^2 + y^2}\right)E_0 \qquad (x^2 + y^2 \geq b^2). \qquad (26)$$

The condition $x^2 + y^2 \geq b^2$ restricts (26) to points on the surface of the hole and points outside it. In cylindrical coordinates,

$$V(\rho, \phi) = -bE_0\left(\frac{\rho}{b} + \frac{b}{\rho}\right)\cos\phi \qquad (\rho \geq b). \qquad (27)$$

The equipotential surfaces are shown edge-on in Fig. 146. The lines drawn *inside* the hole follow from Equation (31) below.

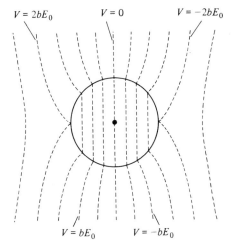

Fig. 146. Equipotential surfaces, shown edge-on for the case of positive charge flowing from left to right past a cylindrical hole in a conductor.

The electric field inside the cylindrical hole. The circle in Fig. 147 is the outline of the hole in the copper block. The points P and Q lie on the surface of the hole; their respective (x, y) coordinates are $(-c, h)$ and (c, h), where $c^2 + h^2 = b^2$. According to (26), the respective potentials of these points are

$$V(P) = 2cE_0, \qquad V(Q) = -2cE_0. \qquad (28)$$

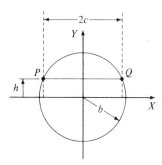

Figure 147

Next we write **E** for the electric intensity inside the hole and compute $V(Q)$ in terms of $V(P)$, using (10) and the path leading from P straight to Q. We then have the equation

$$V(Q) = -\int_{-c}^{c} E_t\, dx + V(P), \tag{29}$$

where E_t is an as yet unknown function of x and y. Combining (28) and (29), and rearranging terms, we get

$$\int_{-c}^{c} E_t\, dx = 2c \cdot 2E_0. \tag{30}$$

Thus, the integral of E_t over any horizontal chord of the circle in Fig. 147 is equal to the length of this chord multiplied by a constant, namely $2E_0$. This result suggests that E_t is constant, and is equal to $2E_0$. The simplest procedure for satisfying this condition—a procedure consistent with the symmetry of the potential (26) about the x axis and the y axis—is to assume that **E** is uniform throughout the hole, is directed from left to right in Fig. 147, and has the magnitude $2E_0$. Therefore, we write

$$\mathbf{E}_{\text{inside the hole}} = 2E_0 \mathbf{1}_x, \tag{31}$$

where E_0 is the magnitude of **E** far from the hole. Our arguments make (31) plausible; a proof is given in §71. In the meantime, the reader should check (31) as suggested in Exercise 3.

A line map of **E** both outside and inside the hole is shown in Fig. 148, together with the equipotentials. In the copper, the lines of **E** begin on positive charges at the far left and end on negative charges at the far right. (One might think of these charges as located on the perfectly conducting end plates of the copper block, connected to the battery.) In the hole, the lines of **E** begin and end on the surface charges first pictured in Fig. 53[14].

The equipotential surfaces inside the hole are planes similar to those in Fig. 144.

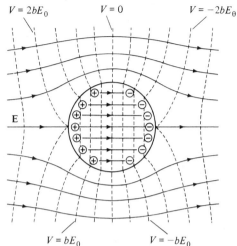

Fig. 148. Lines of **E**, outside and inside the cylindrical hole, superimposed upon the equipotentials of Fig. 146.

EXERCISES

1. Assign an arbitrary value to V_b in Fig. 142 and define V_a and V_c by the process described above. Choose V_b in such a way that the values of V_b and V_a are different from those in (4) and (5), although (7) will still hold, of course. Then use such paths as α, β, and γ in Fig. 142 to show that (8) will also hold. (In the text, the value of V_a was chosen arbitrarily.)

2. Derive (25) and (26), using the following path in Fig. 145: from $P(b, 0)$ straight to $P(b, y_0)$, and then straight to $P(x_0, y_0)$.

3. The respective (x, y) coordinates of the points P and Q in Fig. 149 are (c_1, h_1)

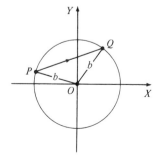

Figure 149

and (c_2, h_2), where $c_1^2 + h_1^2 = c_2^2 + h_2^2 = b^2$. Compute $V(P) - V(Q)$ using (31) and evaluating the integral in (10) over the straight path PQ in Fig. 149. Then verify that the result agrees with the value of $V(P) - V(Q)$ obtained directly from (26).

4. Verify that Equation (31) for **E** inside the hole can be written as

$$E_\rho = 2E_0 \cos\phi, \quad E_\phi = -2E_0 \sin\phi, \quad E_z = 0. \tag{32}$$

5. The insulation of an armored cable can be tested by raising the core to a potential V_1 and the armor to a lower potential V_2 and measuring the leakage current i. It is then found that $i = (V_1 - V_2)/R$, where R is a constant, called the *leakage resistance* of the cable.

   Recall the last sentence of §39, compute **E**, and show that $V_1 - V_2 = (i/2\pi lg) \ln(\rho_2/\rho_1)$ and that

$$R = \frac{1}{2\pi lg} \ln \frac{\rho_2}{\rho_1}. \tag{33}$$

Here ρ_2 and ρ_1 are the outer and inner radii of the insulation, g is its conductivity, and l is the length of the sample.

Next use some version of (7^{45}) to compute the rate at which heat is generated in a length l of the insulation; to check, use (33) and the formula $i^2 R$, taken from circuit theory.

6. As an extension of Exercise 5, show that, at any point in the insulation,

$$E = \frac{V_1 - V_2}{\ln(\rho_2/\rho_1)} \frac{1}{\rho} \mathbf{1}_\rho \tag{34}$$

and

$$V = \frac{V_1 \ln(\rho/\rho_2) - V_2 \ln(\rho/\rho_1)}{\ln(\rho_1/\rho_2)}. \tag{35}$$

These equations do not involve g; therefore, they hold even if two coaxial circular metal cylinders are separated by a vacuum, provided that end effects can be ignored.

7. Consider cylindrical layers of insulation of thickness $d\rho$; note that, so far as leakage current is concerned, they are connected in series; derive (33) on this basis.

8. Show that if the insulation is sufficiently thin, then (33) reduces to $R = (\rho_2 - \rho_1)/2\pi gl\rho_1$; verify this formula on elementary grounds, "without calculus."

57. THE GRADIENT

In §56, we considered conservative electric fields and defined the scalar potential V in terms of the electric intensity **E**. In this section, which is also restricted to conservative fields, we take up the converse problem and show how **E** can be computed when V is given. Our first step is to introduce the concept of the *gradient*. This concept has to do with the differential of a scalar field, say f, associated with the step from a point P to a point Q, which is not necessarily a neighboring point of P. In one-dimensional problems we write x and $x + dx$ for the respective coordinates of P and Q, and define df as

(10⁶) $$df = \frac{df}{dx} dx. \tag{1}$$

Similarly, in three dimensions,

(2⁷) $$df = \frac{\partial f}{\partial s} ds, \tag{2}$$

where $\partial f/\partial s$ is the directional derivative of f taken at P toward Q, and ds is the distance from P to Q. The notation in (2) is quite standard, and, therefore, in a part of this section, we will denote the vector step from P to Q by $d\mathbf{s}$, instead of our usual $d\mathbf{l}$.

In one-dimensional problems the word "gradient" is often used for the derivative with respect to a coordinate variable. For example, in the study of heat conduction in a rod parallel to the x axis, one says "temperature gradient" for "derivative of temperature with respect to x." Let us set

$$g = \frac{df}{dx}, \tag{3}$$

where df/dx is to be evaluated at P. The function g may then be called the gradient of f at P, and (1) may be written as

$$df = g\, dx. \tag{4}$$

To generalize (4) to three dimensions, we associate with the point P a vector, say **G**, defined in terms of df and $d\mathbf{s}$ by the equation

$$df = \mathbf{G} \cdot d\mathbf{s}, \tag{5}$$

which is patterned after (4) and is required to hold for every choice of Q. The vector **G** is called the *gradient* of f at the point **P** and is written as "grad f." In this notation, the definition (5) reads

$$df = (\operatorname{grad} f) \cdot d\mathbf{s}, \tag{6}$$

but we will continue for one more paragraph to write **G** for grad f. The del notation for gradients is described in Appendix D.

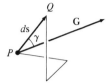

Figure 150

Let the vector **G** in Fig. 150 be the gradient of a scalar field f at P. (The thin lines are intended to help the perspective.) To find the direction of **G** and its magnitude G, we write (5) as

$$df = G \, ds \cos \gamma, \tag{7}$$

where γ is the angle shown in the figure. Comparing (7) with (2), we get

$$\frac{\partial f}{\partial s} = G \cos \gamma. \tag{8}$$

If we now keep P fixed and let Q wander about, we find from (8) that $\partial f/\partial s$ has its maximum value when $\gamma = 0$, that is, when the line PQ has the same direction as **G**. Consequently, **G** points in the direction in which f increases most rapidly at P, and

$$G = \left(\frac{\partial f}{\partial s}\right)_{\max} \tag{9}$$

Thus *the gradient of a scalar field f at a point P is a vector pointing in the direction in which f increases most rapidly at P; the magnitude of this vector is equal to the directional derivative of f in this direction, taken at P.* This description helps to visualize a gradient, but the most important single thing to remember about gradients is the definition (6).

When a scalar field f is given, the vector grad f can be constructed at every point, except perhaps at some singular points. Accordingly, *the gradient of a scalar field is itself a vector field.* It then follows that the gradient of an invariant scalar field is an invariant vector field.

Once a scalar field f has been described and the points P and Q have been chosen, the quantities df and ds in (6) become defined without reference to any coordinate frame; consequently, *the concept of a gradient is coordinate-free.* Furthermore, Equation (8), which in a standard notation reads

$$\frac{\partial f}{\partial s} = |\operatorname{grad} f| \cos \gamma, \tag{10}$$

implies that *the directional derivative of a scalar field f, taken at a point P in an arbitrarily chosen direction, is the component, taken in this direction, of the gradient of f at P.*

Example. The value of a certain scalar field f at a point P is s^2, where s is the distance from P to a straight line A that passes through the point O in Fig. 151(a) at right angles to the page; the circular arcs in this figure show where the cylindrical surfaces of constant f intersect the page. It is perhaps obvious that f increases most rapidly at P if we go from P directly away from the line A at right angles to A, and that the rate of increase of f in this direction is the derivative of s^2 with respect to s, namely $2s$. Consequently, the field grad f at P is directed away from the line A and its magnitude is $2s$; this field is pictured in Fig. 151(b), where the length of a pointer is proportional to the distance of its tail from O.

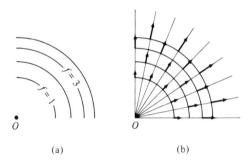

Fig. 151. Level surfaces of the scalar field $f(\rho) = \rho^2$, and a pointer map of grad ρ^2.

The example above was worked out without reference to any coordinate frame. We can summarize the result more succinctly, however, if we let the line A be the z axis of a coordinate frame and use the symbols ρ and $\mathbf{1}_\rho$ in our usual way. Then $f = \rho^2$, and the conclusion reached in the preceding paragraph can be written as

$$\text{grad } \rho^2 = 2\rho \mathbf{1}_\rho. \tag{11}$$

If we let $f = \rho$, we find in a similar way that

$$\text{grad } \rho = \mathbf{1}_\rho. \tag{12}$$

The direction of the vector $\mathbf{1}_\rho$ is not defined at points on the z axis, and hence (12) is meaningless at these points, which are the *singular points* of grad ρ.

Electric fields. Let V be the electric potential, and let us shift from the symbol $d\mathbf{s}$ back to $d\mathbf{l}$. We then have, from (6),

$$dV = (\text{grad } V) \cdot d\mathbf{l} \tag{13}$$

and, from (12^{56}),

$$dV = -\mathbf{E} \cdot d\mathbf{l}. \tag{14}$$

Consequently,

$$(\operatorname{grad} V) \cdot d\mathbf{l} = -\mathbf{E} \cdot d\mathbf{l}. \tag{15}$$

The step $d\mathbf{l}$ can be chosen arbitrarily and, therefore, in view of Exercise 6^{10}, grad $V = -\mathbf{E}$. Conversely,

$$\mathbf{E} = -\operatorname{grad} V. \tag{16}$$

We must stress the fact that (16) makes sense only for *electrostatic fields*. In view of (16), these fields are said to be *derivable from a potential*. The negative sign in (16) implies that a positive electric charge placed at a point P in an electrostatic field is urged by this field in the direction in which the potential V decreases most rapidly at P.

Is every vector field the gradient of some scalar field? The answer is no, as illustrated in Exercise 5.

The gradient as an integral. The gradient was defined above in terms of the coordinate-free concepts of differentials and directional derivatives. An alternative way is to define it as the coordinate-free integral (43^A). That definition is more abstract, but it fits in more neatly with the definition of the divergence; it is also a more convenient starting point for deriving some theorems. A proof of the equivalence of the two definitions is suggested in Exercise 4^{58}.

EXERCISES

1. Let f, f_1, and f_2 be scalar fields and a a numerical constant. Deduce the following equations from (6):

$$\operatorname{grad} af = a \operatorname{grad} f, \tag{17}$$

$$\operatorname{grad}(f_1 + f_2) = \operatorname{grad} f_1 + \operatorname{grad} f_2, \tag{18}$$

and

$$\operatorname{grad}(f_1 f_2) = f_1 \operatorname{grad} f_2 + f_2 \operatorname{grad} f_1. \tag{19}$$

2. Use (12) and (19) to check (11).

3. Show that (except perhaps at singular points) the gradient of a scalar field f at a point P is perpendicular to the surface of constant f passing through P. Ignore any singular points and state the condition under which the follow-

ing statement makes sense: "The lines of **E** are perpendicular to the equipotentials."

4. Consider a scalar field f and the surfaces of constant f, labeled $f = a, a + b$, $a + 2b$, and so on; show that, if b is sufficiently small, then $|\text{grad } f|$ is largest in regions where these surfaces lie closest together.

5. Show graphically that there is no scalar field f whose gradient is $y\mathbf{1}_x$. (Assume that there is such a field and try to draw in the xy plane the lines on which $f = a, f = a + b, f = a + 2b$, and so on, where a and b are any constants of your choice, except that $b \neq 0$. A line map of the field $y\mathbf{1}_x$ is shown in Fig. 76[21].)

58. COORDINATE FORMULAS FOR GRADIENTS

In §57 we emphasized the fact that the concept of the gradient is coordinate-free. Usually, however, the most convenient way to *compute* gradients is to work not from a coordinate-free definition, but from the coordinate formulas presented in this section.

Cartesian coordinates. We replace ds by the equivalent symbol dl, begin with the definition

$$(6^{57}) \qquad df = (\text{grad } f) \cdot dl, \qquad (1)$$

and proceed to compute the cartesian components of grad f, which we denote by $\text{grad}_x f$, $\text{grad}_y f$, and $\text{grad}_z f$, so that

$$\text{grad } f = \mathbf{1}_x(\text{grad}_x f) + \mathbf{1}_y(\text{grad}_y f) + \mathbf{1}_z(\text{grad}_z f). \qquad (2)$$

Now,

$$(10^4) \qquad dl = \mathbf{1}_x dx + \mathbf{1}_y dy + \mathbf{1}_z dz, \qquad (3)$$

and consequently (1) can be written as

$$df = (\text{grad}_x f)\, dx + (\text{grad}_y f)\, dy + (\text{grad}_z f)\, dz. \qquad (4)$$

Comparison with the formula

$$(3^7) \qquad df = \frac{\partial f}{\partial x} dx + \frac{\partial f}{\partial y} dy + \frac{\partial f}{\partial z} dz \qquad (5)$$

shows that, since each of the displacements dx, dy, and dz is arbitrary,

$$\text{grad}_x f = \frac{\partial f}{\partial x}, \qquad \text{grad}_y f = \frac{\partial f}{\partial y}, \qquad \text{grad}_z f = \frac{\partial f}{\partial z}. \qquad (6)$$

Substituting into (2), we get the very useful formula

$$\operatorname{grad} f = \mathbf{1}_x \frac{\partial f}{\partial x} + \mathbf{1}_y \frac{\partial f}{\partial y} + \mathbf{1}_z \frac{\partial f}{\partial z}. \tag{7}$$

For example, if

$$f = x^2 + y^2, \tag{8}$$

we have

$$\frac{\partial f}{\partial x} = 2x, \quad \frac{\partial f}{\partial y} = 2y, \quad \frac{\partial f}{\partial z} = 0, \tag{9}$$

and consequently

$$\operatorname{grad}(x^2 + y^2) = 2(x\mathbf{1}_x + y\mathbf{1}_y). \tag{10}$$

In cylindrical coordinates, (10) becomes $\operatorname{grad} \rho^2 = 2\rho \mathbf{1}_\rho$, in agreement with (11[57]).

Cylindrical and spherical coordinates. To identify the cylindrical components of $\operatorname{grad} f$, we use (12[A]) for $d\mathbf{l}$, so that the right-hand side of (1) becomes $(\operatorname{grad}_\rho f)\,d\rho + (\operatorname{grad}_\phi f)\rho\,d\phi + (\operatorname{grad}_z f)\,dz$. Another formula for df in cylindrical coordinates can be inferred from (5[7]) and (7[7]). A comparison then shows that, since $d\rho$, $d\phi$, and dz can each be chosen arbitrarily, we have $\operatorname{grad}_\rho f = \partial f/\partial \rho$, $\operatorname{grad}_\phi f = \rho^{-1} \partial f/\partial \phi$, and $\operatorname{grad}_z f = \partial f/\partial z$. Consequently,

$$\operatorname{grad} f = \mathbf{1}_\rho \frac{\partial f}{\partial \rho} + \mathbf{1}_\phi \frac{1}{\rho} \frac{\partial f}{\partial \phi} + \mathbf{1}_z \frac{\partial f}{\partial z}. \tag{11}$$

Similarly, in terms of spherical coordinates,

$$\operatorname{grad} f = \mathbf{1}_r \frac{\partial f}{\partial r} + \mathbf{1}_\theta \frac{1}{r} \frac{\partial f}{\partial \theta} + \mathbf{1}_\phi \frac{1}{r \sin \theta} \frac{\partial f}{\partial \phi}. \tag{12}$$

Equations (7), (11), and (12) reflect not only the properties of a gradient, but also the peculiarities of the cartesian, cylindrical, and spherical coordinate frames. Therefore, the significance of a gradient is inferred best not from these equations, but from the definition and the properties discussed in §57.

EXERCISES

1. Verify that the equation $\mathbf{E} = -\operatorname{grad} V$ holds for the fields (17[56]) and (20[56]); the fields (21[56]) and (26[56]); the fields (9[44]) and (27[56]).
2. Use (7) to check the formulas of Exercise 1[57].

3. Show that
$$\operatorname{grad} \frac{\cos \theta}{r^2} = -2\frac{\cos \theta}{r^3} \mathbf{1}_r - \frac{\sin \theta}{r^3} \mathbf{1}_\theta. \tag{13}$$

4. An alternative coordinate-free definition of the gradient is (43^A). Derive (7) from (43^A) and thus prove that that definition is equivalent to (1). [The procedure is similar to that used in deriving the equation $\operatorname{div} \mathbf{F} = \partial F_x/\partial x + \partial F_y/\partial y + \partial F_z/\partial z$ from the coordinate-free definition of divergence. Use the brick-shaped region of Fig. 88[30]; the area of the right-hand face is $dy\,dz$; on this face $\mathbf{1}_n = \mathbf{1}_x$ and $f_Q = f_P + \frac{1}{2}(\partial f/\partial x)_P\,dx + \cdots$; and so on.]

5. Derive (12) as outlined in the text for the case of (11).

6. Derive (12) from (43^A).

CHAPTER 14

Coulomb's Law in Rationalized Form

We now turn to the box in the chart in the preface labeled "Coulomb's law." In this chapter we consider in some detail several specific electric fields of the electrostatic type—first of all, the field of a stationary point charge. The concept of the capacitivity or permittivity of a vacuum is introduced in §59, together with remarks on "rationalized" equations. Next come the electric dipole and some simple continuous distributions of static charges that form straight lines or plane sheets. The concepts of the capacitivity and the dielectric constant of material mediums are discussed in §62 from the macroscopic viewpoint, except for brief remarks on polarization. The electric displacement density **D** is introduced in §§61 and 62.

59. COULOMB'S LAW IN FREE SPACE

Coulomb's law states that the force of interaction between two point charges is directly proportional to the product of the magnitudes of the charges and inversely proportional to the square of the distance between them. The experimental value of the constant of proportionality in *free space* is very nearly 9×10^9 meters per farad, and therefore the equation

$$(7^{42}) \qquad \mathbf{E} = 9 \times 10^9 \frac{q}{r^2} \mathbf{1}_r, \qquad (1)$$

is a good approximation to the electric intensity r meters away from a charge of q coulombs located in free space at the origin of a coordinate frame.

When the MKS-Giorgi units are used and Coulomb's law is written with a literal rather than a numerical factor, this factor, say ϵ_0, is put in the denominator and multiplied by 4π, so that (1) takes the form

$$\mathbf{E} = \frac{q}{4\pi\epsilon_0 r^2} \mathbf{1}_r, \tag{2}$$

and the Coulomb potential (20^{56}) becomes

$$V = \frac{q}{4\pi\epsilon_0 r}. \tag{3}$$

Since (1) is nearly exact, we have, at least approximately,

$$\epsilon_0 = 8.85 \times 10^{-12} \; \frac{\text{farad}}{\text{meter}}. \tag{4}$$

The constant ϵ_0 is called the (rationalized) *capacitivity* or *permittivity* of free space. Its numerical value can be found experimentally by using an evacuated parallel-plate capacitor, whose plates each have the area of a square meters, whose plate separation is s meters, whose guard rings permit ignoring the edge effects, and whose capacitance C (farads) has been measured, say by a calibrated ballistic galvanometer. The value of ϵ_0 can then be found from the formula for the capacitance of the capacitor, namely

$$C = \frac{\epsilon_0 a}{s}. \tag{5}$$

Rationalization. If we want to conceal the factor 4π, (2) can be written more simply as

$$\mathbf{E} = \frac{q}{\delta_0 r^2} \mathbf{1}_r, \tag{6}$$

where $\delta_0 = 4\pi\epsilon_0 = 1.112 \times 10^{-10}$ farad/meter. In either case, the numerical coefficient and the units are the same as in (1); that is,

$$E = \frac{q}{4\pi\epsilon_0 r^2} = \frac{q}{\delta_0 r^2} = 9 \times 10^9 \frac{q}{r^2} \; \frac{\text{volts}}{\text{meter}}. \tag{7}$$

If we begin with (6) and derive the capacitor formula, the ϵ_0 in (5) becomes replaced by $\delta_0/4\pi$; that is,

$$C = \frac{\epsilon_0 a}{s} = \frac{\delta_0 a}{4\pi s} = 8.85 \times 10^{-12} \frac{a}{s} \;\; \text{farads}. \tag{8}$$

Equations (7) and (8) show that, if the factor 4π is concealed in one formula by adjusting the numerical value of a parameter, it may crop up in another formula. In particular, if we adopt the form (6) for Coulomb's law, then the

4π shows up in Maxwell's equations, a fact that some people consider to be somewhat inconvenient.[1]

The form (6) of Coulomb's equation is the older form, usually written with an ϵ instead of our δ. The process of changing from (6) to (2) is called *rationalization* of Maxwell's equations because it removes the factor 4π from these equations. The choice between (2) and (6) is a matter of individual preference and convenience. [Note that in (8) rationalization simplifies the appearance of the literal formula, but in (7) it makes the literal formula look more complicated.] In writings that use the MKS-Giorgi units, the form (2) has become standard. The constant ϵ_0, given by (4), is the "rationalized" capacitivity of free space, but in this book we usually omit the word "rationalized."

EXERCISE

1. A positive charge that repels with the force of one dyne a similar charge placed in free space one centimeter away is called the electrostatic unit of charge or the esu of charge. Show that, within a fraction of one percent, 1 esu of charge is equal to $\frac{1}{3} \times 10^{-9}$ coulomb.

60. EXAMPLES OF ELECTROSTATIC FIELDS

We will now compute E for a few mathematically simple charge distributions, some quite idealized. Our symbols are

q = electric charge (coulombs),
q_l = charge per unit length of a line or a rod (coulombs per meter of length),
q_a = charge per unit area of the surface of a body or of an imagined plane of zero thickness (coulombs per square meter of area).

Our words and figures will imply that the symbols q, q_l, and q_a stand for positive numbers, but our formulas are general.

A short electric dipole. An electric dipole (length l, charges q and $-q$) is located in free space and is oriented as in Figs. 61[17] and 152. The magnitude of its electric dipole moment, defined in Exercise 4[42], is

$$p_e = ql. \tag{1}$$

What are the electric potential and intensity at the point P in the figure if the dipole is so short that $l^2 \ll r^2$?

---

[1] In 1895 Oliver Heaviside "characterized the 4π eruption as absurd, arbitrary, injurious, irrational, mischievous, obtrusive, perverse, preposterous, ridiculous, stupid, unfortunate, unnecessary and unspeakable." G. A. Campbell, *Bull. Nat. Res. Council*, Vol, 93, p. 74 (1933).

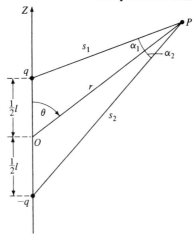

Fig. 152. An electric dipole located on the z axis.

According to (3^{59}), the respective potentials produced at P by the charges q and $-q$ are $q/(4\pi\epsilon_0 s_1)$ and $-q/(4\pi\epsilon_0 s_2)$, and hence the total potential at P is

$$V = \frac{q}{4\pi\epsilon_0}\left(\frac{1}{s_1} - \frac{1}{s_2}\right). \tag{2}$$

Since $l^2 \ll r^2$, we may use (25^7) and reduce (2) to

$$V = \frac{p_e}{4\pi\epsilon_0}\frac{\cos\theta}{r^2}. \tag{3}$$

Next, we put (3) into the equation $\mathbf{E} = -\text{grad } V$ and recall that

(13^{58}) $$\text{grad}\frac{\cos\theta}{r^2} = -2\frac{\cos\theta}{r^3}\mathbf{1}_r - \frac{\sin\theta}{r^3}\mathbf{1}_\theta. \tag{4}$$

Consequently, the spherical components of $\mathbf{E}$ at P are

$$E_r = \frac{p_e}{4\pi\epsilon_0}\frac{2\cos\theta}{r^3}, \quad E_\theta = \frac{p_e}{4\pi\epsilon_0}\frac{\sin\theta}{r^3}, \quad E_\phi = 0. \tag{5}$$

The direction lines of this field are discussed in Exercises 8 and 9.

Straight-line charge. Electric charge is placed all along the z axis with a uniform linear density of q_l coulombs per meter of length. We will compute $\mathbf{E}$ at the distance p from the z axis. The origin O can be put without loss of generality at the foot of the perpendicular drawn from P to the z axis, as in Fig. 153.

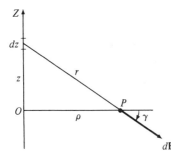

Fig. 153. The field $d\mathbf{E}$ caused at P by an element of a straight-line charge.

The charge on a small segment dz of the z axis is $q_l\, dz$ coulombs, and becomes a point charge when $dz \to 0$. We write $d\mathbf{E}$ for the contribution of this charge to the total intensity $\mathbf{E}$ at P. According to Coulomb's law and Fig. 153, the magnitude of $d\mathbf{E}$ is

$$dE = \frac{1}{4\pi\epsilon_0} \frac{q_l\, dz}{z^2 + \rho^2}. \tag{6}$$

Therefore, the cylindrical components of $d\mathbf{E}$ are

$$(d\mathbf{E})_\rho = (dE)\cos\gamma = \frac{q_l}{4\pi\epsilon_0} \frac{\rho\, dz}{(z^2 + \rho^2)^{3/2}}, \tag{7}$$

$$(d\mathbf{E})_\phi = 0, \tag{8}$$

$$(d\mathbf{E})_z = -(dE)\sin\gamma = -\frac{q_l}{4\pi\epsilon_0} \frac{z\, dz}{(z^2 + \rho^2)^{3/2}}. \tag{9}$$

To include the contributions of *all* the charges, we integrate each component with respect to z from $-\infty$ to ∞. The results are

$$E_\rho = \frac{1}{4\pi\epsilon_0} \frac{2q_l}{\rho}, \qquad E_\phi = 0, \qquad E_z = 0, \tag{10}$$

and, consequently,

$$\mathbf{E} = \frac{1}{2\pi\epsilon_0} \frac{q_l}{\rho} \mathbf{1}_\rho. \tag{11}$$

Thus, $\mathbf{E}$ is a two-dimensional radial vector field of the inverse first-power type. A neater proof is given in §63.

Uniformly charged nonconducting plane sheets of zero thickness. Suppose that electric charge is distributed over the entire zx plane with the uniform density of q_a coulombs per square meter (this statement implies, for example, that the total charge located on the two sides of a square portion of the plane, 10 meters by 10 meters in size, is $100q_a$ coulombs). We will compute **E** at the point P in Fig. 154(a).

The strip of small width dx shown in the figure is, in effect, an infinitely long straight-line charge of linear density q_l, equal to $q_a\,dx$ coulombs per meter of length. According to (11) and Fig. 154(a), the contribution $d\mathbf{E}$ made by this strip to the total intensity at P has the magnitude

$$dE = \frac{1}{2\pi\epsilon_0} \frac{q_a\,dx}{\sqrt{x^2 + y^2}}. \tag{12}$$

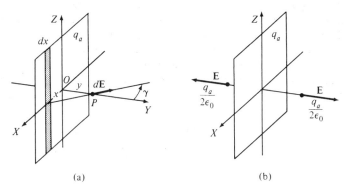

Fig. 154. If a plane carries a uniform charge of q_a coulombs/meter², the field **E** at points in vacuum outside the plane has the magnitude $q_a/2\epsilon_0$ and is directed away from the plane.

Consequently, the cartesian components of $d\mathbf{E}$ are

$$(dE)_x = -(dE)\sin\gamma = -\frac{q_a}{2\pi\epsilon_0}\frac{x\,dx}{x^2+y^2}, \tag{13}$$

$$(dE)_y = (dE)\cos\gamma = \frac{q_a}{2\pi\epsilon_0}\frac{y\,dx}{x^2+y^2}, \tag{14}$$

$$(dE)_z = 0. \tag{15}$$

Integrating these components with respect to x from $-\infty$ to ∞, we finally get

$$E_x = 0, \quad E_y = \frac{1}{2}\frac{q_a}{\epsilon_0}, \quad E_z = 0, \quad (y > 0). \tag{16}$$

If a test charge should be placed *in* the zx plane, it would be pushed equally in all directions in this plane; the net force on it would be zero, and consequently

$$E_x = 0, \quad E_y = 0, \quad E_z = 0 \quad (y = 0). \tag{17}$$

Had we placed P at the left of the charged plane, we would have found that

$$E_x = 0, \quad E_y = -\frac{1}{2}\frac{q_a}{\epsilon_0}, \quad E_z = 0 \quad (y < 0). \tag{18}$$

The fields (16) and (18) are pictured in Fig. 154(b).

Equations (1^{59}) and (11), and the three equations just above, bring out the following sequence, provided that P lies outside the charge distribution: point charge, inverse second-power law; line charge, inverse first-power law; plane charge, "inverse zeroth-power law"—that is, distance does not matter.

Figure 155(a) is a map of the field $\mathbf{E}$ produced by an infinite plane sheet of charge of uniform density q_a; according to (16) and (18), the magnitude of $\mathbf{E}$ at points outside this sheet is $q_a/2\epsilon_0$. Figure 155(b) shows a similar sheet of density $-q_a$; and Fig. 155(c), the two sheets taken together. We conclude that

$$E = \begin{cases} q_a/\epsilon_0 & \text{at points } inside \text{ the pair of sheets,} \\ 0 & \text{at points } outside \text{ the pair of sheets.} \end{cases} \tag{19}$$

(a)

(b)

(c)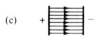

Fig. 155. Superposition of the $\mathbf{E}$'s of two uniformly charged sheets; the charge densities have equal magnitudes but opposite signs.

Finally, suppose that P lies *on* the right-hand sheet in Fig. 155(c). We then have $E = q_a/2\epsilon_0$ at P, because in view of (17), the entire electric field at P is then caused by the charge on the left-hand sheet. (Note Exercise 5.)

Parallel-plate vacuum capacitor. When a slab of a nonconducting substance such as glass or mica is placed in an electric field, the molecules of this substance become distorted and acquire electric moments, so that every portion of the slab becomes "polarized." As a result, the constant ϵ_0, which appears in Coulomb's law for charges in a vacuum, must be replaced inside glass or mica by a larger constant (§62).

By contrast, let a *metal* plate be placed in a moderately strong static electric field. To begin with, this "external" field pervades the interior of the metal and causes a flow of conduction electrons. This flow continues until a number of conduction electrons and of uncovered positive charges become distributed on the surface of the metal in such a way as to cancel the external field that had penetrated into the metal (§72). This process does not cause any effects inside the metal similar to polarization, and does not call for a change from the constant ϵ_0 in Coulomb's law to another constant. For this reason, the field E for a parallel-plate capacitor can be inferred directly from (19), provided we ignore fringing at the edges of an actual capacitor. Indeed, if equal and opposite charges are located on *metal* plates, their mutual attraction causes them to stay on the inner faces as in Fig. 156, so that we have once again the situation pictured in Fig. 155(c), and (19) can be rewritten as

$$E = \begin{cases} q_a/\epsilon_0 & \text{between the plates} \\ 0 & \text{outside the pair of inner faces.} \end{cases} \quad (20)$$

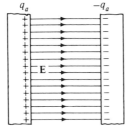

Fig. 156. The field E between the plates of a capacitor; its magnitude is q_a/ϵ_0.

Here q_a is the magnitude of the charge density on one of the two inner faces; the charge density on the other inner face is $-q_a$. In the idealized case of a capacitor having infinite lateral dimensions, Equation (20) covers all space, except for points lying *on* an inner face of a metal plate. The field E at these exceptional points, and hence also the forces acting on the plates, can be found as in Exercise 5.

We remarked above that, when dealing with the interior of a metal in the case of *static* fields, we need not replace ϵ_0 by another constant. This is no longer true if the fields are varying rapidly with time, especially at optical frequencies. For another remark about ϵ_0 and metals, see the end of § 73.

Charges on a single metal plate. When charges of the same sign are placed on a metal plate, their mutual repulsion causes them to recede in equal amounts to the two faces, as in Fig. 157(a), and to form two equally charged plane sheets of charge (we ignore edge effects). The surface charges are shown in the figure by the black dots, with two lines of E beginning on each dot. The lines of E show that the fields of the charged faces cancel inside the plate and that hence $\mathbf{E} = 0$ there; by contrast, outside the plate the two fields reinforce rather than cancel each other.

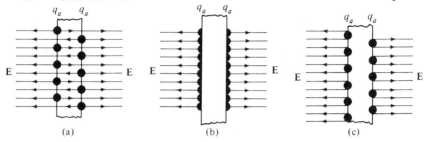

Fig. 157. Three equivalent ways of picturing the electric field of a positively charged metal plate.

Another rough map of **E** is shown in Fig. 157(b), where the dots representing the charges are split into halves and one field line begins on each half-dot. Still another equivalent map appears in Fig. 157(c), where two lines begin on each full dot and go outward from the surface.

Let a point P lie in Fig. 157(a) to the right of the right-hand face of the plate; write q_a for the surface charge density on each of the two faces, and, to stress the fact that **E** is normal to the plate, write E_n for the magnitude of **E** at P. According to (16), the charge on the left-hand face contributes the amount $q_a/2\epsilon_0$ to E_n at P, and so does the charge on the right-hand face. Therefore,

$$E_n = \frac{q_a}{\epsilon_0}. \tag{21}$$

We will see in §72 that (21) holds also in a vacuum just outside a charged, smooth *curved* conductor.

A charged spherical balloon. A spherical surface, say the surface (S) of the balloon shown in Fig. 158(a), carries a charge of uniform density q_a. We will be interested in the magnitude E of the field **E** produced by this charge distribution at points inside (S), at points outside (S), and at points on (S), say the point P in the figure. The simplest way to compute the fields inside and outside (S) is described in §63, where we will find that

$$E_{\text{inside}} = 0, \qquad E_{\text{outside}} = \frac{q}{4\pi\epsilon_0}\frac{1}{r^2} = \frac{q_a}{\epsilon_0}\left(\frac{r_0}{r}\right)^2; \tag{22}$$

here q is the total charge on the sphere, equal to $4\pi r_0^2 q_a$. Note that the field outside (S) is the same as the field that would be produced by a point charge q located at the origin O.

The value of E at P can be found by integration. The area of the shaded ring in the figure is $(2\pi r_0 \sin\theta)(r_0\, d\theta)$, and hence the charge on the ring is

$$dq = 2\pi r_0^2 q_a \sin\theta\, d\theta. \tag{23}$$

Also, the distance s and the angle α in the figure satisfy the equations

$$s = 2r_0 \sin \tfrac{1}{2}\theta, \qquad \cos \alpha = \sin \tfrac{1}{2}\theta. \tag{24}$$

A *point* charge dq would produce at the distance s a field of magnitude dE, equal to $dq/(4\pi\epsilon_0 s^2)$. From symmetry, the only nonvanishing component of **E** at P will be E_z; therefore, to get the formula for dE in the case of the *ring* charge, we need only to replace dq by $dq \cos \alpha$. Accordingly,

$$dE = \frac{dq}{4\pi\epsilon_0} \frac{\cos \alpha}{s^2} = \frac{q_a}{8\epsilon_0} \frac{\sin \theta}{\sin \tfrac{1}{2}\theta} d\theta = \frac{\sqrt{2}\, q_a}{8\epsilon_0} \frac{\sin \theta}{\sqrt{1-\cos\theta}} d\theta. \tag{25}$$

To evaluate the effect at P of the entire charge located on (S), we integrate (25) with respect to θ from 0 to π, and find that $E = q_a/2\epsilon_0$, as in the middle line of (26).

To display the discontinuities in E as we cross the charged surface, we consider the field E_{outside} in (22) at a point lying "just outside" (S), which we locate by letting r approach r_0. Then, in summary,

$$E = \begin{cases} 0 & \text{inside } (S) \\ \dfrac{q_a}{2\epsilon_0} & \text{on } (S) \\ \dfrac{q_a}{\epsilon_0} & \text{just outside } (S). \end{cases} \tag{26}$$

The reasons for the discontinuities in (26) can be pictured as follows. Let us make a hole in (S) by removing a small circular portion of it, centered on the z axis; the two separate portions of (S), say S_1 and S_2, are shown in Fig. 158. Let δ denote the angular aperture of the hole, as shown in the figure. The contribution to E at P made by the charge located on S_2 is then approximately $(q_a/4\epsilon_0)\delta$; we assume that the hole is small enough to make this contribution negligible. If we should now move in and out of the hole along the z axis, we

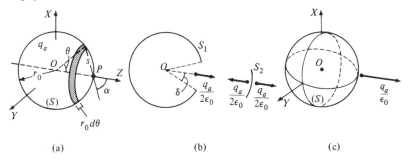

Fig. 158. Computation of **E** for a uniformly charged balloon.

would not cross any charged surfaces; therefore, we would expect E to vary continuously and to have essentially the same value, namely $q_a/2\epsilon_0$, in the neighborhood of P. The field **E** in this neighborhood is pictured by the first pointer in Fig. 158(b).

The rest of Fig. 158(b) deals with the electric field at points on the z axis near the portion S_2 of (S). If we come close enough to S_2, we may regard it as an infinite and practically flat nonconducting charged surface of zero thickness, and may adopt the field pictured in Fig. 154(b)—as we have done in Fig. 158(b). Superposition of the fields shown by the pointers in Fig. 158 then gives $E = 0$ just inside (S), and $E = q_a/\epsilon_0$ just outside, as required by (26).

To summarize: The field **E** at a point P' just inside (S) is a superposition of two fields—the inward field of magnitude $q_a/2\epsilon_0$, produced by a small patch of charge adjacent to P', and the outward field of the same magnitude, produced by the charge on the rest of (S); these fields cancel. At a point P'' just outside (S), the magnitudes of the two fields are the same as at P', but both are directed outward, and hence $E = q_a/\epsilon_0$. This argument implies that, if we move close enough to S_2, we may regard it as practically flat—which may seem wrong, because the radius of curvature of even a small portion of a sphere is the same as that of the whole sphere. However, *Duhamel's theorem*,[2] discussed in books on advanced calculus, justifies the argument we used above.

A coin between charged plates. Suppose that a small, thin, smooth uncharged metal disk—say a flat coin—is placed inside an evacuated charged capacitor, its faces parallel to the plates. The field **E** then has the form indicated in Fig. 159, in which the effect of the coin on the charge distribution on the

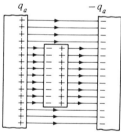

Fig. 159. A coin placed between charged plates.

plates and the fringing of the lines of **E** near the edges of the coin are ignored. When the coin is first put into the capacitor, it is momentarily pervaded by the electric field of the charged plates. This field drives conduction electrons toward its left-hand face and uncovers positive charges on the right-hand face, until the field inside the coin is reduced to zero. When the transient currents have subsided, the interior of the coin will be uncharged, but its faces will carry certain charges, say Q_d and $-Q_d$; here the subscript d suggests that the faces

---

[2] Jean Marie Constant Duhamel (1797–1872).

become charged because of a *displacement* of the charges in the coin. Furthermore, as one can infer from the lines of E in Fig. 159, the charge density on the positive face of the coin is equal to the charge density q_a on the positive plate, while that on the negative face is $-q_a$. Therefore, if the face area of the coin is A, we have

$$\frac{Q_d}{A} = q_a, \qquad (27)$$

at least in the limit of a very small and thin coin.

Next suppose that, as in Fig. 160, the coin is replaced by what we will call a "two-coin probe," which consists of two coins in contact, each with an insulating handle. Using the handles, we can separate the coins while they are inside the capacitor and then take them out. One of them will then carry the charge $-Q_d$ and the other the charge Q_d. If we measure Q_d, say by a ballistic galvanometer, and measure the area A, we can compute q_a from (27). But we can go even further and, using the first line of (20), can write (27) as

$$\frac{Q_d}{A} = \epsilon_0 E. \qquad (28)$$

Fig. 160. A two-coin probe placed between charged plates.

Therefore, once the value of ϵ_0 is known, the electric intensity inside an evacuated capacitor can be measured in principle by a two-coin probe.

EXERCISES

1. Because of a momentary perturbation, the respective charge densities on the two faces of the metal plate in Fig. 157 become $q_a(1 + \delta)$ and $q_a(1 - \delta)$. Show that, as the perturbation subsides, currents will flow in the plate in such directions as to equalize the charge densities.

2. The faces of two parallel metal plates, located in vacuum, are labeled 1, 2, 3, and 4, as in Fig. 161. The initial charge densities on these faces are $q'_a(1)$, $q'_a(2)$, and so on. Suppose that

$$q'_a(1) = 0, \qquad q'_a(2) = 2q_a, \qquad q'_a(3) = -q_a, \qquad q'_a(4) = 0, \qquad (29)$$

where q_a is specified. The initial electric fields do not vanish in the plates and cause currents. Show, relying on rough diagrams, that the final densities will be

$$q_a(1) = \frac{1}{2} q_a, \qquad q_a(2) = \frac{3}{2} q_a, \qquad q_a(3) = -\frac{3}{2} q_a, \qquad q_a(4) = \frac{1}{2} q_a.$$
(30)

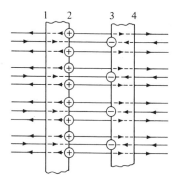

Fig. 161. Diagram for Exercise 2.

3. One face of the copper plate in Fig. 162 lies in the zx plane and carries a surface charge of uniform density q_a. Assume that the dimensions of the plate in the $\pm x$ and $\pm z$ directions are infinite, and find the flaw in the following computation of **E** at the point P in the figure: "Divide the zx plane into strips of width dx, as in Fig. 162. The magnitude of the contribution of a strip to **E** at the point P is then given by (12), and integration gives $E_y = q_a/2\epsilon_0$, as in (16), in contradiction to (21)."

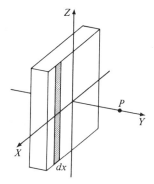

Fig. 162. Diagram for Exercise 3.

4. The face of a charged metal plate carries some 55 "extra" electrons per mm². Show that, just outside the metal, E is about 1 volt/meter.

5. A portion of the right-hand sheet in Fig. 155(c) has the area a. Show that the force of attraction exerted on this portion by the entire left-hand sheet

is $aq_a^2/2\epsilon_0$ newtons. (State clearly the reason for the factor $\frac{1}{2}$ in this formula.) Then consider a parallel-plate vacuum capacitor (area of each plate $= a$) carrying charges $\pm q$ and provided with guard rings, so that fringing can be ignored; show that the right-hand plate is pulled leftward with the force of $q^2/2\epsilon_0 a$ newtons.

6. A uniform line charge of density q_l lies along the z axis of Fig. 153. Show that

$$V(\rho) = -\frac{q_l}{2\pi\epsilon_0} \ln \frac{\rho}{\rho_0}, \qquad (31)$$

where ρ_0 is an arbitrary nonzero length (because of the logarithm, the choice of ρ_0 amounts to a choice of an additive constant in V). Sketch graphs of $V(\rho)$ as a function of ρ for two values of ρ_0, one twice as large as the other.

7. Derive (5) without referring to potentials, as follows: Use (2^{59}) and the necessary coordinate shifts to derive the *exact* formula for the field **E** produced at P by the dipole of Fig. 152. Then let $l^2 \ll r^2$ and recall the pertinent exercises in §7.

8. Show that, if $r \neq 0$, the field (5) satisfies the equation div $\mathbf{E} = 0$, which suggests that the direction lines of **E** can begin and end only at the origin. Reconcile this conclusion with Fig. 61^{17}.

9. Show that the direction lines of the field (5) satisfy the equation $r^{-1} \sin^2 \theta$ = constant.

10. Assume that, for a point charge, E is proportional to r^n and show that a uniformly charged infinite plane produces a uniform electric field if and only if $n = -2$.

61. ELECTRIC DISPLACEMENT DENSITY

Electric fields have many properties, some of special interest in one problem, some in another. Their basic property is what we have called the *force-exerting property*: If a charge is located in an electric field, the field exerts a force on it. This property can be explored, in principle, with the aid of a test charge. It is described quantitatively by the field **E**, the electric intensity.

A related property of electric fields—the *torque-exerting property*—was discussed in a preliminary way in §55; it has to do with torques that electric fields may exert on symmetric charge distributions. This property can be explored, in principle, with the aid of an electric pinwheel. It is described quantitatively by the curl of **E**.

We now come to still another property of electric fields, which becomes important in practice when mediums other than a vacuum are involved. This property, which may be called the *charge-displacing property*, was mentioned in §60, in connection with the "two-coin probe." It is a consequence of the force-exerting property, but it proves to be sufficiently important in its own right to warrant introducing, in addition to E and curl E, still another vector field for describing electric fields quantitatively. This field is denoted by **D**, is called *electric displacement density*, and is measured in coulombs per square meter; another name of **D** is *electric flux density*. Our opening remarks on **D** pertain to fields in a vacuum and are patterned after those of Pohl,[3] of Schelkunoff,[4] and of Jefimenko.[5]

First, a remark on E. By definition, the direction of E at a point P is the direction of the force exerted on a (positive) test charge q^* placed at P, and the magnitude of E at P is the magnitude of this force divided by the magnitude of the test charge [strictly, a limit is involved, as in (5^{42})]. Figure 163(a) shows a test charge placed in an electric field and a few lines of E of this field.

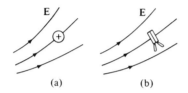

Fig. 163. A test charge for measuring E and a two-coin probe for measuring **D**.

The vector **D** at P can be determined, in principle, with the help of an uncharged two-coin probe. To find the direction of **D**, we place the probe at P and tilt it about P until the magnitudes of the charges displaced to the outer faces of the coins are maximized. The direction of **D** at P is then defined as the direction of the normal to the probe, drawn outward from the coin that has become charged positively. As long as we restrict ourselves to fields in a vacuum, it is perhaps apparent that the properly oriented probe makes right angles with E, as in Fig. 163(b), and that therefore **D** has the same direction as E.

To determine the magnitude D of **D** at P, we orient the probe as described above, separate the coins, and take them out of the electric field. We then measure the charge on the positively charged coin (say Q_d coulombs), divide it by the face area of the coin (say A square meters), and write

$$D = \frac{Q_d}{A} \frac{\text{coulombs}}{\text{meter}^2}. \tag{1}$$

The exact definition of **D** requires that the thickness of the coins and their area

---

[3] R. W. Pohl, *Physical Principles of Electricity and Magnetism* (Blackie and Son, 1930), p. 66.

[4] S. A. Schelkunoff, *Electromagnetic Waves*, (D. Van Nostrand, 1943), p. 62; *Electromagnetic Fields* (Blaisdell, 1963), p. 25.

[5] Oleg D. Jefimenko, *Electricity and Magnetism* (Appleton-Century-Crofts, 1966), pp. 80 and 225.

approach zero, so that $Q_d \to 0$. The methods for measuring **E** and **D** described above are useful for formulating definitions, but in practice the measurements are made with the help of various equations derived from the definitions.

Now, in principle, the electric field at a point P in vacuum (but not necessarily both at P and in the neighborhood of P) can be reproduced at any instant by removing its actual causes and surrounding P by a properly charged and properly oriented parallel-plate vacuum capacitor. If the probe procedure is then used at P, the ratio Q_d/A will be equal to $\epsilon_0 E$, as required by (28^{60}). Therefore, in view of (1), we conclude that $D = \epsilon_0 E$. Since in free space **D** has the same direction as **E**, it follows that, in free space,

$$\mathbf{D} = \epsilon_0 \mathbf{E}. \tag{2}$$

One can, of course, reverse our order of presentation, take (2) as the definition of **D**, and then speak of two-coin probes to illustrate the physical significance of **D**.

We have defined the magnitude and direction of **D**, but have not yet considered the parallelogram rule and hence have not yet proved that **D** is indeed a *vector* field. The proof can be based on (2) and on the fact that we have already shown that **E** is a vector field.

In view of (2) and (2^{59}), the electric displacement density r meters away from a charge of q coulombs located in empty space at the origin of a coordinate frame is

$$\mathbf{D} = \frac{q}{4\pi r^2} \mathbf{1}_r. \tag{3}$$

Similarly, the magnitude of **D** between the plates of a parallel-plate vacuum capacitor is

(20^{60}) $$D = q_a \frac{\text{coulomb}}{\text{meter}^2}, \tag{4}$$

where q_a is the surface charge density on the positive plate.

62. DIELECTRICS

A "dielectric" is a good electrical insulator—solid, liquid, or gas; a vacuum, too, is sometimes called a dielectric. We are concerned primarily with fields in vacuum; therefore, we consider material dielectrics only briefly and assume them to be isotropic.

The molecules of some dielectrics are *nonpolar*: The respective average "electric centers" of their negative and positive charge coincide, unless the molecules are distorted by external fields. The molecules of other dielectrics are *polar*: The two centers do not coincide, and each molecule is, in effect, a small electric dipole even in the absence of external fields. An example of a polar molecule is HCl. Here the electric center of the positive charges lies much nearer to the Cl nucleus than to the H nucleus, because the Cl nucleus has 17 protons and the H nucleus only one. But the average center of the 18 electrons lies even nearer to the Cl nucleus.

When an uncharged slab of dielectric is put between the charged plates of a vacuum capacitor, as in Fig. 164, the molecules become affected by the new electric field in which they find themselves. If they are nonpolar, the average center of the electrons in a molecule shifts relative to the average electric center of the protons contained in the atomic nuclei. If they are polar, they tend, in addition, to align themselves along the electric field. In either case, positive charges, bound to the slab, become uncovered on the face nearest to the negative plate, bound negative charges appear on the other face, and, *if the slab is homogeneous*, its interior remains uncharged. In short, a homogeneous slab becomes "polarized" as suggested in the figure, where q_a and $-q_a$ are the surface densities (coulombs/meter^2) of the charges on the plates, and q'_a and $-q'_a$ are the densities of the "polarization charges" bound to the faces of the slab. Experiment shows that the ratio q'_a/q_a is less than unity and that, for a large class of isotropic homogeneous dielectrics and for moderate electric fields, this ratio is independent of the electric intensity in the capacitor and depends only on the chemical composition and temperature of the dielectric. Dielectrics of this kind are said to be "linear."

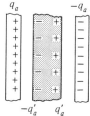

Fig. 164. An uncharged dielectric slab becomes polarized when placed in an electric field.

Let $\mathbf{E}_{\text{dielectric}}$ denote the electric intensity in the dielectric slab in Fig. 164. Let $\mathbf{E}_{\text{vacuum}}$ denote the electric intensity in the space between the plates *before* the slab is inserted. The foregoing remarks imply that $\mathbf{E}_{\text{dielectric}}$ is proportional to $\mathbf{E}_{\text{vacuum}}$. We write the constant of proportionality as ϵ_0/ϵ, and then rewrite the equation $\mathbf{E}_{\text{dielectric}} = (\epsilon_0/\epsilon)\mathbf{E}_{\text{vacuum}}$ as

$$\epsilon \mathbf{E}_{\text{dielectric}} = \epsilon_0 \mathbf{E}_{\text{vacuum}}. \tag{1}$$

Here

$$\epsilon > \epsilon_0 \tag{2}$$

and $\epsilon_0 = 8.85 \times 10^{-12}$ farad/meter, as in (4^{59}). The factor ϵ is the (rationalized) *capacitivity* or *permittivity* of the dielectric. The ratio ϵ/ϵ_0 is the "relative capacitivity" or "relative permittivity," or simply the *dielectric constant* of the material of which the slab is made. Typical values of ϵ/ϵ_0 are given in Table 4. Since the value for air is nearly unity, we will treat it as unity. Dielectric "constants" are actually not constant; they depend on the temperature and, in the case of alternating fields, on the frequency.

TABLE 4. Dielectric Constants

Substance	$\dfrac{\epsilon}{\epsilon_0}$
Air	1.0006
Petroleum oil	2.2
Fused quartz	4.
Pyrex glass	4.8
Mica	5.
Methyl alcohol	31.2
Distilled water	81.1

As shown in Fig. 165, the field **E** in the space occupied by the slab is weaker than it was before the slab was put into this space. Indeed, in the absence of the slab, the lines of **E** begin on the left-hand plate and end on the right-hand plate. After the slab is put in, however, some of these lines do not continue into the slab but end on the negative polarization charges. In the right-hand gap in Fig. 165(b), the density of the lines is built up to its original value by the lines of **E** that begin on the positive polarization charges. As one can tell by counting lines, we have $\epsilon/\epsilon_0 = 2$ in Fig. 165.

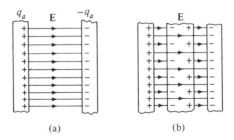

Fig. 165. The effect on a uniform field **E** caused by placing in this field an uncharged dielectric slab.

We have been discussing only *uncharged* dielectrics. Note, for example, that if the dielectric is a slab of glass, one face of which was rubbed with silk before the slab was inserted into the capacitor, then such figures as 164 and 165(b) are no longer applicable.

Coulomb's law. Next suppose that a positive point charge q is placed inside a homogeneous and previously uncharged dielectric, pervading all space. The radial electric field of this charge will affect the molecules of the dielectric. For example, if q is placed in hydrochloric acid, the HCl molecules will tend to turn their more negative parts toward q. As a result, the net positive charge contained in any imagined sphere centered on q will be less than q, and the magnitude of **E** at the distance r from q will be less than $q/4\pi\epsilon_0 r^2$. A detailed argument, which we omit, shows that the reduction factor is ϵ_0/ϵ; that is, instead of the equation $E = q/4\pi\epsilon_0 r^2$ we have $E = q/4\pi\epsilon r^2$. Accordingly, if q lies at the origin of a coordinate frame, the expression for **E** takes the form

$$\mathbf{E} = \frac{q}{4\pi\epsilon r^2} \mathbf{1}_r. \tag{3}$$

Equation (3) is Coulomb's law for a homogeneous dielectric whose capacitivity is ϵ and whose extent is *infinite*. This equation holds also for a *homogeneous dielectric sphere* centered on the origin, when polarization effects are spherically symmetric.

Displacement density. If the dielectric is a gas, the displacement density **D** can be defined with the aid of two-coin probes we used in a vacuum. The relation between **D** and **E** then proves to be

$$\mathbf{D} = \epsilon \mathbf{E}, \tag{4}$$

as one might expect from (2^{61}). Combining (4) and (3), we get

$$\mathbf{D} = \frac{q}{4\pi r^2} \mathbf{1}_r. \tag{5}$$

This equation does not involve ϵ and is identical with (3^{61}), which holds for a vacuum.

If the dielectric is a solid, the vector **D** at a point P inside it can still be measured, in principle, by a two-coin probe inserted into a *stubby* cylindrical cavity surrounding P. The vector **E** at P can be measured, in principle, with the help of a test charge placed in an *elongated* cylindrical cavity surrounding P—see the cavities in Fig. 166, which shows in perspective a portion of the dielectric slab drawn edge-on in Fig. 167. In this case, the conceptual measurements lead again to the equation $\mathbf{D} = \epsilon \mathbf{E}$. In other cases, **D** and **E** may have different directions, as in Fig. 168. In books that treat dielectric materials in greater detail from the microscopic viewpoint, the field **D** at any point inside a dielectric is defined by Equation (7), namely, $\mathbf{D} = \epsilon_0 \mathbf{E} + \mathbf{P}$, where the field $\mathbf{P}$, called "intensity of polarization," is the electric moment (of the polarized dielectric) per unit volume at the point.

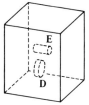

Fig. 166. Cavities for conceptual measurements of **E** and **D** in a solid dielectric.

It is instructive to compare the lines of **E** with the lines of **D**, as in Fig. 167, where the capacitivity of the dielectric is taken to be $2\epsilon_0$, and the densities of the lines of **D** and **E** are scaled in such a way that the factor 8.85×10^{-12} in the equations $\mathbf{D} = \epsilon_0 \mathbf{E}$ and $\mathbf{D} = 2\epsilon_0 \mathbf{E}$ is replaced by unity. Figure 167(a), a replica of Fig. 165(b), describes the field **E**. To get the corresponding field **D** in the gaps between the slab and the plates, we must multiply the **E** in the gaps by ϵ_0. To get the **D** in the slab, we must multiply the **E** in the slab by ϵ, which in this example is $2\epsilon_0$. But the **E** in the slab is half as strong as in the gaps, and consequently **D** is as strong in the slab as in the gaps. Accordingly, all lines of **D** run from the left-hand plate, through the slab, to the right-hand plate, as in Fig. 167(b). We should stress the fact that no lines of **D** begin or end on polarization charges.

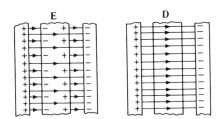

Fig. 167. The fields **E** and **D** between uniformly charged plates when the capacitivity of the dielectric slab is $2\epsilon_0$.

Electrets. If certain melted waxy substances are placed in strong electric fields, are allowed to solidify, and are then taken out of the field, they may stay polarized for weeks. Discs or cylinders of such permanently polarized materials are called *electrets*.[6] Lines of **E** and **D** for a stubby cylindrical electret with flat ends are shown in Fig. 168. Since only polarization charges are involved, the lines of **D** have no beginnings or ends. By contrast, the lines of **E** begin on

---

[6] F. Gutman, "The Electret," *Reviews of Modern Physics*, Vol. 20 (1948), p. 457.

the positive and end on the negative polarization charges, which are confined in the figure to the flat ends of the electret; the shapes of these lines are discussed in Exercise 3. The value of ϵ_0 in the equation $\mathbf{D} = \epsilon_0 \mathbf{E}$ is replaced in the figure by unity, so that *outside* the electret the line map of $\mathbf{D}$ is the same as that of $\mathbf{E}$. Incidentally, if an electret is cut into two parts, each part retains the properties of an electret, but has a smaller total electric moment.

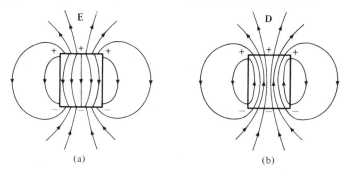

Fig. 168. The fields $\mathbf{E}$ and $\mathbf{D}$ for an electret.

A charge near a dielectric slab. The positive point charge q in Fig. 169 is held in a vacuum near a homogeneous dielectric filling the half-space to the right of the plane interface, which is perpendicular to the page. The capacitivity of the dielectric is $2\epsilon_0$. Figure 169(a) gives an idea of the lines of $\mathbf{E}$.

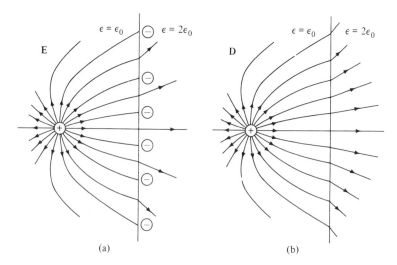

Fig. 169. A positive point charge held in a vacuum near the flat face of an uncharged dielectric slab.

All lines begin on the charge q; some of them end on the negative polarization charges on the surface of the dielectric, and some go out to infinity. Figure 169(b) is a similar picture of the lines of **D**.

Were it not for the dielectric, the lines of **E** and **D** in Fig. 169 would be straight and go out radially from the charge q. (The lines prove to be straight only in the dielectric.) The figure also shows that, in contrast to the lines of **E**, the lines of **D** do not begin or end on polarization charges. The boundary conditions responsible for the bending or "refraction" of the field lines at the surface of the dielectric are given in §70.

EXERCISES

1. A dielectric slab (capacitivity ϵ) is inserted into a vacuum capacitor, as in Fig. 165. Use (1) to show that the surface density of the polarization charges satisfies the equation

$$q'_a = \left(\frac{\epsilon - \epsilon_0}{\epsilon}\right) q_a, \tag{6}$$

where q_a is the charge density on the plates.

2. Redraw Fig. 167 for the case when $\epsilon = 1.5\epsilon_0$; mark charges by pluses and minuses. Note that if we keep the rule that lines of **E** begin on positive and end on negative charges, then a uniform field **E** cannot always be pictured by equally spaced lines. Use (6) to check the density of polarization charges.

3. In the electret of Fig. 168, the lines of **E** and **D** have opposite curvatures. One can show the reason for this graphically, using an idealized electret and the equation

$$\mathbf{D} = \epsilon_0 \mathbf{E} + \mathbf{P}. \tag{7}$$

(Since **E** and **D** are not parallel in the electret, the formula $\mathbf{D} = \epsilon \mathbf{E}$ does not hold there, unless the scalar ϵ is replaced by a "tensor.")

Assume that the field **P** (electric moment per unit volume of the electret) points up the page and is uniform throughout the electret. Guided by Fig. 168(a), draw vectors describing **E** at several points in the electret, and add to each a sufficiently long vector pointing up the page. Note the slopes of the vector sums, and sketch a few lines of the field described by these sums.

CHAPTER 15

Integral and Differential Forms of Coulomb's Law

It is convenient to *introduce* Coulomb's law by stating it in terms of the interaction between two point charges, as we did in Chapter 14. In most computations, however, other ways of stating the same law are more useful. Several of the alternatives are discussed in this chapter. One is an integral formula involving the electric displacement **D** and named after Gauss. Next come Poisson's and Laplace's formulas, which involve the electrostatic potential V and are thus restricted to conservative fields. The chapter ends with the form of Coulomb's law used in a different notation by Maxwell, and listed in the chart in the preface.

63. GAUSS'S FORM OF COULOMB'S LAW

The flux of **D** across a surface S is called the *electric flux* across S. We will now evaluate this flux across a *closed* surface for the special case of point charges located in free space, but our main result—Equations (4) and (5)—is very general.

By definition of the term "flux,"

$$(7^{22}) \qquad \text{flux of } \mathbf{D} \text{ across a closed surface } (S) = \iint_{(S)} D_n \, da. \qquad (1)$$

We begin with a single point charge q_1 and an irregular but smooth surface (S), say the surface shown in Fig. 170(a). A small and sufficiently flat element of (S), having the area da and lying at the distance r from q_1, is pictured in Fig. 170(b). Since $D = q_1/4\pi r^2$, we have

$$D_n \, da = \frac{q_1}{4\pi} \frac{\cos \gamma_n}{r^2} \, da. \tag{2}$$

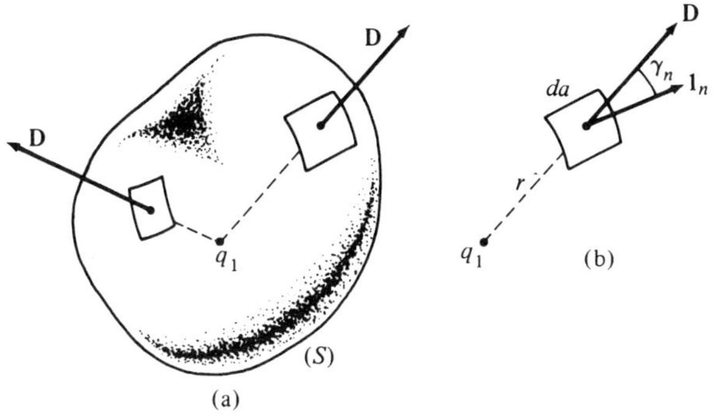

Fig. 170. Derivation of Gauss's formula.

Apart from the factor $q_1/4\pi$, the right-hand side of (2) is the solid angle subtended at the charge q_1 by the element of (S). Consequently, according to (13^9),

$$\iint_{(S)} D_n \, da = \begin{cases} q_1 & \text{if } q_1 \text{ lies inside } (S) \\ 0 & \text{if } q_1 \text{ lies outside } (S). \end{cases} \tag{3}$$

A similar argument shows that (3) holds also if q_1 is negative.

Next suppose that several point charges, say $q_1, q_2, \ldots,$ and q_n coulombs, are located in empty space. Equation (3) holds for each of them, and hence

$$\iint_{(S)} D_n \, da = q, \tag{4}$$

where

$$q = \begin{cases} \text{algebraic sum of the electric} \\ \text{charges, other than polarization} \\ \text{charges, located inside } (S). \end{cases} \tag{5}$$

Equation (4) is *Gauss's form of Coulomb's law* or the *integral form of Coulomb's law*; it is also called "Gauss's formula" and "Gauss's law." The reason for excluding polarization charges in (5) is that, as we will now indicate by pictorial arguments, Equation (4) holds even in the presence of conductors and dielectrics, provided the polarization charges are ignored in evaluating q. The imagined surface (S) on the left-hand side of (4) is sometimes called a "Gaussian surface."

First, consider charges located in vacuum. Equations (4) and (5) then mean pictorially that the net number of lines of **D** emerging from a region R is equal to the net charge (coulombs) contained in R; in other words, one line of **D** originates on each coulomb of positive charge and one line of **D** ends on each coulomb of negative charge. (A coulomb is a very large charge by laboratory standards; therefore, in line maps of **D**—such as Fig. 167[62]—the lines of **D** are understood to begin and end on minute fractions of a coulomb.)

We have shown in Chapter 14, at least pictorially, that lines of **D** *do not* begin or end on polarization charges, but *do* begin and end on charges other than polarization charges. Consequently, (4) will hold even if the region bounded by (S) contains dielectrics and conductors, provided that in evaluating q we ignore any polarization charges that may be present. We will be content with this pictorial justification of (4) and (5) for the general case. Rigorous proofs of (4) and (5) involve a vector field that describes the polarization of the dielectric, a field that we will not need in this book. An analytic rather than pictorial proof of (4) and (5) for a simple special case is outlined in Exercises 10 and 11.

"Charges" and "polarization charges." From now on we will usually say, for short, "charges" instead of "charges other than polarization charges." If polarization charges are to be included, we will state this explicitly. In these terms, the sources and sinks of **D** are charges (such as charges placed on or in various bodies by conduction currents, convection currents, or friction, and charges rearranged on conductors by conduction currents). By contrast, the sources and sinks of **E** are these charges and also any polarization charges. We say that a dielectric is "uncharged" or "neutral" if it carries no charges other than polarization charges; we write q, q_a, and so on for charges, and q', q'_a, and so on for polarization charges.

Continuous charge distributions. Many problems involve continuous charge distributions rather than the isolated point charges that we used in deriving (4). To adapt (4) to this case, we consider the region R bounded by (S), write ρ or q_v for the volume charge density in R, subdivide R into differential elements, treat the charge spread over each element as a point charge, and write the right-hand side of (4) as

(4³)
$$q = \iiint_R q_v \, dv. \tag{6}$$

Gauss's formula then reads

$$\iint_{(S)} D_n \, da = \iiint_R q_v \, dv. \tag{7}$$

If only *surface* charges are involved, the right-hand side of (7) should be replaced by the appropriate surface integrals that yield the total charge located in R.

Uniformly charged nonconducting rod. To illustrate the use of Gauss's formula for computing **D**, we consider an infinitely long nonconducting rod, coaxial with the z axis in Fig. 171; the radius of its circular cross section

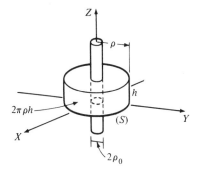

Fig. 171. The field (11^{60}) of a uniformly charged long rod derived from Gauss's form of Coulomb's law.

is ρ_0. A positive charge (not a polarization charge) is spread throughout the interior of the rod with a constant density q_v. We write q_l for the amount of this charge per meter of length of the rod. The imagined surface (S) in the figure has the radius ρ and the height h. By symmetry, we have $\mathbf{D} = f(\rho)\mathbf{1}_\rho$, and hence $D_n = 0$ on the top and bottom of (S), while on the curved part D_n is as yet an undetermined constant, say D. The total charge inside (S) can be evaluated without explicit integration—it is simply hq_l, and hence (4) becomes

$$D \times \text{(area of curved part of } S) = hq_l. \tag{8}$$

Therefore, $2\pi \rho h D = hq_l$, $D = q_l/2\pi\rho$, and

$$\mathbf{D} = \frac{q_l}{2\pi\rho}\mathbf{1}_\rho \quad \text{if} \quad \rho \geq \rho_0. \tag{9}$$

If the rod is surrounded by vacuum, we can find **E** outside the rod by dividing (9) by ϵ_0. The result is (11^{60})—as though the charge distributed in the rod were concentrated on the z axis.

Comparison with our proof of (11^{60}) in §60 illustrates the fact that, if the charge distribution is simple and highly symmetric, Gauss's formula may save quite a few mathematical steps. The problem is to find a surface (S) for which the surface integral of D_n can be easily expressed in terms of an undetermined constant; in the example above, this integral proved to be $2\pi\rho h D$, where D was eventually found from the equation $2\pi\rho h D = hq_l$.

Note that if the surface (S) in Fig. 171 were not coaxial with the charged rod, or if the rod had a noncircular cross section, then Gauss's formula, while still true, would not be useful for computing **D**. Note also that the problems where Gauss's formula can be used for computing **D** are precisely the problems where the field-line picture can be used to get exact results.

Finally we let ρ in Fig. 171 be smaller than the radius ρ_0 of the rod. The charge inside (S) will then be $(\rho^2/\rho_0^2)hq_l$ rather than hq_l; Equation (8) will change to $2\pi\rho hD = (\rho^2/\rho_0^2)hq_l$, and D will become $q_l\rho/2\pi\rho_0^2$. Consequently, (9) can be generalized to read

$$\mathbf{D} = \begin{cases} \dfrac{q_l}{2\pi\rho}\mathbf{1}_\rho & \text{if } \rho \geq \rho_0 \\ \dfrac{q_l}{2\pi\rho_0^2}\rho\mathbf{1}_\rho & \text{if } \rho \leq \rho_0. \end{cases} \tag{10}$$

Note that **D** changes continuously as we cross the surface of the rod.

EXERCISES

1. Use Gauss's formula to check (21[60]) and to analyze the false argument of Exercise 3[60].

2. A nonconducting rod has a circular cross section and is oriented as in Fig. 171. The volume density of charge in it is $q_v = k\rho$, where k is a constant. Show that

$$\mathbf{D} = \begin{cases} \dfrac{1}{3}k\rho_0^3\dfrac{1}{\rho}\mathbf{1}_\rho & \text{if } \rho \geq \rho_0 \\ \dfrac{1}{3}k\rho^2\mathbf{1}_\rho & \text{if } \rho \leq \rho_0. \end{cases} \tag{11}$$

Recall Exercise 5[3], rewrite (11) in terms of q_l, and compare the result with (10).

3. The volume density of charge in the rod of Fig. 171 is $q_v = f(\rho)$. Show that *outside the rod*,

$$\mathbf{D} = \frac{q_l}{2\pi}\frac{1}{\rho}\mathbf{1}_\rho, \tag{12}$$

irrespective of the form of the function $f(\rho)$.

4. An isolated sphere (radius r_0), centered on the origin of a coordinate frame, is charged uniformly throughout its interior; that is, $q_v = a =$ constant inside the sphere. Show that

$$\mathbf{D} = \begin{cases} \dfrac{1}{3}ar_0^3\dfrac{1}{r^2}\mathbf{1}_r & \text{if } r \geq r_0 \\ \dfrac{1}{3}ar\mathbf{1}_r & \text{if } r \leq r_0. \end{cases} \tag{13}$$

5. The volume density of charge inside a sphere centered on the origin is $q_v = br$, where b is a constant. Show that

$$\mathbf{D} = \begin{cases} \dfrac{1}{4} br_0^4 \dfrac{1}{r^2} \mathbf{1}_r, & \text{if } r \geq r_0 \\ \dfrac{1}{4} br^2 \mathbf{1}_r, & \text{if } r \leq r_0. \end{cases} \quad (14)$$

6. Rewrite (13) and (14) in terms of the respective total charges on the spheres, and compare the results.

7. Show that if the magnetic field in a homogeneous conductor carrying a steady current can be ignored, then the interior of the conductor can be regarded as uncharged. Use (6^{44}), (4^{63}), and the result of Exercise 1^{17}, namely

$$\iint_{(S)} J_n \, da = 0. \quad (15)$$

8. Compute **E** inside the charged nonconducting rod whose **D** is given by (10) and show that, apart from an additive constant, the potential inside the rod is

$$V = -\frac{q_l}{4\pi\epsilon \rho_0^2} \rho^2 \quad \text{for} \quad \rho \leq \rho_0; \quad (16)$$

here ϵ is the capacitivity of the rod material. Also, compute **E** and V outside the rod, where the capacitivity ϵ_0 is smaller than ϵ. Note that, unlike **D**, the field **E** changes discontinuously as we cross the surface of the rod. What phenomenon accounts for the discontinuity? Draw rough line maps of **D** and **E**, distributing the charges inside the rod as in Fig. 98^{34} and assuming that $\epsilon = 2\epsilon_0$.

9. The nonconducting plate of Fig. 72^{19} carries a charge (other than polarization charge) distributed throughout the interior of the plate with the constant density q_v. Use Gauss's formula to show that

$$\mathbf{D} = q_v x \mathbf{1}_x \quad \text{for} \quad -\tfrac{1}{2}b \leq x \leq \tfrac{1}{2}b, \quad (17)$$

where b is the thickness of the plate. We anticipated the form of this result in (2^{19}).

10. From the microscopic viewpoint, every electron and every atomic nucleus is located in vacuum. Show on this basis that the D_n in (4) can be replaced by E_n, provided that the q on the right-hand side of (4) is replaced by $(q + q')/\epsilon_0$, where q' is the total polarization charge contained in (S).

11. Let (S) be an imagined drum-shaped surface located in Fig. 164 as follows: One of its flat faces (area A) lies in, and is parallel to, the left-hand plate of the capacitor; the opposite face lies in the dielectric slab, whose capacitivity is ϵ. Write **E** for the electric intensity in the gaps, recall the experimental relation (1^{62}), evaluate the integral of E_n over (S), and, according to the preceding problem, set it equal to $(q + q')/\epsilon_0$. Then recall (6^{62}) and verify that this procedure leads to (4) without pictorial arguments.

64. DELTA FUNCTIONS

In §63 we derived Gauss's formula (4^{63}), which may be written as

$$\iint_{(S)} D_n \, da = q_1 + q_2 + \ldots + q_n, \tag{1}$$

where the q's are point charges enclosed by (S). We then turned to continuous charge distributions and outlined the steps leading from (1) to the equation

(7^{63})
$$\iint_{(S)} D_n \, da = \iiint_R q_v \, dv. \tag{2}$$

Our next task is to develop a formula for q_v that will apply to point charges, so that the single equation (2) could be used not only for continuous but also for discrete charge distributions.

Let a single point charge q be located at the origin O of a coordinate frame, so that

$$q_v = 0 \quad \text{for } r \neq 0. \tag{3}$$

Let (S) be a spherical surface of radius r_0, centered on O, as in Fig. 172. According to (1), the *left-hand side* of (2) should be equal to q, as a trivial com-

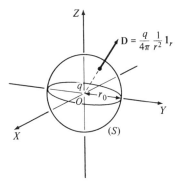

Figure 172

putation will verify. Indeed, the normal component of **D** at any point on (S) is $q/(4\pi r_0^2)$, the area of (S) is $4\pi r_0^2$, and hence

$$\iint_{(S)} D_n \, da = \frac{q}{4\pi r_0^2} \iint_{(S)} da = q. \tag{4}$$

On the *right-hand side* of (2), however, we meet a complication: The triple integral in (2) should be equal to q, but in view of (3) the integrand is zero everywhere except at O, and the standard theory of integrals tells us that, if the integrand is different from zero at only one point, then the integral is zero

—even if the value of the integrand at the exceptional point is greater than any preassigned number.

To avoid the difficulty, one can simply say that we *know* that (1) gives correct results for point charges, and that hence there is no need to force Equation (2) to include the case of point charges. There are, however, at least two ways of extending (2) to point charges. One is to begin with a function q_v that describes a total charge q smeared out smoothly in the neighborhood of O, to make the required computation using this q_v, and then to take the limit approached by the result when the smeared-out charge is assumed to contract upon O. Another way, which we will follow, is to make use of the δ function—an "improper" function brought into prominence by Dirac.[1]

Definitions and comments. The function of x denoted by the symbol $\delta(x - a)$ is defined by the two equations

$$\int_{-\infty}^{\infty} \delta(x - a)\, dx = 1 \tag{5}$$

and

$$\delta(x - a) = 0 \quad \text{for } x \neq a. \tag{6}$$

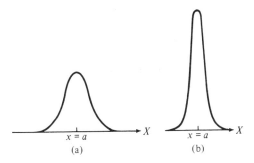

Fig. 173. Bell-shaped curves whose limit is a delta function.

To get a picture of $\delta(x - a)$, consider Fig. 173. The bell-shaped curve in Fig. 173(a) has a maximum at $x = a$; the area under it is presumably unity. If its "width" is cut in half—without changing the area under it, or the position of the maximum, or the general shape—we get Fig. 173(b). The function $\delta(x - a)$ can be visualized as the limit of the functions pictured in Fig. 173 when the width of the curves shrinks to zero. According to the rules of standard calculus however, the limit of the integral of a function is not necessarily equal to the integral of the limit of the function; the limit of the functions in Fig. 173 is so degenerate that equations (5) and (6) are actually not consistent with each

---

[1] P. A. M. Dirac, *The Principles of Quantum Mechanics*, 4th ed., p. 58. Oxford (1958).

other. Therefore, the δ functions—also called "peak functions"—should be used with caution.

Let x be expressed in meters; according to (5), the unit to be used for $\delta(x - a)$ is then the reciprocal meter—the "per meter." The δ function whose peak lies at $x = 0$ is written as $\delta(x)$, rather than $\delta(x - 0)$. Because of (6), Equation (5) can be written as

$$\int_{a-b_1}^{a+b_2} \delta(x - a)\, dx = 1, \tag{7}$$

where b_1 and b_2 are any nonvanishing positive numbers.

The most important single property of $\delta(x - a)$ is that, for any $f(x)$,

$$\int_{-\infty}^{\infty} f(x)\, \delta(x - a)\, dx = f(a). \tag{8}$$

Indeed, suppose that, at every stage of the process of shrinking in width, the curve in Fig. 173(a) is multiplied by the same function $f(x)$. When the curve becomes sufficiently narrow, the only values of $f(x)$ that affect the value of the integral in (8) are values lying near $x = a$. In the limit of zero width, the integrand in (8) becomes $f(a)\, \delta(x - a)$, and the factor $f(a)$ can be taken outside the integral sign. Equation (8) then follows from (5). The symbols on the left-hand side of (8) thus signal this message: Write a for x in $f(x)$.

The δ functions are quite "improper," but *integrals* containing them can nevertheless be handled by standard methods. To illustrate, we will now show that the integral

$$\int_{-\infty}^{\infty} f(x) \left[\frac{d}{dx} \delta(x - a)\right] dx \tag{9}$$

can be evaluated "by parts" without leading to inconsistencies. Applying to (9) the standard formula, namely,

$$\int_b^c u \frac{dv}{dx}\, dx = u(c)v(c) - u(b)v(b) - \int_b^c \frac{du}{dx} v\, dx, \tag{10}$$

we get

$$\int_{-\infty}^{\infty} f(x) \left[\frac{d}{dx} \delta(x - a)\right] dx = f(\infty)\delta(\infty) - f(-\infty)\delta(-\infty)$$
$$- \int_{-\infty}^{\infty} \frac{df(x)}{dx} \delta(x - a)\, dx. \tag{11}$$

In view of (6) and (8), this reduces to

$$\int_{-\infty}^{\infty} f(x) \left[\frac{d}{dx} \delta(x - a)\right] dx = -\frac{df(a)}{da}. \tag{12}$$

To check that (12) is consistent with the result of another standard way of handling the integral (9), we differentiate both sides of (8) with respect to a, getting

$$\int_{-\infty}^{\infty} f(x) \left[\frac{d}{da} \delta(x - a)\right] dx = \frac{df(a)}{da}. \tag{13}$$

Now, if a function $g(x - a)$ is differentiable, we have $d[g(x - a)]/da = -d[g(x - a)]/dx$. When we apply this formula in (13)—even though $\delta(x - a)$ is actually not differentiable—we get (12) once again.

An equation involving δ functions need not include any integral signs. But such an equation as, for example,

$$\delta(kx) = \frac{1}{k} \delta(x), \qquad k > 0, \tag{14}$$

implies, nevertheless, that its two sides are equivalent *when used as factors in integrands*.

The respective definitions of the functions $\delta(y - b)$ and $\delta(z - c)$ of y and z are similar to the definition of $\delta(x - a)$. An example of a three-dimensional δ function is $\delta(x) \delta(y) \delta(z)$, which vanishes everywhere except at the origin O. Let us integrate it over a cubical region R whose center lies at O and whose edges (length $2l$) are parallel to the coordinate axes. The result is

$$\iiint_R \delta(x) \delta(y) \delta(z) \, dx \, dy \, dz = \int_{-l}^{l} \delta(x) \, dx \int_{-l}^{l} \delta(y) \, dy \int_{-l}^{l} \delta(z) \, dz$$
$$= 1 \cdot 1 \cdot 1 = 1. \tag{15}$$

A more general formula is

$$\iiint_R \delta(x) \delta(y) \delta(z) \, dx \, dy \, dz = \begin{cases} 1 & \text{if the origin lies inside } R \\ 0 & \text{if the origin lies outside } R; \end{cases} \tag{16}$$

here R is a region of *any* shape (Exercise 1).

Point charge at the origin. Our task in this section has been this: to find a formula for q_v that would be consistent with (3) and yet would make the triple integral in (2) equal to q rather than zero when R is the spherical region of Fig. 172. The answer is perhaps apparent from (16): In the case of a single point charge q at the origin, we may describe the distribution of charge throughout all space by writing

$$q_v = q\,\delta(x)\,\delta(y)\,\delta(z) \ \frac{\text{coulombs}}{\text{meter}^3}. \tag{17}$$

If we do this, Gauss's formula (2) will hold, even though the charge distribution is actually not continuous.

Divergence of the field $r^{-2}\mathbf{1}_r$. To get a slightly different illustration of the usefulness of δ functions, we return to the equation

$$(6^{37}) \qquad \operatorname{div} \frac{1}{r^2}\mathbf{1}_r = 0 \qquad \text{for } r > 0, \tag{18}$$

in which the origin is excluded. With the aid of δ functions, however, we may replace (18) by the equation

$$\operatorname{div} \frac{1}{r^2}\mathbf{1}_r = 4\pi\,\delta(x)\,\delta(y)\,\delta(z), \tag{19}$$

in which the origin is no longer excluded (Exercise 3).

EXERCISES

1. Verify that (16) holds for a region R of any shape.
2. Check the units in (17).
3. Verify that (19) is consistent with the divergence theorem (2^{31}) and avoids the complications at $r = 0$ that arise in Exercise 2^{37}.
4. Use the divergence theorem to verify that Equation (8^{34}), namely,

$$\operatorname{div} \frac{1}{\rho}\mathbf{1}_\rho = 0 \qquad \text{for } \rho > 0 \tag{20}$$

   can be generalized to read

$$\operatorname{div} \frac{1}{\rho}\mathbf{1}_\rho = 2\pi\,\delta(x)\,\delta(y). \tag{21}$$

5. Sketch the derivatives of the curves in Fig. 173 and account graphically for the minus sign on the right-hand side of (12).
6. Verify (14) in detail (integrate both sides, change variables in the left-hand side, and so on). Also, justify (14) on the basis of Fig. 173.

65. THE DIVERGENCE OF A GRADIENT

One of the important fields of the electromagnetic theory is div **D**, where $\mathbf{D} = \epsilon \mathbf{E}$. An important field that arises in *electrostatics* is the potential V. If we write

$$\mathbf{E} = -\text{grad } V, \tag{1}$$

(16[57])

then div **D** involves the divergence of a gradient. We will now introduce this type of a field for the general case.

The gradient of a scalar field f is the vector field defined in §57 and written as grad f. The divergence of a vector field **F** is the scalar field defined in §29 and written as div **F**. If **F** is in fact the gradient of a field f, we get the scalar field

$$\text{div (grad } f), \tag{2}$$

which can be written without ambiguity as

$$\text{div grad } f. \tag{3}$$

The usual abbreviation for the field div grad f is $\nabla^2 f$, which is read "the Laplacian of f" or "del squared f" and is defined by the equation[2]

$$\nabla^2 f = \text{div grad } f. \tag{4}$$

The field $\nabla^2 f$ has a clear-cut meaning whenever f and its derivatives are sufficiently continuous for the right-hand side of (4) to have a meaning, or if the discontinuities can be handled with the help of δ functions.

In cartesian coordinates,

(7[58])
$$\text{grad } f = \mathbf{1}_x \frac{\partial f}{\partial x} + \mathbf{1}_y \frac{\partial f}{\partial y} + \mathbf{1}_z \frac{\partial f}{\partial z} \tag{5}$$

and

(10[30])
$$\text{div } \mathbf{F} = \frac{\partial F_x}{\partial x} + \frac{\partial F_y}{\partial y} + \frac{\partial F_z}{\partial z}. \tag{6}$$

When we write $\partial f/\partial x$ for F_x in (6), and make similar substitutions for F_y and F_z, we find that

$$\text{div grad } f = \frac{\partial^2 f}{\partial x^2} + \frac{\partial^2 f}{\partial y^2} + \frac{\partial^2 f}{\partial z^2}, \tag{7}$$

---

[2] The Laplacian is named after Marquis Pierre Simon de Laplace (1749–1827).

or, in shorthand,
$$\nabla^2 f = \frac{\partial^2 f}{\partial x^2} + \frac{\partial^2 f}{\partial y^2} + \frac{\partial^2 f}{\partial z^2}. \tag{8}$$

Explicit formulas for $\nabla^2 f$ in terms of cylindrical and spherical coordinates are given in (58^A) and (66^A).

EXERCISES

1. Use (56^A) and (57^A) to check the formula (58^A) for the Laplacian of f in cylindrical coordinates.
2. Verify (66^A) in the manner of Exercise 1.
3. Show that
$$\nabla^2 \frac{1}{r} = -4\pi\, \delta(x)\, \delta(y)\, \delta(z). \tag{9}$$

66. POISSON'S AND LAPLACE'S FORMS OF COULOMB'S LAW

This section is confined to conservative fields, so that **E** can be written as
$$\mathbf{E} = -\operatorname{grad} V. \tag{1}$$

Also, except for Exercise 5, we consider only homogeneous mediums and set
$$\epsilon = \text{constant}. \tag{2}$$

Since $\mathbf{D} = \epsilon \mathbf{E}$, we have $\mathbf{D} = -\epsilon \operatorname{grad} V$; replacing D_n in Gauss's formula by $-\epsilon(\operatorname{grad} V)_n$, we get
$$-\epsilon \iint_{(S)} (\operatorname{grad} V)_n\, da = \iiint_R q_v\, dv. \tag{3}$$

The divergence theorem states that

(2^{31})
$$\iint_{(S)} F_n\, da = \iiint_R \operatorname{div} \mathbf{F}\, dv. \tag{4}$$

The field **F** in (4) corresponds to the field grad V in (3), and consequently the surface integral in (3) can be transformed into a volume integral whose integrand is div grad V, which can be written as $\nabla^2 V$. Equation (3) then becomes

$$-\epsilon \iiint_R \nabla^2 V \, dv = \iiint_R q_v \, dv, \tag{5}$$

and can be written as

$$\iiint_R \left(\nabla^2 V + \frac{1}{\epsilon} q_v \right) dv = 0. \tag{6}$$

Now that the δ function is available for taking care of point charges, we can claim that (6) holds for *every* region R. It then follows from (45[5]) and (46[5]) that the integrand in (6) vanishes, and that

$$\nabla^2 V = -\frac{1}{\epsilon} q_v. \tag{7}$$

Equation (7), generalized in Exercise 5, is *Poisson's equation*, named after Siméon Denis Poisson (1781–1840). In charge-free regions, (7) reduces to

$$\nabla^2 V = 0, \tag{8}$$

which is *Laplace's equation*; we will return to it in §68. Since both (7) and (8) involve the potential V, both are restricted to conservative electric fields.

EXERCISES

1. Verify that Poisson's equation holds inside and outside the charged rod of Exercise 8[63].

2. Verify after the pattern of Exercise 1 that Poisson's equation holds inside and outside (a) the plate of Exercise 9[63], (b) the rod of Exercise 2[63], and (c) the sphere of Exercise 4[63].

3. Verify Poisson's equation for the case of a point charge located in free space at the point (x_0, y_0, z_0).

4. The field V outside the cylindrical hole in our current-carrying conductor is given by (27[56]). Take ϵ to be constant and use Poisson's equation to show that the conducting material is uncharged.

5. Show that, if ϵ is not a constant, then (7) must be replaced by

$$\epsilon \nabla^2 V + (\text{grad } \epsilon) \cdot (\text{grad } V) = -q_v. \tag{9}$$

6. If the c's are constants, the function $c_1 f_1 + c_2 f_2 + \ldots + c_n f_n$ is called a *superposition* or a *linear combination* of the functions $f_1, f_2, \ldots, f_n$. Show that if each of these functions satisfies Laplace's equation, then every linear combination of them also satisfies it.

67. MAXWELL'S FORM OF COULOMB'S LAW

We now come to the field equation written in the chart in the preface as

$$\text{div } \mathbf{D} = \rho. \tag{1}$$

This equation is *Maxwell's form of Coulomb's law*; it is one of the four equations underlying Maxwell's theory of electromagnetic fields. In terms of the alternative symbol q_v for volume charge density, we have

$$\text{div } \mathbf{D} = q_v. \tag{2}$$

To deduce (2), we apply the divergence theorem to the left-hand side of Gauss's formula (7^{63}), get the equation

$$\iiint_R (\text{div } \mathbf{D} - q_v) \, dv = 0, \tag{3}$$

claim that (3) holds for *every* region R, and end up with (2) and (1).

By now we have drifted quite far from the unsophisticated form of Coulomb's inverse-square law. Note that the steps that led us to (1) do not involve assuming that $\mathbf{E} = -\text{grad } V$, and consequently (1) holds whether or not the electric field is conservative.

EXERCISES

1. Show that (2) gives the correct charge densities both inside and outside the spheres whose **D**'s are given by (13^{63}) and (14^{63}).
2. Extend Exercise 1 to the rods whose **D**'s are given by (10^{63}) and (11^{63}).
3. Consider the sphere of Exercise 1, take the capacitivity to be the constant ϵ_0 outside and a constant ϵ inside, compute the potential V inside and outside, and show that Poisson's equation gives the correct charge densities irrespective of the choice of the arbitrary constant in V. (Adjust V to be continuous at the surface of the sphere.)
4. Derive Poisson's equation from Maxwell's equation (2).
5. Had we not used the factor 4π in (2^{59}), but otherwise proceeded as before, Equation (1) would have its nonrationalized form, namely $\text{div } \mathbf{D} = 4\pi\rho$. Verify this statement.

CHAPTER 16

Examples of Solutions of Laplace's Equation

Laplace's equation, $\nabla^2 f = 0$, is one of the most important partial differential equations of theoretical physics. Just now we are interested in the case when f is the electrostatic potential V, so that the explicit form of the equation, say in cartesian coordinates, is

$$\frac{\partial^2 V}{\partial x^2} + \frac{\partial^2 V}{\partial y^2} + \frac{\partial^2 V}{\partial z^2} = 0.$$

The general solution of an *ordinary* differential equation involves *arbitrary constants*; by contrast, the general solution of a *partial* differential equation involves *arbitrary functions*. Studies of Laplace's equation have been concerned mostly with methods for finding *particular solutions* that are useful in particular physical situations. One of these methods consists in specifying a restricted class of problems (say problems with axial symmetry), specifying a particular coordinate frame, and then computing once and for all a set of special solutions of Laplace's equation that may be useful for this class of problems. The explicit particular solution pertaining to a specific problem of this class is then inferred from the special solutions with the help of the conditions that V and its derivatives must satisfy at the boundaries of the region of interest in the problem. In this chapter we illustrate this procedure for the case of a current-carrying copper block with a cylindrical hole in it, and prove the formulas that we took for granted in §14.

Solutions of Laplace's equation are often called *harmonic functions* of the coordinate variables, or simply *harmonics*.

68. HARMONIC FUNCTIONS OF PAIRS OF COORDINATE VARIABLES

In this section we work out in detail the two-dimensional harmonics that are listed in Table 5 and involve only the cylindrical coordinates (ρ, ϕ). In Exercise 2 they are expressed in terms of the cartesian coordinates (x, y); a different set of harmonic functions of x and y is introduced in Exercise 3. Examples of axially symmetric harmonics involving only the spherical coordinates r and θ are given in Exercise 4.

If the scalar potential V does not depend on z, the last term in (58^A) is zero and Laplace's equation, $\nabla^2 V = 0$, becomes

$$\frac{1}{\rho}\frac{\partial}{\partial \rho}\left(\rho \frac{\partial V}{\partial \rho}\right) + \frac{1}{\rho^2}\frac{\partial^2 V}{\partial \phi^2} = 0. \tag{1}$$

To avoid complications on the z axis, we let

$$\rho > 0, \tag{2}$$

when (1) can be written as

$$\rho\frac{\partial}{\partial \rho}\left(\rho \frac{\partial V}{\partial \rho}\right) + \frac{\partial^2 V}{\partial \phi^2} = 0. \tag{3}$$

A standard way of handling a partial differential equation of type (1) is to assume temporarily that V has the special form

$$V = R(\rho)\Phi(\phi), \tag{4}$$

where the factor R depends only on ρ, and the factor Φ only on ϕ. Substituting (4) into (3), we get

$$\Phi \rho \frac{d}{d\rho}\left(\rho \frac{dR}{d\rho}\right) + R\frac{d^2\Phi}{d\phi^2} = 0 \tag{5}$$

and then

$$\frac{1}{R}\rho \frac{d}{d\rho}\left(\rho \frac{dR}{d\rho}\right) = -\frac{1}{\Phi}\frac{d^2\Phi}{d\phi^2}. \tag{6}$$

Next, suppose that we keep ρ fixed and vary the angular coordinate ϕ. The left-hand side of (6) does not depend on ϕ and will stay constant while we vary ϕ. It then follows from (6) that the right-hand side must also stay constant and that consequently even this side does not depend on ϕ. A similar argument shows that the left-hand side of (6) is, in fact, independent of ρ. In other words, each side of (6) is simply some numerical constant, which we will denote by n^2. Equation (6) then separates into two *ordinary* differential equations, namely

$$\frac{d^2\Phi}{d\phi^2} = -n^2\Phi \tag{7}$$

and

$$\rho\frac{d}{d\rho}\left(\rho\frac{dR}{d\rho}\right) = n^2 R. \tag{8}$$

The procedure that leads from (3) to (7) and (8) is called *separation of variables*.

If $n \neq 0$, the general solution of (7) is

$$\Phi = A\cos n\phi + B\sin n\phi, \tag{9}$$

where A and B are arbitrary constants. The form of the general solution of (8) also depends on the value of n. It is

$$R = C\rho^n + D\rho^{-n} \quad \text{if } n > 0, \tag{10}$$

where C and D are arbitrary constants; if $n = 0$, it is a linear combination of the two terms listed in (14). In view of (4), the product of (9) and (10) is a particular solution of Laplace's equation.

If we set $A = 1$, $B = 0$, $C = 1$, and $D = 0$, we get a particularly simple solution of (3), namely

$$\rho^n \cos n\phi. \tag{11}$$

Other simple choices of A, B, C, and D yield such solutions as

$$\rho^n \sin n\phi, \quad \rho^{-n} \cos n\phi, \quad \rho^{-n} \sin n\phi. \tag{12}$$

Laplace's equation holds only for conservative electric fields, when we have a potential function V that has a fixed value at every point P in the field. This fact puts a limitation on the values on the parameter n. For example, if the coordinates of P are $(\rho, \phi = 0, z)$, the solution (11) is ρ^n. However, the coordinates of *the same* point P can also be given as $(\rho, \phi = 2\pi, z)$, when (11) becomes $\rho^n \cos 2\pi n$. Consequently, since V must have a unique value at P, the parameter n must satisfy the condition $\rho^n = \rho^n \cos 2\pi n$, or simply $\cos 2\pi n = 1$. Accordingly, the physically permissible values of n are

$$n = 0, 1, 2, 3, \ldots, \tag{13}$$

and $n = -1, -2, -3, \ldots$.

The arguments above refer specifically to (11), but they apply equally well to any other product of (9) and (10), so the condition (13) holds in general. Furthermore, the negative values of n need not be displayed: in the first term of (9), the sign of n does not matter; in the second term, reversing the sign of n amounts to reversing the sign of the arbitrary constant B.

If $n = 0$, the simplest pair of linearly independent solutions of (8)—that is, solutions which do not differ merely by a constant factor—is

$$1, \quad \ln \frac{\rho}{\rho_0}, \tag{14}$$

where ρ_0 is an arbitrary nonzero length. These solutions are listed in the top row of Table 5; the other entries in that row are zeros because $\sin n\phi = 0$ when $n = 0$. The rest of the table is obtained by writing $n = 1, 2,$ and 3 in (11) and (12). The four formulas in (11) and (12) could be used as headings for the columns in Table 5, except that the top term in the third column does not have the form $\rho^{-n} \cos n\phi$ when $n = 0$.

TABLE 5. Two-dimensional Harmonics; Cylindrical Coordinates

1	0	$\ln (\rho/\rho_0)$	0
$\rho \cos \phi$	$\rho \sin \phi$	$\rho^{-1} \cos \phi$	$\rho^{-1} \sin \phi$
$\rho^2 \cos 2\phi$	$\rho^2 \sin 2\phi$	$\rho^{-2} \cos 2\phi$	$\rho^{-2} \sin 2\phi$
$\rho^3 \cos 3\phi$	$\rho^3 \sin 3\phi$	$\rho^{-3} \cos 3\phi$	$\rho^{-3} \sin 3\phi$
...	...	...	...
...	...	...	...

A harmonic listed in Table 5 must, of course, be multiplied by a constant of the proper dimensionality before we can use it to describe a potential V, which we express in volts. In the case of the entries 0 and 1 in the table, V is constant, and therefore $\mathbf{E} = 0$. The entry $\ln (\rho/\rho_0)$ is not new to us—it appears in the formula for V when a uniform line charge is spread along the z axis, so that

(31[60]) $$V(\rho) = -\frac{q_l}{2\pi\epsilon_0} \ln \frac{\rho}{\rho_0}. \tag{15}$$

Since the entry $\rho \cos \phi$ is equal to x, the corresponding potential is a constant times x—as in Equation (16[56]), which holds when $\mathbf{E} = E_0 \mathbf{1}_x$.

Every entry in Table 5 is a solution of Laplace's equation and has the rather special form $R(\rho)\Phi(\phi)$, to which we have restricted ourselves so far. According to Exercise 6[66], however, every linear combination of the entries is also a solution of Laplace's equation, and these linear combinations do not necessarily have the form $R(\rho)\Phi(\phi)$. For example, the function $5 \ln (\rho/\rho_0) + 7\rho \cos \phi$, put together from two entries in the table, satisfies Laplace's equation but does not have this form. We will illustrate in §69 how a solution of Laplace's equation pertaining to a definite physical problem can be found by combining some entries in Table 5.

Next this question: Can *every* solution of Laplace's equation of interest in a two-dimensional physical problem be expressed as a linear combination of the entries in Table 5? In more technical terms: Is the set of functions listed in the table *complete* for scalar potentials of the form $V(\rho, \phi)$? The answer is yes, provided that the problem does not involve certain complicated discontinuities that arise only seldom in physical problems. A detailed answer is given by the theory of polynomials and Fourier series.

EXERCISES

1. Verify that the functions (11) and (12) satisfy Laplace's equation if $n > 0$, and that the functions (14) satisfy it if $n = 0$. The function $\Phi = \phi$ satisfies (7) if $n = 0$; explain why this Φ must be excluded in the case of the potential V.

2. (a) Show that, in cartesian coordinates, the entries in the first column of Table 5 are

$$1, \quad x, \quad x^2 - y^2, \quad x^3 - 3xy^2. \tag{16}$$

The functions (16) are examples of functions called *two-dimensional rational integral harmonics*. (b) Transform to cartesian coordinates the second column of Table 5.

3. If V does not depend on z, Laplace's equation in cartesian coordinates is

$$\frac{\partial^2 V}{\partial x^2} + \frac{\partial^2 V}{\partial y^2} = 0. \tag{17}$$

Write $X(x)$ for a function of x alone and $Y(y)$ for a function of y alone, assume that $V = X(x)Y(y)$, separate the variables in (17), and deduce the following special solutions of (17):

$$e^{kx} \sin ky, \quad e^{kx} \cos ky, \quad e^{-kx} \sin ky, \quad e^{-kx} \cos ky, \tag{18}$$

$$e^{ky} \sin kx, \quad e^{ky} \cos kx, \quad e^{-ky} \sin kx, \quad e^{-ky} \cos kx, \tag{19}$$

where k is an arbitrary constant. These harmonics are useful in two-dimensional problems involving rectangular regions.

4. The functions listed in Table 6 are examples of harmonics that are symmetric about the z axis. The factors 1, $\cos \theta$, and $\frac{1}{2}(3 \cos^2 \theta - 1)$ in the table are examples of polynomial functions of $\cos \theta$, called *Legendre polynomials*.

258 Examples of Solutions of Laplace's Equation

TABLE 6. Six Axially-symmetric Harmonics; Spherical Coordinates

1	r^{-1}
$r \cos \theta$	$r^{-2} \cos \theta$
$\tfrac{1}{2} r^2 (3 \cos^2 \theta - 1)$	$\tfrac{1}{2} r^{-3} (3 \cos^2 \theta - 1)$

(a) Verify that the entries in the table satisfy Laplace's equation. (b) The V's of three nonvanishing electric fields that we considered earlier in this book involve harmonic functions listed in Table 6. Identify these fields.

69. THE POTENTIAL V IN A CURRENT-CARRYING COPPER BLOCK WITH A CYLINDRICAL HOLE

The potential in question is given in (27^{56}). The derivation of that equation, however, was based on formulas that we had accepted without proof—the formulas (2^{14}) and (3^{14}) for the current density in the copper block. We will now show that (27^{56}) does indeed follow from the fundamental equation of Laplace for electrostatic fields in uncharged regions.

Our problem is described in the first paragraph of §14, but we now focus our attention on V and **E**, rather than **J**. The electric field in the copper block is steady and the block is homogeneous, so its interior remains uncharged. Consequently, inside the block,

$$\nabla^2 V = 0. \tag{1}$$

The electric field is parallel to the xy plane, so V does not depend on z; therefore, the solution of (1) that we need can be put together from the harmonics listed in Table 5^{68}, and our task is to find *which* of them pertain to the present problem. The search is based on two *boundary conditions*:

- Far from the hole, $\mathbf{E} = E_0 \mathbf{1}_x$.
- At the surface of the hole, the current density **J**—and hence also the electric intensity **E**—is tangential to the surface.

In symbols,
$$\mathbf{E} = E_0 \mathbf{1}_x \quad \text{for } \rho \gg b, \tag{2}$$

and
$$E_\rho = 0 \quad \text{for } \rho = b, \tag{3}$$

where b is the radius of the hole.

Now, if $\mathbf{E} = E_0 \mathbf{1}_x$, then, apart from an arbitrary additive constant, $V = -E_0 x = -E_0 \rho \cos \phi$. Also, since $\mathbf{E} = -\text{grad } V$, it follows from ($56^A$) that the condition $E_\rho = 0$ implies that $\partial V/\partial \rho = 0$. Therefore, (2) and (3) lead to the following conditions for V:

$$V = -E_0 \rho \cos \phi \quad \text{for } \rho \gg b \tag{4}$$

and

$$\frac{\partial V}{\partial \rho} = 0 \qquad \text{for } \rho = b. \tag{5}$$

The potential $-E_0 \rho \cos \phi$ in (4) cannot be the complete solution of our problem, because its derivative with respect to ρ is $-E_0 \cos \phi$ and does not satisfy (5). Therefore, we set

$$V = -E_0 \rho \cos \phi + V_1 \tag{6}$$

and proceed to identify the extra term V_1. Substituting (6) into (5), we get $\partial V/\partial \rho = -E_0 \cos \phi + \partial V_1/\partial \rho = 0$ and then

$$\frac{\partial V_1}{\partial \rho} = E_0 \cos \phi \qquad \text{for } \rho = b, \tag{7}$$

so V_1 should have the factor $\cos \phi$. The only entries in Table 5[68] that have this factor are $\rho \cos \phi$ and $\rho^{-1} \cos \phi$. We have already included $\rho \cos \phi$ in V, and consequently V_1 must have the form $C\rho^{-1} \cos \phi$, where C is a constant. Substituting into (6), we get

$$V = -E_0 \rho \cos \phi + C\rho^{-1} \cos \phi, \tag{8}$$

and then

$$\frac{\partial V}{\partial \rho} = -E_0 \cos \phi - C\rho^{-2} \cos \phi. \tag{9}$$

Therefore, (5) will hold if $C = -E_0 b^2$. The term in $\rho^{-1} \cos \phi$ vanishes when $\rho \to \infty$, and does not upset the condition (4). Accordingly—apart from a trivial additive constant—the complete solution of our problem is

$$V = -E_0 \rho \cos \phi - E_0 \frac{b^2}{\rho} \cos \phi, \tag{10}$$

namely (27[56]). The equation $\mathbf{E} = -\text{grad } V$ now leads to the field $E_z = 0$,

$$E_\rho = E_0 \left(1 - \frac{b^2}{\rho^2}\right) \cos \phi, \qquad E_\phi = -E_0 \left(1 + \frac{b^2}{\rho^2}\right) \sin \phi, \tag{11}$$

and then to (8[44]); and Ohm's law finally brings us back to our assumptions (2[14]) and (3[14]), which we have now derived from first principles.

EXERCISES

1. Consider the entries in the *first column* of Table 5[68] and explain without computation why 1 and $\rho \cos \phi$ are the only entries that can appear in V in the problem studied in this section. Why are the nonzero entries in the fourth column excluded from V?

2. The potential V in the insulation of a simple armored cable is (35[56]). Deduce it from Laplace's equation and the boundary conditions stated in Exercise 5[56].

3. Compute V in the copper block for the following situation involving a cylindrical hole: The coordinate axes are oriented as in Fig. 54[14], but far from the hole $\mathbf{J}$ has the form

$$\frac{1}{\sqrt{2}} J_0 \mathbf{1}_x + \frac{1}{\sqrt{2}} J_0 \mathbf{1}_y, \tag{12}$$

rather than $J_0 \mathbf{1}_x$.

4. Electric charge flows past a spherical cavity in a copper block, as in Exercise 2[16]. The boundary conditions are: Far from the cavity, $\mathbf{E} = E_0 \mathbf{1}_z$; $\mathbf{E}$ is tangential to the surface of the cavity, so $\partial V / \partial r = 0$ for $r = r_0$. Use these conditions and Table 6 to show that, outside and at the surface of the cavity, and apart from an additive constant,

$$V_{\substack{\text{outside} \\ \text{cavity}}} = -\frac{1}{2} E_0 \frac{r_0^3}{r^2} \cos \theta - E_0 r \cos \theta. \tag{13}$$

Then find $\mathbf{E}$ and verify Equations (16[36]) for $\mathbf{J}$. (This will prove the formulas for $\mathbf{J}$ that we took for granted in Exercise 2[16].)

5. If $r = r_0$, the formula for V inside the cavity must give the values of V on the boundary of the cavity, which are given by (13) for $r = r_0$. Use this boundary condition and Table 6, and (remembering that $r = 0$ at the center of the cavity) show that, inside and on the surface of the cavity we have $V = -\frac{3}{2} E_0 r \cos \theta$. Then show that

$$\mathbf{E}_{\substack{\text{inside} \\ \text{cavity}}} = \tfrac{3}{2} E_0 \mathbf{1}_z. \tag{14}$$

6. A homogeneous conductor of unspecified shape carries a current whose density varies from point to point but does not depend on t. Assume that a potential V exists and use Ohm's law, the law of conservation of charge, and the equation $\mathbf{E} = -\operatorname{grad} V$ to show that $\nabla^2 V = 0$.

70. BOUNDARY CONDITIONS FOR D AND E

We will now derive the boundary conditions (2) and (4) below, which account for such phenomena as the "refraction" of the field lines illustrated in Fig. 169[62].

***The normal component of* D.** The shaded surface in Fig. 174(a) is the uncharged interface of two different—and in general polarized—mediums, which we distinguish below by a prime and a double prime. The closed surface (S) that we will use in the equation

$$\iint_{(S)} D_n \, da = 0 \tag{1}$$

is shown in Fig. 174(a) and, edge-on, in Fig. 174(b). It has the shape of a circular drum of radius r and depth h, where $h \ll r$, and where r is so small that the portion of the interface contained in (S) can be taken as flat. We will be concerned with the limiting situation, when (S) contracts upon the point P on the interface. The vectors $\mathbf{D}'$ and $\mathbf{D}''$ in Fig. 174(b) pertain to a pair of points on opposite flat portions of (S); they do not necessarily lie in the plane of the page.

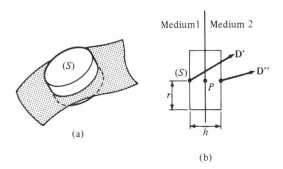

Fig. 174. Drum-shaped surface (S) used for deriving the boundary condition for **D**.

We will need two sets of normal components of **D**: components taken relative to (S), which we write for the moment as D'_{ns} and D''_{ns}, and components relative to the interface, which we write as D'_n and D''_n. If we assume, for definiteness, that the interface in Fig. 174(b) looks rightward, then $D''_{ns} = D''_n$, because the right-hand flat part of (S) also looks rightward. But the other flat part of (S) looks leftward, and hence $D'_{ns} = -D'_n$.

Since r is small, the respective integrals of D_n over the left-hand and the right-hand flat part of (S) are $\pi r^2 D'_{ns}$ and $\pi r^2 D''_{ns}$; since $h \ll r$, the integral of

D_n over the curved part of (S) can be ignored. Therefore, (1) becomes $\pi r^2(D'_{ns} + D''_{ns}) = 0$, so $D''_{ns} = -D'_{ns}$ and

$$D''_n = D'_n. \tag{2}$$

In words: *As we cross the interface, the normal component of* **D** *relative to the interface does not change, provided that no charges other than polarization charges are located on the interface.* This result is generalized in Exercise 2.

The tangential component of E. Gauss's formula is not suited for dealing with tangential components, so we shift our attention from **D** to **E**, and to the work involved in carrying a test charge around a closed path. The interface of the two mediums in Fig. 175 is perpendicular to the plane of the page;

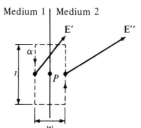

Fig. 175. Rectangular path α used for deriving the boundary condition for **E**.

the rectangular contour α lies in this plane, and the length l and the width w satisfy the condition $w \ll l$. We will be concerned with the limiting situation, when α contracts upon the point P on the interface. We can then ignore any emf's induced in α by time-varying magnetic fields, because such emf's depend on the magnetic flux linking α and vanish when α shrinks to a point. Furthermore, since in this book we avoid regions where there is chemical or thermal action, we can confine ourselves to electrostatic fields. But these fields are conservative, and hence we use the equation

$$(6^{51}) \qquad \oint E_t \, dl = 0 \tag{3}$$

when the area wl tends to zero.

The vectors **E**' and **E**'' in the figure do not necessarily lie in the plane of the page. We will need two sets of components of these vectors. One set, which we write for the moment as $E'_{t\alpha}$ and $E''_{t\alpha}$, consists of tangential components taken relative to α. (Note, for example, that in the figure $E'_{t\alpha}$ is negative, but $E''_{t\alpha}$ is positive.) The second set, which we write as $E'_{\tau u}$ and $E''_{\tau u}$, is the pair of "up-the-page" components of the **E**'s. In Fig. 175, both $E'_{\tau u}$ and $E''_{\tau u}$ are positive; furthermore, $E'_{t\alpha} = -E'_{\tau u}$ and $E''_{t\alpha} = E''_{\tau u}$.

If we ignore the two short straight parts of α and if l is short enough, Equation (3) implies that $lE'_{t\alpha} + lE''_{t\alpha} = 0$, so that $E''_{t\alpha} = -E'_{t\alpha}$, and hence $E''_{\tau u} = E'_{\tau u}$. Thus, as we cross the interface, the up-the-page component of **E** does not change.

Next we turn the path α about its horizontal midline until its plane is perpendicular to the page. Reasoning as before, we then find that the out-of-page component of **E** has no discontinuity at the boundary. Combining this with the previous result, we conclude that *as we cross the interface, the tangential component of* **E** *relative to the interface does not change*. In symbols,

$$E''_\tau = E'_\tau, \tag{4}$$

where the subscript τ means that we are dealing here with tangential components relative to a *surface*, rather than a line.

If the capacitivities of the two mediums are ϵ' and ϵ'', we have $D' = \epsilon' E'$ and $D'' = \epsilon'' E''$. Equation (2) can then be written as

$$\epsilon'' E''_n = \epsilon' E'_n, \tag{5}$$

and Equation (4) as

$$\frac{1}{\epsilon''} D''_\tau = \frac{1}{\epsilon'} D'_\tau. \tag{6}$$

Our derivation of (4) and (6) did not presume that the dielectrics are uncharged, but (2) and (5) hold only if the interface is free from any charges other than polarization charges—see Exercise 2 below.

EXERCISES

1. The line of **D** shown in Fig. 176 lies in the plane of the page. The interface of the two dielectrics is normal to the page and is uncharged. Show that

$$\epsilon'' \cot \theta'' = \epsilon' \cot \theta'. \tag{7}$$

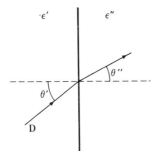

Fig. 176. "Refraction" of an electric field at an interface of two mediums.

264   Examples of Solutions of Laplace's Equation

2. One face of a glass plate is rubbed with silk until it acquires a surface charge density q_a. This face is then clamped to the face of an uncharged plate made of a different glass, and the two plates are placed in an electric field. Show that (2) must then be replaced by

$$D_n'' - D_n' = q_a, \qquad (8)$$

and show by a diagram the direction in which the normal components in (8) are taken as positive.

3. Derive the relation between θ' and θ'' in Fig. 176 if the interface carries a surface charge (other than a polarization charge) of density q_a.

4. A steady current flows from a medium of conductivity g' to one of conductivity g'', and $\mathbf{J}$ is not necessarily normal to the interface. Show that

$$J_n'' = J_n' \qquad (9)$$

and that consequently

$$g'' E_n'' = g' E_n'. \qquad (10)$$

Exercise 3[44] is a simpler version of this problem.

71. FIELDS INSIDE THE CYLINDRICAL HOLE AND CHARGES ON ITS SURFACE

The hole in the copper block includes the z axis, where $\rho = 0$, and where $\ln(\rho/\rho_0)$ and the negative powers of ρ become infinite. Consequently, only the first two columns of Table 5[68] are of physical interest inside the hole. Our task is to find the entries in these columns that provide for the continuity of the tangential component of $\mathbf{E}$ across the surface of the hole.

The field $\mathbf{E}$ in the copper is (11[69]). To evaluate the tangential component of $\mathbf{E}$ just *outside* the hole, we let $\rho = b$ in the circular component and get

$$E_\phi = -2E_0 \sin \phi \qquad \text{for } \rho = b. \qquad (1)$$

To compute E_ϕ just *inside* the hole, we write the equation $\mathbf{E} = -\text{grad } V$ in cylindrical coordinates and find from (56[A]) that

$$E_\phi = -\frac{1}{\rho} \frac{\partial V}{\partial \phi}. \qquad (2)$$

Therefore, V must be so chosen that, when $\rho = b$, the right-hand sides of (2) and (1) are equal; in particular, the function $\partial V/\partial \phi$ in (2) must have the factor $\sin \phi$ that appears in (1), and consequently V must have the factor $\cos \phi$. Hence, we have no choice other than the entry $\rho \cos \phi$ in Table 5[68]. Accordingly, we set $V = C\rho \cos \phi$, get $E_\phi = C \sin \phi$ from (2), and then $C = -2E_0$ from (1). Thus

$$V = -2E_0 \rho \cos \phi = -2E_0 x \tag{3}$$

and, as we anticipated in (31[56]),

$$\mathbf{E} = 2E_0 \mathbf{1}_x. \tag{4}$$

Surface charges. When surface charges (other than polarization charges) of density q_a are located on the interface of copper and a vacuum, the boundary condition for **D** is

(8[70]) $$D_n'' - D_n' = q_a, \tag{5}$$

where D' pertains to the copper and D'' to the hole. Polarization is not involved in the present problem, because the conductor is a metal. Consequently, the complete charge density on the surface can be computed from (5) and from our knowledge of E. It is convenient to have the normal component of E directed *into* the hole, and therefore we will reverse the signs of the ρ components. Since E_ρ vanishes as we approach the hole, we have $D_n' = 0$. According to (4), $E_\rho = 2E_0 \cos \phi$, and hence $D_n'' = \epsilon_0(-E_\rho) = -2\epsilon_0 E_0 \cos \phi$. Therefore, (5) gives the following surface charge density:

$$q_a = -2\epsilon_0 E_0 \cos \phi \; \frac{\text{coulombs}}{\text{meter}^2}. \tag{6}$$

Note that q_a is positive on the left-hand side of the hole, as in Fig. 53[14].

EXERCISES

1. Compute the normal component of **E** just *inside* the cylindrical hole in two ways: first, using the field $2E_0 \mathbf{1}_x$ inside the hole; second, using (6) and (21[60]).

2. An "infinitely long" cylindrical surface, made of thin paper, has a circular cross section of radius b; it is located in vacuum, is coaxial with the z axis, and carries a surface charge given by (6). Show that $\mathbf{E} = E_0 \mathbf{1}_x$ on the axis of the cylinder. (Divide the surface into straight strips of angular width $d\phi$, and so on.)

3. According to (4), $\mathbf{E} = 2E_0 \mathbf{1}_x$ on the z axis and throughout the hole in the current-carrying conductor. Explain why the E in Exercise 2 is half as strong. (A similar situation arose in Exercise 3[60].) Also, explain without any computations why we would expect the field E inside the hole to be stronger than the field E_0 far from the hole.

4. The point a lies *on* the paper cylinder of Exercise 2; its coordinates are $\rho = b$

and $\phi = 180$ deg. (The point a in Fig. 83^{25} has these coordinates.) Show that the intensity of the electric field produced by the charge distribution (6) is zero at a.

5. (a) Show that the electric field produced by the charge distribution (6) cancels the "original" field $E_0 \mathbf{1}_x$ just to the left of the point a in Fig. 83^{25} and doubles it just to the right of a. (b) Show that one would not expect this cancellation and doubling, were it not for the result of Exercise 4. (c) Compare the present case with that of the uniformly charged spherical balloon of Fig. 158^{60}.

6. According to (14^{69}), we have $\mathbf{E} = \tfrac{3}{2} E_0 \mathbf{1}_z$ inside the cavity.[1] Use (5) to show that the density of charge on the surface of the cavity is

$$q_a = -\tfrac{3}{2}\epsilon_0 E_0 \cos\theta. \tag{7}$$

7. Take (S) in Fig. 158^{60} to be the surface of the cavity[1] and show that the charge distribution (7) produces at the point P the field $\mathbf{E}(P) = -\tfrac{1}{4} E_0 \mathbf{1}_z$. (The vectors in Figs. 158(b) and 158(c) apply to a uniformly charged balloon, but not to the present case.)

8. Let (S) in Fig. 158^{60} be the surface of the cavity.[1] According to (21^{16}), (22^{16}), and (23^{16}), the current density on the z axis just to the right of P is zero, and hence $\mathbf{E}$ is also zero there. In view of (14^{69}), we then have, on the z axis,

$$\mathbf{E} = \begin{cases} 0 & \text{just to the right of } P \\ \tfrac{3}{2} E_0 \mathbf{1}_z & \text{just to the left of } P. \end{cases} \tag{8}$$

The top line of (8) can be checked by an argument similar to that we used for a charged balloon. The field $\mathbf{E}$ at a point just to the right of P is a superposition of three fields: the "original" field, equal to $E_0 \mathbf{1}_z$; the field produced at P (and hence also near P) by the bulk of the surface charge, and found in Exercise 7 to be $-\tfrac{1}{4} E_0 \mathbf{1}_z$; and the extra field $q_a/2\epsilon_0$, produced just outside P by the patch of surface charge adjacent to P—a field equal in the present case to $-\tfrac{3}{4} E_0 \mathbf{1}_z$. These fields add to zero.

Account in a similar way for the second line of (8) and also for the electric intensities just inside and just outside the cavity on the negative part of the z axis.

9. Show without integration that the charge density (7) produces the field $\tfrac{1}{2} E_0 \mathbf{1}_z$ everywhere *inside* the cavity. Verify by integration that this conclusion holds at the center of the cavity.

10. A long uncharged *nonconducting rod* of circular cross section (radius b) is put into a uniform electric field. The intensity of the "original" field is $\mathbf{E} = E_0 \mathbf{1}_x$ with $E_0 > 0$, but the polarization of the rod distorts this field. Assume that the axis of the rod is the z axis and derive (not merely verify) the following

---

[1] The spherical cavity in a current-carrying copper block, described in Exercise 2^{16}.

formulas, where the arbitrary constant in V is omitted, and where ϵ and ϵ_0 are the capacitivities of the rod and the surrounding space:

$$V_{\substack{\text{inside}\\\text{the rod}}} = -\frac{2E_0}{(\epsilon/\epsilon_0)+1}\,\rho\cos\phi, \tag{9}$$

$$V_{\substack{\text{outside}\\\text{the rod}}} = -E_0\rho\cos\phi + E_0 b^2\frac{(\epsilon/\epsilon_0)-1}{(\epsilon/\epsilon_0)+1}\frac{1}{\rho}\cos\phi. \tag{10}$$

Compute **E** both outside and inside the rod, and verify the formula

$$\mathbf{E}_{\substack{\text{inside}\\\text{the rod}}} = \frac{2}{(\epsilon/\epsilon_0)+1}\,E_0\mathbf{1}_x. \tag{11}$$

Check the general features of the field **D** in Fig. 177, where $\epsilon = 3\epsilon_0$, and plot the corresponding map of **E**.

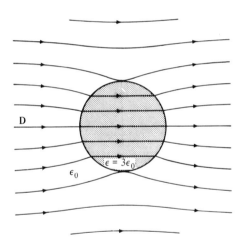

Fig. 177. Distortion of a uniform electric field by a long dielectric rod.

11. Consider the space outside the rod of Fig. 177, compute **E**, and derive the equation of the lines of **E**.

12. Replace the nonconducting rod of Exercise 10 by a *metal rod*, set $V = 0$ inside the rod, and show that

$$V_{\substack{\text{outside}\\\text{the rod}}} = -E_0 b\left(\frac{\rho}{b}-\frac{b}{\rho}\right)\cos\phi. \tag{12}$$

Then compute **E** and check the general features of Fig. 178. Finally, compute the charge density q_a on the surface of the metal rod and compare it with (6).

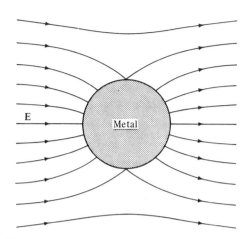

Fig. 178. Distortion of a uniform electric field by a long metal rod.

13. The field **E** vanishes inside the metal rod of Exercise 12. Therefore, we must conclude that the "original" field, $E_0 \mathbf{1}_x$, pervading the rod is canceled everywhere inside the rod by the field produced inside the rod by the charges that have developed on its surface. Verify this conclusion for points on the z axis.

14. A dielectric sphere (capacitivity ϵ, radius r_0, center at the origin) is located in a vacuum in an electric field whose intensity in the absence of the sphere was $E_0 \mathbf{1}_z$. Use Table 6[68] and the boundary conditions to compute V and **E** in the presence of the sphere; in particular, show that

$$\mathbf{E}_{\substack{\text{inside the}\\ \text{sphere}}} = \frac{3}{(\epsilon/\epsilon_0) + 2} E_0 \mathbf{1}_z. \tag{11}$$

Sketch line maps of **E** and **D**.

15. Consider along the lines of Exercise 14 a *metal sphere* placed in an originally uniform electric field.

CHAPTER 17

Charged Conductors and Capacitors

Our remarks on charged conductors were confined so far to rather simple examples, such as the metal plates discussed in §60. In this chapter we present a few theorems that apply to isolated conductors of any shape. If a conductor is initially charged throughout its interior, the charges will eventually migrate to the surface; in §72, we work out the rate of this migration. In §73, where we turn to capacitors, our main aim is to prepare for the concepts of electrostatic energy and the density of this energy, to be introduced in Chapter 18.

When a piece of copper is brought into an electric field—or when a charge is put on it by an electrostatic machine—it becomes pervaded by an "external" electric field. In response to this field, the conduction electrons of the copper atoms form conduction currents in the metal. But no molecular distortion is involved in this process, and the copper does not become polarized in the sense of §62. This suggests letting $\epsilon = \epsilon_0$ in a metal, as we do in this book. Actually, the interior of a metal resembles a highly ionized gas (a "plasma"), and a more careful study would call for statistical computations.

72. CHARGED CONDUCTORS

When an electric charge is put on an isolated conductor—or when an isolated conductor is placed in an "external" electric field—the field produced inside the conductor drives currents in the conductor until the charges become redistributed in such a way that E vanishes at all points lying in the conducting material; the currents then subside and the situation becomes static. In a good conductor this process is completed almost instantly; in a poor conductor it may take days.

Suppose that an isolated conductor containing no cavities is located in an electric field and that the currents in it have subsided (which is not so in Fig. 148[56]). As we will see, the following conditions then hold:

(a) the fields **E** and **D** vanish in the conducting material;
(b) the potential V has the same value at all points inside and on the surface of the conductor;
(c) the entire charge resides on the surface of the conductor;
(d) the field **E** at a point P just outside a conductor is normal to the surface of the conductor; and
(e) the formula for the normal component of **E** is

$$E_n = \frac{q_a}{\epsilon}, \tag{1}$$

where q_a is the surface charge density on the conductor near P, and ϵ is the capacitivity of the (homogeneous) medium surrounding the conductor.

The statements (a) and (b) are implied in the assumption that the currents have subsided; and so is (d), because, according to the boundary condition (4[70]), a nonvanishing tangential component of **E** just outside the conductor implies such a component—and hence a tangential current—just inside.

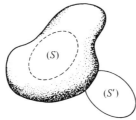

Fig. 179. The imagined closed surface (S) lies completely inside a conductor. The closed surface (S') lies partly inside and partly outside the conductor.

To verify (c), consider the potato-shaped conductor of Fig. 179 and the imagined closed surface (S) lying entirely inside it. According to (a), we have $\mathbf{D} = 0$ at all points on (S), so that

$$\iint_{(S)} D_n \, da = 0. \tag{2}$$

Therefore, according to Gauss's formula (4[63]), the net charge contained in (S) is zero. This conclusion holds for every closed surface lying completely inside the conductor, and consequently no charges can stay *inside* the conductor. To show that they can stay on the *surface*, we turn to the closed surface (S') in Fig. 179. The integral of D_n over the portion of (S') lying outside the conductor is not necessarily zero. Therefore, the integral of D_n over the entire surface (S') is not necessarily zero, and hence the charge contained in (S') is not necessarily zero. Thus, the theory excludes charges from the interior of the conductor, but permits them to stay on the surface.

As we have shown, statement (c) follows from Gauss's formula, which in turn follows from Coulomb's inverse-square law. Conversely, one can show that if (c) is correct, then the force of interaction between two point charges, a distance r apart, must be proportional to $1/r^2$. Modern experimental tests of Coulomb's law are based on (c).

In §60 we considered a flat metal plate surrounded by vacuum and carrying a uniform charge density q_a on each face; we found that the normal component of **E** just outside a face is given by Equation (21^{60}), namely $E_n = q_a/\epsilon_0$. Equation (1) generalizes that result when the conductor is surrounded by a dielectric other than a vacuum and when the surface of the conductor is not necessarily flat—provided it is smooth. To prove (1), we use Gauss's formula and a drum-shaped surface (S) that includes a portion of the interface of a charged conductor and a dielectric as in Fig. 174(a). The field **D** in the conductor is zero, and, therefore, for a sufficiently small and thin surface (S), the integral in Gauss's formula is aD_n, where a is the area of one flat face of (S). The q in Gauss's formula is now aq_a, where q_a is the density of charge on the portion of the surface of the conductor contained in (S). Hence, $aD_n = aq_a$ and $E_n = q_a/\epsilon$, as in (1).

Image charges. One face of a grounded metal plate is the xy plane. A positive point charge q is held on the z axis, in vacuum, at the distance s from this face—as in Fig. 180(a). The lines of **E** begin on q and end on the negative charges that come up from the earth to the left-hand face of the plate. The field **E** at the left of the plate satisfies the following conditions:

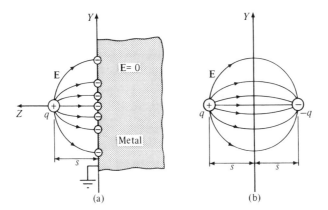

Fig. 180. A positive charge q outside a metal plate and its image, $-q$, inside the plate.

- at a sufficiently small distance d from q, E is directed radially away from q, and $E = q/(4\pi\epsilon_0 d^2)$;
- at sufficiently small distances from the plate, E is normal to the plate.

The two conditions given above imply the corresponding boundary conditions on the potential V, and the theory of Laplace's equation then shows that these conditions determine the field E at the left of the plate *uniquely*; that is, if in some way—say by guessing—we can find an E satisfying these conditions, then this E is the correct E. The standard way for identifying E in the present problem is to suppose that the grounded plate is removed and a charge $-q$ is placed at the position of the mirror image of q, as in Fig. 180(b). It is then perhaps obvious that the field E at the left of the xy plane in Fig. 180(b) satisfies the two conditions given above. Since these conditions determine E uniquely, we conclude that the electric field in the left-hand half of Fig. 180(a) is the same as that in the left-hand half of Fig. 180(b), and is, therefore, the field of an electric dipole of length $2s$. The value of E_n just outside the metal face is given in Exercise 3. The field E vanishes everywhere at the right of the xy plane; in other words, the face of the grounded metal plate "shields" the interior of the metal from the field of the charge q.

One can think of electrostatic shielding in either of two ways:

- the electric field of the charge q cannot penetrate into the metal;
- the electric field of the charge q does extend into the metal, and so does the field of the surface charges, but in the metal these fields are equal and opposite, and their resultant is zero.

The second alternative is illustrated in Exercise 4. In electrostatics it does not matter which view one takes, but in nonstatic situations one must allow for the penetration of external fields into metals.

Flow of charge to the surface. If the interior of a conductor is charged to begin with, the entire charge will eventually move to the surface. To see how this comes about, we consider a point P in the conductor, imagine a closed surface (S) surrounding P and lying entirely in the conductor, write q for the charge (other than any polarization charge) contained in (S), and assume that the region R enclosed by (S) is so small that in it the conductivity g and the capacitivity ϵ can be treated as constants. By Gauss's formula,

$$\iint_{(S)} D_n \, da = q. \tag{3}$$

Now, $\mathbf{D} = \epsilon \mathbf{E}$ and, by Ohm's law, $\mathbf{E} = \mathbf{J}/g$, Therefore (3) can be written as

$$\iint_{(S)} J_n \, da = \frac{g}{\epsilon} q. \tag{4}$$

The integral in (4) is the current leaving R. Since charge is conserved, this current is equal to the time rate at which the charge in R is decreasing, namely, $-dq/dt$. Consequently,

$$-\frac{dq}{dt} = \frac{g}{\epsilon}q. \tag{5}$$

Dividing both sides of (5) by the volume of R, we get

$$-\frac{dq_v}{dt} = \frac{g}{\epsilon}q_v. \tag{6}$$

Therefore,

$$q_v(t) = q_v(0)e^{-(g/\epsilon)t}, \tag{7}$$

where $q_v(0)$ is the volume charge density at P when $t = 0$. Thus, the magnitude of the charge density at every point in the conductor decreases exponentially with time and approaches zero. Our proof does not apply if P lies *on* the surface of the conductor; therefore, the charges that move to the surface may stay there indefinitely (Exercise 9).

The ratio ϵ/g is called the (rationalized) *relaxation time* of the conducting material. It is long for a good insulator (about 10^6 seconds for fused quartz) and very short for even a moderately good conductor (about 10^{-10} second for sea water).

EXERCISES

1. Equations (12^{63}) and (15^{63}) hold, respectively, outside a charged nonconducting rod and a nonconducting sphere. Verify that they hold also outside a metal rod and a metal sphere.

2. The potato-shaped object in Fig. 181 is an isolated uncharged conductor that completely encloses a cavity. A metal pellet hangs in the cavity from a nonconducting thread. Suppose that the pellet is taken out through a trap door, is charged positively, and then put back without touching the conductor. Sketch the resulting charge distribution and the field **D** in the cavity, in the conducting material, and outside the conductor. Also, make a similar sketch

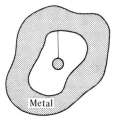

Fig. 181. A pellet suspended in a cavity in a piece of metal.

on the assumption that the pellet touched the surface of the cavity after the trap door was closed. Faraday investigated phenomena of this kind in his celebrated ice-pail experiments.

3. Verify that, at points on the xy plane, the field $\mathbf{E}$ produced jointly by the charges q and $-q$ in Fig. 180(b) is normal to this plane, and show that

$$E_n = \frac{sq}{2\pi\epsilon_0(s^2 + \rho^2)^{3/2}}, \tag{8}$$

where $\rho^2 = x^2 + y^2$. Then show that the charge density on the metal face in Fig. 180(a) is

$$q_a = -\frac{sq}{2\pi(s^2 + \rho^2)^{3/2}}. \tag{9}$$

4. The net electric field in Fig. 180(a)—not only in the vacuum but also in the metal—can be regarded as a superposition of the fields produced by the charge q and by the charge distributed on the surface with the density (9). Using the arguments of this section and setting $\epsilon = \epsilon_0$ in the metal, prove that the superposition of these fields does indeed give $\mathbf{E} = 0$ in the metal.

5. Granted that $\mathbf{E} = 0$ in the metal in Fig. 180, show, without using formulas or boundary conditions, that the $\mathbf{E}$ outside the metal due to the surface charges is the same as would be produced by an image charge $-q$.

6. Figure 180 suggests that the charge q is attracted to the metal plate as though it were pulled toward the right by its image (a charge $-q$) located at the distance $2s$. If so, the force acting on q should be

$$\frac{q^2}{4\pi\epsilon_0(2s)^2}\mathbf{1}_z. \tag{10}$$

To verify (10), compute the force exerted on q by the distribution (9) of the induced surface charge. (Divide the surface charge into rings of radii ρ and $\rho + d\rho$, and so on.)

7. If the charge q in Fig. 180(a) should be moving toward the plate, it would constitute an element of current flowing from left to right. What would be the direction of flow of the image of this current? Consider in a similar way the case when q moves parallel to the plate, say up the page in Fig. 180.

8. Our discussion of the image charge in Fig. 180 was based on the properties of $\mathbf{E}$. Derive (8) by first computing the scalar potential V.

9. Show that our proof of (7) breaks down if the point P lies *on* the surface of the conductor.

73. CAPACITORS

Suppose that we first discharge the two plates of a capacitor by connecting them to ground, and then connect them across a battery. A positive charge, say q coulombs, will move from one plate to the other. (Actually, electrons will carry $-q$ coulombs in the opposite direction.) The charge q proves to be proportional to the emf of the battery, say V volts; that is,

$$q = CV, \qquad (1)$$

where C is a constant—the *capacitance* (farads) of the capacitor. The charge q is said to be "stored" in the capacitor, even though it has only moved from one plate to the other. The energy, say U_e joules, required to charge the capacitor (apart from losses in the charging circuit) is said to be "stored" in the capacitor. The formulas for U_e are

$$U_e = \frac{1}{2}qV = \frac{1}{2}CV^2 = \frac{1}{2}\frac{q^2}{C}, \qquad (2)$$

provided we ignore such perturbations as temperature changes and the distortion of dielectrics caused by electric fields (electrostriction). Equation (2) can be proved by referring to the following idealized experiment: Imagine that we transfer the charge q from one plate to the other in small portions and at a constant rate. According to (1), the potential difference between the plates will then rise linearly from 0 to V volts, and its average value during the transfer will be $\frac{1}{2}V$ volts. This means that the total work we will have done is (q coulombs) $\times$ ($\frac{1}{2}V$ volts), and consequently $U_e = \frac{1}{2}qV$ joules. The other parts of (2) follow from (1).

When a capacitor is charged by a battery, the battery provides $\frac{1}{2}qV$ joules to be stored in the capacitor, but an equal amount of energy is dissipated as heat, so this process is only 50 percent efficient (Exercise 3).

Parallel-plate capacitors. We now turn to the parallel-plate capacitor of Fig. 182, under the assumption that edge effects are eliminated by

Fig. 182. A parallel-plate capacitor. (Fringing is eliminated by guard rings; see Fig. 63[17].)

guard rings, as in Fig. 63[17]. The area of each plate (excluding the guard rings) is a, the plate separation is s, and hence the (effective) volume of the capacitor is

$$v = as. \qquad (3)$$

The capacitivity of the dielectric between the plates is ϵ, and the charge stored in the capacitor (excluding the guard rings) is q, so the charge density on the positive plate is q/a. According to (21^{60}), with ϵ replacing ϵ_0, the magnitude of E in the space between the plates is

$$E = \frac{q}{\epsilon a}. \tag{4}$$

The work we would have to do to carry a test charge q^* from the negative to the positive plate is q^*Es; the work we would have to do per unit charge is Es; and consequently the potential drop across the capacitor is

$$V = Es = \frac{qs}{\epsilon a}. \tag{5}$$

Combining (5) and (1), we find that

(5^{59})
$$C = \frac{\epsilon a}{s}, \tag{6}$$

and hence (2) can be written as

$$U_e = \frac{1}{2}\frac{q^2 s}{\epsilon a}. \tag{7}$$

Now, in the space between the plates,

$$E = \frac{q}{\epsilon a}, \quad D = \epsilon E = \frac{q}{a}, \tag{8}$$

and, therefore, for a parallel-plate capacitor,

$$U_e = \frac{1}{2}EDv, \tag{9}$$

where v is the volume of the capacitor.

Spherical capacitor. Another capacitor that leads to simple formulas is made of two concentric spherical metal shells, as in Fig. 183. The positive lead of a battery is momentarily connected to the inner shell through a hole in the outer shell; it is then withdrawn and the hole is closed. If the charge on the inner shell is q, the electric intensity in the space between the shells points radially toward the outer shell and has the magnitude

(15^{63})
$$E = \frac{q}{4\pi\epsilon r^2}, \tag{10}$$

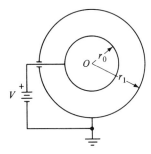

Fig. 183. A spherical capacitor.

where r is measured from the center (Exercise 1[72]). The potential of the grounded outer shell is zero, and that of the inner shell is V volts, say. To express V in terms of q, we integrate the formula

(11[56]) $$dV = -E_t\,dl \tag{11}$$

along a radial path extending from the outer to the inner shell. For such a path $dl = -dr$ and E_t is the negative of (10), so

$$V = -\int_{r_1}^{r_0} \frac{q}{4\pi\epsilon r^2}\,dr = \frac{q}{4\pi\epsilon}\left(\frac{1}{r_0} - \frac{1}{r_1}\right) = \frac{q}{4\pi\epsilon}\frac{r_1 - r_0}{r_0 r_1}. \tag{12}$$

Comparison with (1) shows that, for a spherical capacitor,

$$C = \frac{4\pi\epsilon r_0 r_1}{r_1 - r_0}. \tag{13}$$

Therefore, from (2),

$$U_e = \frac{1}{2}\frac{(r_1 - r_0)q^2}{4\pi\epsilon r_0 r_1}. \tag{14}$$

Capacitivity of metals. If a parallel-plate capacitor is connected across a constant-voltage generator, a current will flow until the charge stored in the capacitor reaches the value $q = \epsilon a V/s$, given by (5). Suppose, however, that a metal plate of thickness s is inserted snugly between the plates of the capacitor. The generator will then keep "charging" the shorted capacitor indefinitely, as though the capacitance has become infinite. In this context, one would have to set $\epsilon = \infty$ for a metal. But in computing the fields of static charges located near, in, or on the surface of a metal object, we may write $\epsilon = \epsilon_0$ inside the metal, as in §60.

EXERCISES

1. This problem outlines an alternative proof of (2). Let Q be the initial charge stored in a capacitor. When $t = 0$, the capacitor is shorted by a resistor of

R ohms; the self-inductance of the circuit is negligible. Recall footnote 1[40] and use Kirchhoff's voltage law to show that the charge $q = q(t)$ remaining in the capacitor at the instant t satisfies the equation

$$R\frac{dq}{dt} + \frac{1}{C}q = 0. \tag{15}$$

Solve (15) for q; compute the current from the equation $i = dq/dt$; show that the rate of generation of heat in the resistor is $(Q^2/RC^2)\exp(-2t/RC)$ watts, and compute the energy converted into heat during the discharge (which, in principle, takes infinite time). Then verify (2) by arguing that this energy must have been originally stored in the capacitor.

2. This exercise, restricted to a parallel-plate vacuum capacitor equipped with guard rings, outlines still another proof of (2). According to Exercise 5[60], the right-hand plate (area a) is pulled leftward with the force of $q^2/2\epsilon_0 a$ newtons. Suppose that the left-hand plate and guard ring are held fixed, but the right-hand plate is moved without acceleration through the distance s' by a force of magnitude $q^2/2\epsilon_0 a$, and that the corresponding guard ring is moved through the same distance by a separate mechanism. The plate separation then changes from s to $s + s'$, and U_e in (7) changes accordingly. Verify that the increase of U_e is equal to the work done in moving the plate of the capacitor.

3. When $t = 0$, an "empty" capacitor is connected across a battery whose emf is V volts; the total resistance of the circuit, including the internal resistance of the battery, is R ohms; the self-inductance of the circuit is negligible. Show that the equation for the charge $q = q(t)$ delivered to the capacitor is (15), with the zero on the right-hand side replaced by V. Compute q and the charging current i; find the energy lost as heat during the charging process (which, in principle, takes infinite time), and show that this energy is equal to that stored in the fully charged capacitor.

4. Two infinitely long cylindrical metal shells of circular cross section and negligible thickness are centered on the z axis; their radii are ρ_1 and ρ_2, where $\rho_1 < \rho_2$; the inner shell carries q_l coulombs per meter of length. Show that the electric intensity in the vacuum between the shells is given by (11[60]). Then show that the capacitor formed by the coaxial cylinders has the following capacitance per unit length:

$$C_l = \frac{2\pi\epsilon_0}{\ln(\rho_2/\rho_1)} \quad \frac{\text{farads}}{\text{meter}}. \tag{16}$$

5. The single slab of dielectric located between the plates of the capacitor of Fig. 182 is replaced by two slabs, each $\frac{1}{2}s$ meters thick; their respective capacitivities are ϵ_1 and ϵ_2. Show that, for the modified capacitor,

$$C = \frac{2\epsilon_1\epsilon_2 a}{(\epsilon_1 + \epsilon_2)s} \tag{17}$$

and

$$U_e = \frac{(\epsilon_1 + \epsilon_2)s}{4\epsilon_1\epsilon_2 a} q^2; \tag{18}$$

here q is the charge stored in the capacitor (that is, the charge that has been transferred—say, by a battery—from one metal plate to the other).

6. Let $r_1 \to \infty$ in (13); show that the capacitance of an isolated conducting sphere (radius r_0) in free space is

$$C = 4\pi\epsilon_0 r_0, \tag{19}$$

and compute the capacitance of the earth in farads.

7. Small charges, dq coulombs each, are brought in turn from infinity to the surface of an isolated conducting sphere (radius r_0) in free space, until the total charge on the sphere is q. On its way, each dq of charge is repelled by the charge that is already on the sphere. Show that the work required to charge the sphere is

$$U_e = \frac{q^2}{8\pi\epsilon_0 r_0}, \tag{20}$$

and verify that (20) agrees with a limiting case of (14). Note that $U_e \to \infty$ if $r_0 \to 0$.

8. The "classical radius of the electron," say r_0, is computed by equating (20) to the mass energy mc^2 of the electron and letting q be the electronic charge. Compute r_0 in fermis. (1 fermi = 10^{-15} meter.)

CHAPTER 18

Electric Energy Density and Displacement Current

The formulas of §73 for the electric energy U_e stored in simple capacitors are based on straightforward electromechanical arguments, and we derived them without making any assumptions as to *where* this energy is located. In circuit theory, the value of U_e is all that one needs to know, but in the field theory the question of the location of this energy was raised even long before there was any tangible evidence that electromagnetic energy can flow from place to place outside of conductors—as demonstrated today by the radio. The present view is the view of Faraday and Maxwell: *Electric energy is distributed with a definite density throughout every region that is pervaded by an electric field.* (We will extend this statement in §100 to include magnetic fields and magnetic energy.) Faraday pictured this energy as stored in the electric "lines of force" —which he imagined to be under tensile stress. Maxwell pictured it as the stress energy of an elastic ether. Since then, Maxwell's formula for the electric energy density has become dissociated from any ether. It can be inferred from simple examples, as in this chapter, and shown to fit in with relativistic ideas. Today electric energy is thought to be contained in any part of space (even empty space) pervaded by an electric field.

After the definition of electric energy density, this chapter continues with remarks on the differentiation of time-varying vector fields and ends with Maxwell's formula for the density of the electric "displacement current." This "current" may "flow" in empty space, devoid of any charged particles—a flow of this current in a region R does not imply either a presence or a motion of any electrons or ions in R. The displacement current is in some ways the most abstract concept mentioned in this book so far.

74. ELECTRIC ENERGY DENSITY

We write u_e for the density of electric energy and express it in joules per cubic meter. Accordingly, our formula for the electric energy (joules) stored in a region R is

$$U_e = \iiint_R u_e \, dv. \tag{1}$$

To derive an explicit expression for u_e, we assume outright that the value of u_e at a point P depends only on the electric field and the properties of the medium at P. This assumption implies that, if the medium is homogeneous and if the vector $\mathbf{E}$ at P is equal to the vector $\mathbf{E}$ at another point P', then the values of u_e at P and P' are also equal. In particular, if the medium is homogeneous and the field $\mathbf{E}$ is uniform throughout a region R, then u_e is constant throughout R. This special case arises in the space between the plates of the parallel-plate capacitor of the preceding section, and consequently the value of u_e at any point between the plates is simply the electric energy U_e stored in the capacitor divided by the volume of the capacitor. It then follows from (9^{73}) that

$$u_e = \tfrac{1}{2}ED. \tag{2}$$

Equation (2) is simple and coordinate-free; yet we have deduced it only for a rather special case, and our assumption that the value of u_e at P depends only on the conditions at P is open to question. One must, therefore, check that (2) applies in more complicated cases, say the cases described below in the exercises. We will discuss the validity of (2) in a more general context when we come to the theory of radiation.

Note that, in view of (2), Equation (1) becomes

$$U_e = \tfrac{1}{2} \iiint_R ED \, dv, \tag{3}$$

and that (2) can be written as

$$u_e = \tfrac{1}{2} \mathbf{E} \cdot \mathbf{D}, \tag{4}$$

provided that the medium is isotropic.

EXERCISES

1. Use the energy density formula (2) and the field (10^{73}) to compute by integration the energy (14^{73}) stored in a spherical capacitor.

282   Electric Energy Density and Displacement Current

2. Use (2) to compute the energy (18^{73}) stored in the parallel-plate capacitor of Exercise 5^{73}.
3. Compute in two ways the energy stored per meter of length in a cylindrical capacitor. First, use (2^{73}) and (16^{73}). Next, use (2) and (11^{60}) in the manner of Exercise 1.
4. The energy (20^{73}) of a charged sphere can be deduced from (14^{73}) or computed as in Exercise 7^{73}. Derive (20^{73}) in still another way, based on (2).

75. TIME-VARYING VECTOR FIELDS

If the magnitude of a vector field **F** or its direction (or both) change with the time t at some or at all points in space, this field is called *time-dependent* or *time-varying*. The current density **J** in Fig. 81^{24} is an example. If **F** depends on t, this fact can be emphasized by such symbols as $\mathbf{F}(x, y, z, t)$, $\mathbf{F}(\rho, \phi, z, t)$, or $\mathbf{F}(P, t)$, where P denotes a point in space.

Let the vectors $\mathbf{F}(P, t + \Delta t)$ and $\mathbf{F}(P, t)$ pertain to the same point but to two different instants of time. Their difference is then a vector pertaining to P. The limit, as $\Delta t \to 0$, of the ratio of this vector to Δt is called the partial derivative of **F** at P with respect to t; it may be written as $\partial \mathbf{F}(P, t)/\partial t$ or $\partial \mathbf{F}/\partial t$. That is,

$$\frac{\partial \mathbf{F}(P, t)}{\partial t} = \frac{\partial \mathbf{F}}{\partial t} = \lim_{\Delta t \to 0} \frac{\mathbf{F}(P, t + \Delta t) - \mathbf{F}(P, t)}{\Delta t}. \tag{1}$$

Note that $\partial \mathbf{F}/\partial t$ is itself a vector field; it depends, in general, on the time and can be pictured at any instant by a pointer map or a line map pertaining to this instant.

The components of $\partial \mathbf{F}/\partial t$ are partial time derivatives of the corresponding components of **F**. To illustrate, let

$$\mathbf{F} = F_x \mathbf{1}_x + F_y \mathbf{1}_y + F_z \mathbf{1}_z, \tag{2}$$

where $\mathbf{F} = \mathbf{F}(P, t)$, $F_x = F_x(P, t)$, and so on, and where the cartesian basevectors $\mathbf{1}_x, \mathbf{1}_y$, and $\mathbf{1}_z$ are constant vectors. When we substitute (2) into the right-hand side of (1), the coefficient of $\mathbf{1}_x$ proves to be

$$\lim_{\Delta t \to 0} \frac{F_x(P, t + \Delta t) - F_x(P, t)}{\Delta t}, \tag{3}$$

which is just the partial derivative $\partial F_x/\partial t$ evaluated at P. The coefficients of $\mathbf{1}_y$ and $\mathbf{1}_z$ are similar, and the end result is

$$\frac{\partial \mathbf{F}}{\partial t} = \frac{\partial F_x}{\partial t} \mathbf{1}_x + \frac{\partial F_y}{\partial t} \mathbf{1}_y + \frac{\partial F_z}{\partial t} \mathbf{1}_z. \tag{4}$$

We may denote the cartesian components of the field $\partial \mathbf{F}/\partial t$ by $(\partial \mathbf{F}/\partial t)_x$, and so on. It then follows from (4) that

$$\left(\frac{\partial \mathbf{F}}{\partial t}\right)_x = \frac{\partial F_x}{\partial t}, \qquad \left(\frac{\partial \mathbf{F}}{\partial t}\right)_y = \frac{\partial F_y}{\partial t}, \qquad \left(\frac{\partial \mathbf{F}}{\partial t}\right)_z = \frac{\partial F_z}{\partial t}. \tag{5}$$

The results are similar for other coordinate frames that are at rest relative to the observer. Also, in the case of a normal component relative to a stationary surface element we have

$$\left(\frac{\partial \mathbf{F}}{\partial t}\right)_n = \frac{\partial F_n}{\partial t}, \tag{6}$$

with a similar result for the tangential component relative to a stationary line element.

Under certain continuity conditions, which are usually satisfied in physical problems, we have such coordinate-free equations as

$$\frac{\partial}{\partial t} \operatorname{div} \mathbf{F} = \operatorname{div} \frac{\partial \mathbf{F}}{\partial t}. \tag{7}$$

EXERCISES

1. Write out (7) in detail for the cartesian case and state the conditions under which it is strictly true.

2. Let $\mathbf{F} = k(t - t_0)\mathbf{1}_x$, where t is the time, and k and t_0 are positive constants.
   (a) While $t < t_0$, what is the direction of $\mathbf{F}$? Is F increasing or decreasing?
   (b) When $t > t_0$ what is the direction of $\mathbf{F}$? Is F increasing or decreasing?
   (c) Compute $\partial \mathbf{F}/\partial t$ and describe its direction and magnitude.

76. ELECTRIC DISPLACEMENT CURRENT

The stage is now set for introducing the electric "displacement current." The law of conservation of charge and Maxwell's form of Coulomb's law are

$$(8^{40}) \qquad \operatorname{div} \mathbf{J} = -\frac{\partial \rho}{\partial t} \tag{1}$$

and

$$(1^{67}) \qquad \operatorname{div} \mathbf{D} = \rho, \tag{2}$$

where ρ is the volume charge density, which we often denote by q_v. If we differentiate (2) partially with respect to t and use (7^{75}), we get the equation

$$\operatorname{div} \frac{\partial \mathbf{D}}{\partial t} = \frac{\partial \rho}{\partial t}, \tag{3}$$

and consequently

$$\operatorname{div} \left(\mathbf{J} + \frac{\partial \mathbf{D}}{\partial t} \right) = 0. \tag{4}$$

The vector field $\partial \mathbf{D}/\partial t$ is called *displacement current density*; for the moment we will write it as $\mathbf{J}_d$, so that

$$\mathbf{J}_d = \frac{\partial \mathbf{D}}{\partial t}, \tag{5}$$

and we will denote *conduction current density* by $\mathbf{J}_c$ rather than by $\mathbf{J}$. Equation (4) then becomes

$$\operatorname{div} (\mathbf{J}_c + \mathbf{J}_d) = 0. \tag{6}$$

Since "divergence" is but another name for source density, Equation (6) may be stated as follows: *The source density of the field* $\mathbf{J}_c + \mathbf{J}_d$ *is zero*. In our pictorial terms, *the direction lines of the field* $\mathbf{J}_c + \mathbf{J}_d$ *have neither beginnings nor ends*.

The flux of $\mathbf{J}_d$ across a surface S is called the *displacement current* "flowing" across S. This terminology was introduced by Maxwell and is still standard; but the interpretation of displacement current used by Maxwell is no longer in fashion, because it implied what may be called an "electric elasticity" of dielectrics, even of a vacuum. Our unit for both i_c and i_d is the ampere.

Example. To illustrate the validity of (4) and its pictorial representation, we consider, ignoring edge effects, a parallel-plate capacitor that is being

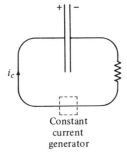

Figure 184

charged by a constant-current generator, as in Fig. 184. The two regions of special interest to us are

- the space between the plates,
- the space inside the wire leads.

Write a for the area of either plate of the capacitor, write i_c for the constant conduction current produced by the generator, and assume that the generator is turned on when $t = 0$. The surface charge density on the positive plate is then

$$q_a = \frac{i_c}{a} t. \tag{7}$$

The field **D** between the plates is directed in Fig. 184 from left to right, and its magnitude everywhere between the plates is

(4^{62})
$$D = q_a = \frac{i_c}{a} t. \tag{8}$$

Accordingly, the displacement current density at all points between the plates has the magnitude

$$J_d = \frac{\partial D}{\partial t} = \frac{i_c}{a} \tag{9}$$

and is directed from left to right. Multiplying J_d by the plate area, we find that the magnitude of the displacement current that flows between the plates from left to right is

$$i_d = J_d a = i_c. \tag{10}$$

Thus, the displacement current flowing in Fig. 184 from the positive to the negative plate in the space between the plates is equal in magnitude to the conduction current *driven by the generator* through the wire leads from the negative to the positive plate.

The plates are separated by an insulator, and, therefore, $i_c = 0$ in the space between them. The electric field in the wires does not vary with t; therefore, $\partial D/\partial t$ is zero in the wires, and so is i_d. In the space outside the plates and outside the wires we have $i_c = 0$; as long as we can ignore the fringing of the electric field at the edges of the capacitor, the electric field in this space does not vary with time, and hence $i_d = 0$.

The conduction current density $\mathbf{J}_c$ has its sources on the negative plate and its sinks on the positive plate. By contrast, the displacement current density $\mathbf{J}_d$ has its sinks on the negative plate and its sources on the positive plate. Furthermore, according to (10), the totality of sources of $\mathbf{J}_d$ on the positive plate is equal to the totality of the sinks of $\mathbf{J}_c$ on this plate. The situation on the other plate is reversed, and hence we may regard a source of $\mathbf{J}_d$ as a sink of $\mathbf{J}_c$, and vice versa. In other words, the "total" current density, $\mathbf{J}_c + \mathbf{J}_d$, has neither sources nor sinks, in agreement with (6). A line map of the field $\mathbf{J}_c + \mathbf{J}_d$ is shown in Fig. 185, which should be compared with Fig. 80[24].

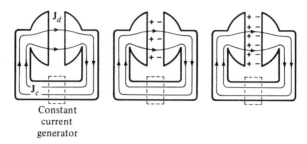

Fig. 185. A line map of a conduction current (in the constant-current generator and the leads) and of a displacement current (in the space between the plates of the capacitor). This figure is a sequel to Fig. 80[24].

Equation (6) holds even when our simplifying assumptions are not tenable, but then the problem of verifying it in detail becomes complicated. For example, if the charging conduction current i_c changes with time, the field **D** in the wires will depend on t, and a displacement current will also "flow" *in the wires*, in addition to i_c. Again, if edge effects can not be ignored, the time-varying field **D** produced between the plates by the accumulating charges will spread outside the circuit, and so will the displacement current.

This has been a convenient place to introduce the concept of displacement current, but its usefulness will show up only after we study magnetic fields.

CHAPTER 19

Cross Products and Curls

We showed in §55 that conservative electric fields exert no torques on electric pinwheels. A nonrigorous discussion of small pinwheels then led us to the tentative conclusion that, if an electric field is conservative, then each of the three quantities

$$\frac{\partial E_z}{\partial y} - \frac{\partial E_y}{\partial z}, \quad \frac{\partial E_x}{\partial z} - \frac{\partial E_z}{\partial x}, \quad \frac{\partial E_y}{\partial x} - \frac{\partial E_x}{\partial y}$$

is zero. In anticipation of definitions that were yet to come, we called these quantities the cartesian components of the "curl" of **E**.

In this chapter we state a coordinate-free definition of the curl of a general vector field **F**, and show how this definition leads to coordinate formulas for the components of a curl. We begin with remarks on torques and on "vector products" or "cross products" of vectors, which are useful in defining both torques and curls. Then come proofs of several identities involving curls, one of which implies pictorially that the direction lines of a curl of a vector field have no beginnings and no ends.

77. TORQUES AND CROSS PRODUCTS OF VECTORS

The points O and Q in Fig. 186(a), and the force **F** acting on the body shown in the figure, all lie in the xy plane, which is the plane of the page. The distance from O to the line of action of **F** is called "the lever arm of **F** about O"; in terms of the symbols used in the figure, this distance is $r \cos \alpha$.

The magnitude, say T, of the torque produced by **F** about O is, by definition, the magnitude of **F** times the lever arm; accordingly, in Fig. 186(a) we have $T = rF \cos \alpha$. The vector **r** pertaining to Q is shown in Fig. 186(b), which leads to the formula

$$T = rF \sin \gamma, \tag{1}$$

where γ is the angle between **r** and **F**. Note that T is *not* the dot product of **r** and **F**, which is $rF \cos \gamma$.

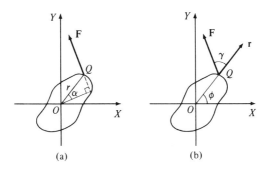

Fig. 186. In planar mechanics, the torque applied by the force **F** about the point O is the product of F and the lever arm, equal to $r \sin \alpha$.

In three-dimensional mechanics the torque produced by a force **F** about a point P, such as the origin O in Fig. 186, is a vector, say **T**, defined as follows: T is F times the lever arm; **T** lies along a line passing through P and perpendicular to both **F** and the lever arm; the direction of **T** along this line is inferred by the right-hand rule from the sense of the rotation about this line that the force **F** would cause. In Fig. 186, for example, the torque about O points toward the reader, so

$$T_x = 0, \quad T_y = 0, \quad T_z = rF \sin \gamma. \tag{2}$$

Vectors whose direction is perpendicular to two given vectors and whose magnitude is proportional to the sine of the angle between them arise often enough to warrant a special name and notation, which we will now describe. Torques are defined in this notation in Equation (18) below.

Cross products. The vector called the "cross product" of **A** and **B** is denoted by the symbol **A** × **B**, which is read "**A** cross **B**." By definition, the magnitude of this vector is $AB \sin \gamma$, where γ is the angle between **A** and **B**; that is,

$$|\mathbf{A} \times \mathbf{B}| = AB \sin \gamma. \tag{3}$$

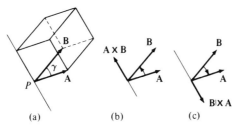

Fig. 187. Two vectors, **A** and **B**, and their two cross products, **A** × **B** and **B** × **A**.

Next consider Fig. 187(a), where the thin line passing through P is perpendicular to both **A** and **B**. (The parallelepiped is drawn to help the perspective.) If a right-handed screw is placed along this line and the vector **A** is turned about P toward **B**, turning the screw with it, the screw will advance *upward* along this line. The direction of the vector **A** × **B** is defined to be the direction of the advance of the screw, as shown in Fig. 187(b).

To identify the vector **B** × **A**, we interchange in the preceding paragraph the letters **A** and **B**, as well as A and B, and get the equation

$$|\mathbf{B} \times \mathbf{A}| = BA \sin \gamma. \tag{4}$$

It follows that the magnitudes of the vectors **B** × **A** and **A** × **B** are the same. However, when **B** is turned toward **A** in Fig. 187, and turns the imagined screw with it, the screw will advance *downward* along the thin line, and therefore the vector **B** × **A** will point as in Fig. 187(c). Consequently,

$$\mathbf{B} \times \mathbf{A} = -\mathbf{A} \times \mathbf{B}. \tag{5}$$

Since **B** × **A** is not equal to **A** × **B**, cross multiplication is said to be *noncommutative*. One can show by a geometrical construction that this kind of multiplication is, nevertheless, distributive; for example,

$$(\mathbf{A} + \mathbf{B}) \times (\mathbf{C} + \mathbf{D}) = \mathbf{A} \times \mathbf{C} + \mathbf{A} \times \mathbf{D} + \mathbf{B} \times \mathbf{C} + \mathbf{B} \times \mathbf{D}. \tag{6}$$

When we are given the left-hand side of (6), however, we must be careful to write the **A**'s and **B**'s at the left of the **C**'s and **D**'s on the right-hand side.

The respective alternative names for dot products and cross products are "scalar products" and "vector products." We recall that

(6^{10}) $\qquad$ if $\quad \mathbf{A} \cdot \mathbf{B} = 0, \quad$ then $\quad \gamma = 90°,$ $\qquad$ (7)

and conversely. On the other hand, according to (3),

$$\text{if} \quad \mathbf{A} \times \mathbf{B} = 0, \quad \text{then} \quad \gamma = 0 \text{ or } 180°, \tag{8}$$

and conversely. In particular, the equation

$$\mathbf{A} \times \mathbf{A} = 0 \tag{9}$$

should be contrasted with the equation

(7[10]) $$\mathbf{A} \cdot \mathbf{A} = A^2. \tag{10}$$

The cross product of two *vector fields*, say **F** and **G**, is itself a vector field. It is denoted by $\mathbf{F} \times \mathbf{G}$ and is constructed by computing at every point P the cross product of the vectors **F** and **G** pertaining to P.

Cross products in terms of components. As seen from Fig. 58[16] and from the coordinate-free definition of cross products given above, the "cross" multiplication table for the cartesian base-vectors is

$$\mathbf{1}_x \times \mathbf{1}_x = 0, \quad \mathbf{1}_y \times \mathbf{1}_x = -\mathbf{1}_z, \quad \mathbf{1}_z \times \mathbf{1}_x = \mathbf{1}_y, \tag{11a}$$

$$\mathbf{1}_x \times \mathbf{1}_y = \mathbf{1}_z, \quad \mathbf{1}_y \times \mathbf{1}_y = 0, \quad \mathbf{1}_z \times \mathbf{1}_y = -\mathbf{1}_x, \tag{11b}$$

$$\mathbf{1}_x \times \mathbf{1}_z = -\mathbf{1}_y, \quad \mathbf{1}_y \times \mathbf{1}_z = \mathbf{1}_x, \quad \mathbf{1}_z \times \mathbf{1}_z = 0. \tag{11c}$$

When we write

$$\mathbf{A} \times \mathbf{B} = (A_x \mathbf{1}_x + A_y \mathbf{1}_y + A_z \mathbf{1}_z) \times (B_x \mathbf{1}_x + B_y \mathbf{1}_y + B_z \mathbf{1}_z), \tag{12}$$

recall (6), and open the parentheses, we get a sum of nine terms of the form $A_x B_x (\mathbf{1}_x \times \mathbf{1}_x)$, $A_x B_y (\mathbf{1}_x \times \mathbf{1}_y)$, and so on. Three of them vanish, and the result is

$$\begin{aligned}\mathbf{A} \times \mathbf{B} = \mathbf{1}_x (A_y B_z - A_z B_y) \\ + \mathbf{1}_y (A_z B_x - A_x B_z) \\ + \mathbf{1}_z (A_x B_y - A_y B_x).\end{aligned} \tag{13}$$

Note that, in each of the three parts of the right-hand side of (13), the letters x, y, and z first appear in their cyclic order. According to (13), the formulas for the cartesian components of the vector $\mathbf{A} \times \mathbf{B}$ are

$$(\mathbf{A} \times \mathbf{B})_x = A_y B_z - A_z B_y, \tag{14}$$

$$(\mathbf{A} \times \mathbf{B})_y = A_z B_x - A_x B_z, \tag{15}$$

$$(\mathbf{A} \times \mathbf{B})_z = A_x B_y - A_y B_x. \tag{16}$$

EXERCISES

1. Show that if $\mathbf{A} = 2\mathbf{1}_x + 2\mathbf{1}_y + 3\mathbf{1}_z$ and $\mathbf{B} = 4\mathbf{1}_x + 5\mathbf{1}_y + 6\mathbf{1}_z$, then $\mathbf{A} \times \mathbf{B} = -3\mathbf{1}_x + 2\mathbf{1}_z$.

2. The vector $\mathbf{A} \times \mathbf{B}$ is perpendicular to both $\mathbf{A}$ and $\mathbf{B}$; therefore, $(\mathbf{A} \times \mathbf{B}) \cdot \mathbf{A} = 0$ and $(\mathbf{A} \times \mathbf{B}) \cdot \mathbf{B} = 0$. Verify these equations, using cartesian components.

3. Since $\mathbf{A} \times \mathbf{B} = -\mathbf{B} \times \mathbf{A}$, the sine of the angle between the vectors $\mathbf{A} \times \mathbf{B}$ and $\mathbf{B} \times \mathbf{A}$ is zero, and, therefore, $(\mathbf{A} \times \mathbf{B}) \times (\mathbf{B} \times \mathbf{A}) = 0$. Verify this equation, using cartesian components.

4. Describe the direction of the vector $\mathbf{A} \times \mathbf{B}$ in terms of the thumb and fingers of the right hand, without referring to a right-handed screw.

5. Two nonzero vectors, say $\mathbf{A}$ and $\mathbf{B}$, are usually called "parallel" if their directions are either the same or opposite. In more precise terms, $\mathbf{A}$ and $\mathbf{B}$ are *parallel* if they have the same direction, and *antiparallel* if they have opposite directions. Verify that $\mathbf{A}$ and $\mathbf{B}$ are parallel in the latter sense if and only if $\mathbf{A} \cdot \mathbf{B} > 0$ and $\mathbf{A} \times \mathbf{B} = 0$. What conditions do $\mathbf{A}$ and $\mathbf{B}$ satisfy if they are antiparallel?

6. Write in terms of cartesian components the formula for all vectors that are perpendicular to the vector $\mathbf{1}_x + \mathbf{1}_z$ and also to the vector $\mathbf{1}_y + \mathbf{1}_z$.

7. Write a cross multiplication table for (a) the vectors $\mathbf{1}_\rho$, $\mathbf{1}_\phi$, and $\mathbf{1}_z$; (b) the vectors $\mathbf{1}_r$, $\mathbf{1}_\theta$, and $\mathbf{1}_\phi$.

8. Show that at the point $P(r, \theta, \phi)$ we have $\mathbf{1}_z \times \mathbf{1}_r = \mathbf{1}_\phi \sin \theta$.

9. Verify the determinantal formula

$$\mathbf{A} \times \mathbf{B} = \begin{vmatrix} \mathbf{1}_x & \mathbf{1}_y & \mathbf{1}_z \\ A_x & A_y & A_z \\ B_x & B_y & B_z \end{vmatrix}. \tag{17}$$

10. The general formula for the torque produced about the origin O by a force applied at a point P is

$$\mathbf{T} = \mathbf{r} \times \mathbf{F}, \tag{18}$$

where the vector $\mathbf{r}$ pertains to P. Verify that (18) agrees with the definition of torques given earlier in this section.

11. Use (18) and (13) to verify (2).

12. The point of application of the force in Exercise 10 is moved through the distance s in the direction of the force. Show that $\mathbf{r}$ becomes replaced by $\mathbf{r} + (s/F)\mathbf{F}$, but $\mathbf{T}$ does not change.

13. The forces $\mathbf{F}$ and $-\mathbf{F}$ are applied, respectively, at the points P_1 and P_2 of a body. Use (18) to show that the torque produced by this couple about any point P does not depend on the position of P relative to P_1 and P_2.

14. The vectors $\mathbf{A}$, $\mathbf{B}$, and $\mathbf{C}$ have a common origin and form three edges of a parallelepiped. Show that the products (41^A), called "triple scalar products," give the volume of this parallelepiped, except perhaps for sign. Show also that, in terms of cartesian components,

$$\mathbf{A} \cdot \mathbf{B} \times \mathbf{C} = \begin{vmatrix} A_x & A_y & A_z \\ B_x & B_y & B_z \\ C_x & C_y & C_z \end{vmatrix} = \mathbf{A} \times \mathbf{B} \cdot \mathbf{C}. \qquad (19)$$

Here the symbol $\mathbf{A} \cdot \mathbf{B} \times \mathbf{C}$ stands for $\mathbf{A} \cdot (\mathbf{B} \times \mathbf{C})$, and $\mathbf{A} \times \mathbf{B} \cdot \mathbf{C}$ stands for $(\mathbf{A} \times \mathbf{B}) \cdot \mathbf{C}$.

15. Verify the "*bac* minus *cab*" formula (42^A).

16. Recall Exercise (4^{42}) and show that the torque applied to an electric dipole by a uniform electric field can be written as $\mathbf{T} = \mathbf{p}_e \times \mathbf{E}$.

17. Sometimes a vector $\mathbf{A}$ must be resolved into two components, one parallel (or antiparallel) and the other perpendicular to a given unit vector, say $\mathbf{1}$. Verify the following formula for computing these components:

$$\mathbf{A} = \mathbf{1}(\mathbf{A} \cdot \mathbf{1}) + \mathbf{1} \times (\mathbf{A} \times \mathbf{1}). \qquad (20)$$

18. A vector $\mathbf{X}$ satisfies the equations $\mathbf{A} \times \mathbf{X} = \mathbf{B}$ and $\mathbf{C} \times \mathbf{X} = \mathbf{D}$, where $\mathbf{A}$, $\mathbf{B}$, $\mathbf{C}$, and $\mathbf{D}$ are known nonvanishing vectors. What information does *one* of these equations provide about the magnitude and direction of $\mathbf{X}$? Under what conditions do the *two* equations determine $\mathbf{X}$ completely?

19. The boundary of the surface S in Fig. 188 is a parallelogram, whose area has the magnitude da and whose edge-vectors $d\mathbf{s}$ and $d\mathbf{l}$ are directed as shown. Let the unit vector

$$\mathbf{1}_n = \frac{d\mathbf{l} \times d\mathbf{s}}{da} \qquad (21)$$

indicate the positive normal to S, so that the point P sees the front of S. Draw the unit vector $\mathbf{1}_r$ as shown, use a result of Exercise 14, and verify

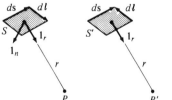

Figure 188

that, if the distance r is large enough compared to $d\mathbf{l}$ and $d\mathbf{s}$, then the solid angle, say $d\Omega$, subtended by S at P can be written in the following equivalent forms:

$$d\Omega = -\frac{\mathbf{1}_r \cdot \mathbf{1}_n \, da}{r^2} = -\frac{\mathbf{1}_r \cdot d\mathbf{l} \times d\mathbf{s}}{r^2}$$
$$= -\left(\frac{\mathbf{1}_r}{r^2} \times d\mathbf{l}\right) \cdot d\mathbf{s}. \qquad (22)$$

Show that (22) gives also the solid angle subtended at the point P' in Fig. 188 by the surface S'. In this case the direction of dl is reversed, so that, according to (21), P' sees the back of S'.

78. THE CURL OF A VECTOR FIELD

The coordinate-free definition of the *divergence* of a vector field **F** at a point P can be written as (1^{29}) or as

$$\text{div } \mathbf{F} = \lim_{v \to 0} \frac{1}{v} \left(\iint_{(S)} \mathbf{1}_n \cdot \mathbf{F} \, da \right). \tag{1}$$

(44^A)

Here (S) is a closed surface bounding a region R that includes the point P and has the volume v.

The coordinate-free definition of the *curl* of a vector field **F** at a point $\dot{P}$ is

$$\text{curl } \mathbf{F} = \lim_{v \to 0} \frac{1}{v} \left(\iint_{(S)} \mathbf{1}_n \times \mathbf{F} \, da \right), \tag{2}$$

where (S) and v have the same meanings as in (1). However, the integrand in (2) involves a cross product rather than a dot product, and therefore—in contrast to the divergence—*the curl of a vector field is itself a vector field* (Exercise 1).

Coordinate formulas. We will now express the curl of **F** in terms of cartesian coordinates. The procedure is similar to that used in §30 for the divergence, and we need not describe it in as much detail. We write

$$\mathbf{F} = F_x \mathbf{1}_x + F_y \mathbf{1}_y + F_z \mathbf{1}_z, \tag{3}$$

consider the brick-shaped region R centered on P as in Fig. 189, confine ourselves to the midpoints of the six faces of R—such as Q—and finally let each

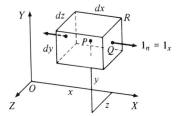

Figure 189

of the edges dx, dy, and dz tend to zero. On the right-hand face of R we have $\mathbf{1}_n = \mathbf{1}_x$, so that, in view of ($13^{77}$),

$$\mathbf{1}_n \times \mathbf{F}(Q) = \mathbf{1}_x \times \mathbf{F}(Q) = \mathbf{1}_z F_y(Q) - \mathbf{1}_y F_z(Q). \tag{4}$$

Next we let

$$F_y(Q) = F_y(P) + \frac{1}{2}\frac{\partial F_y}{\partial x}dx + \ldots, \tag{5}$$

where the derivative pertains to P. When we make a similar substitution for $F_z(Q)$ in (4), we find that, in the limit,

$$\iint_{\substack{\text{right}\\\text{face}}} \mathbf{1}_n \times \mathbf{F}\, da \to \mathbf{1}_z\left(F_y(P) + \frac{1}{2}\frac{\partial F_y}{\partial x}dx\right) dy\, dz$$

$$- \mathbf{1}_y\left(F_z(P) + \frac{1}{2}\frac{\partial F_z}{\partial x}dx\right) dy\, dz. \tag{6}$$

To derive the corresponding formula for the contribution of the left-hand face of R, we let $\mathbf{1}_n = -\mathbf{1}_x$ and replace $\frac{1}{2}dx$ by $-\frac{1}{2}dx$. The combined contribution of the right-hand and left-hand faces then proves to be

$$\left(\mathbf{1}_z \frac{\partial F_y}{\partial x} - \mathbf{1}_y \frac{\partial F_z}{\partial x}\right)v, \tag{7}$$

where v is the volume of R, equal to $dx\, dy\, dz$.

When we treat the remaining four faces in a similar way and substitute the totals into (2), the factor v cancels out and (2) turns into the following very useful formula:

$$\operatorname{curl} \mathbf{F} = \mathbf{1}_x\left(\frac{\partial}{\partial y}F_z - \frac{\partial}{\partial z}F_y\right) + \mathbf{1}_y\left(\frac{\partial}{\partial z}F_x - \frac{\partial}{\partial x}F_z\right) + \mathbf{1}_z\left(\frac{\partial}{\partial x}F_y - \frac{\partial}{\partial y}F_x\right). \tag{8}$$

Note that (8) resembles the formula (13^{77}) for a cross product. In cylindrical coordinates,

$$\operatorname{curl} \mathbf{F} = \mathbf{1}_\rho\left(\frac{1}{\rho}\frac{\partial F_z}{\partial \phi} - \frac{\partial F_\phi}{\partial z}\right) + \mathbf{1}_\phi\left(\frac{\partial F_\rho}{\partial z} - \frac{\partial F_z}{\partial \rho}\right) + \mathbf{1}_z\left(\frac{1}{\rho}\frac{\partial}{\partial \rho}(\rho F_\phi) - \frac{1}{\rho}\frac{\partial F_\rho}{\partial \phi}\right). \tag{9}$$

In spherical coordinates we have (68^A).

EXERCISES

1. Show that $\operatorname{curl}(\mathbf{F} + \mathbf{G}) = \operatorname{curl} \mathbf{F} + \operatorname{curl} \mathbf{G}$, and that the parallelogram rule for addition is applicable to curls.

The Curl of a Vector Field

2. Sketch the respective line maps of the three fields $-y\mathbf{1}_x$, $x\mathbf{1}_y$, and $\rho\mathbf{1}_\phi$, which are related as follows: $\rho\mathbf{1}_\phi = -y\mathbf{1}_x + x\mathbf{1}_y$. What relation among the *curls* of these fields do the maps suggest? Compute the curls of $-y\mathbf{1}_x$ and $x\mathbf{1}_y$, using (8); compute the curl of $\rho\mathbf{1}_\phi$, using (9), and verify formally the relation among the three curls.

3. Write curl_z for the z component of a curl; show that

$$\text{if } n > -1, \quad \text{curl}_z(\rho^n \mathbf{1}_\phi) > 0, \tag{10}$$

$$\text{if } n < -1, \quad \text{curl}_z(\rho^n \mathbf{1}_\phi) < 0, \tag{11}$$

and note that these relationships are consistent with Fig. 140[55]. Show also that

$$\text{if } n = -1, \quad \text{curl}(\rho^n \mathbf{1}_\phi) = 0, \quad \text{for } \rho > 0. \tag{12}$$

Equation (12) is generalized in Exercise 5[81].

4. Show that the curls of radial fields of the form $f(\rho)\mathbf{1}_\rho$ and $f(r)\mathbf{1}_r$ are zero.

79. DIRECTION LINES OF CURLS

An important identity involving the curl of a field $\mathbf{F}$ is div (curl $\mathbf{F}$) = 0, which can be written without ambiguity as

(75[A]) $$\text{div curl } \mathbf{F} = 0. \tag{1}$$

To prove (1), we write

(49[A]) $$\text{div } \mathbf{G} = \frac{\partial}{\partial x} G_x + \frac{\partial}{\partial y} G_y + \frac{\partial}{\partial z} G_z, \tag{2}$$

and then take $\mathbf{G}$ to be the curl of $\mathbf{F}$. When we put into (2) the cartesian components of curl $\mathbf{F}$, given in (8[78]), we get

$$\text{div curl } \mathbf{F} = \frac{\partial}{\partial x}\left(\frac{\partial}{\partial y} F_z - \frac{\partial}{\partial z} F_y\right) + \frac{\partial}{\partial y}\left(\frac{\partial}{\partial z} F_x - \frac{\partial}{\partial x} F_z\right) + \frac{\partial}{\partial z}\left(\frac{\partial}{\partial x} F_y - \frac{\partial}{\partial y} F_x\right). \tag{3}$$

Under continuity conditions that are usually satisfied in physical problems, the order of partial differentiation of a scalar field can be reversed, and such equations as

(22[5]) $$\frac{\partial}{\partial x}\frac{\partial}{\partial y} f = \frac{\partial}{\partial y}\frac{\partial}{\partial x} f \tag{4}$$

are valid. The six derivatives in (3) then cancel in pairs, and we arrive at (1).

The fact that div curl **F** = 0 implies a striking pictorial feature of the curl of a vector field: *The direction lines of a curl have no beginnings or ends.*

The field **E** of an electric dipole can be pictured, as in Fig. 61[17], by lines that begin on the positive charge and end on the negative charge. Can this field be the curl of another vector field? In other words, if **E** is the field pictured in Fig. 61[17], is there a vector field, say **G**, such that **E** = curl **G**? The answer is no, because the direction lines in Fig. 61[71] do have beginnings and ends. A line map gives only approximate information about a field, but in this example it leads to a clear-cut conclusion.

The fact that div curl **F** = 0 suggests a question: If the divergence of a vector field, say **G**, is zero, does it follow that **G** is the curl of some other vector field, say **F**? The answer is yes. That is, if

$$\text{div } \mathbf{G} = 0, \tag{5}$$

then there are fields, say **F**, such that

$$\mathbf{G} = \text{curl } \mathbf{F}. \tag{6}$$

This theorem, which we take for granted, is listed at the end of Appendix A; but note Exercise 4.

EXERCISES

1. Compute the explicit expressions for curl $(z\mathbf{1}_y)$, curl $(x^2\mathbf{1}_y)$, and curl $(x^3\mathbf{1}_y)$, and describe in words the field lines of these curls.

2. Figures 62[17], 68[18], 72[19], 80[24], 81[24], and 185[76] describe certain vector fields. Decide on pictorial grounds which of these fields *cannot* be the curls of some other vector fields.

3. Use formulas expressed in cylindrical coordinates to verify that div curl **F** = 0.

4. If div **G** = 0, the field **F** in (6) is not determined completely. Describe the arbitrary features of **F**.

80. NORMAL COMPONENTS OF CURLS

Suppose that a point P lies on a simple surface S fixed in space and write "normal component" for the component of a vector field that is normal to S at P. The normal component of the field curl **F**, which we denote here by curl$_n$ **F**, can be expressed in a coordinate-free way as follows:

$$\text{curl}_n \mathbf{F} = \lim_{a \to 0} \frac{1}{a} \oint_C F_t \, dl. \tag{1}$$

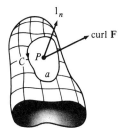

Figure 190

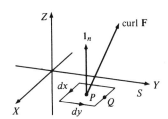

Fig. 191 Computation of the component of curl F normal to a surface element lying in the *xy* plane.

Here C is any simple contour lying on S and surrounding the point P, as in Fig. 190; the sense of C is related by the right-hand rule to the direction of the unit vector $\mathbf{1}_n$, which marks the positive normal to S at P. The symbol F_t denotes the tangential component of the field $\mathbf{F}$ relative to C, and a is the area of the portion of S bounded by C. The condition $a \to 0$ implies, among other mathematical conditions, that every point on C approaches the point P.

To verify (1), we begin with the special case when S is the xy plane, and take C to be the rectangular contour shown in Fig. 191. The integral in (1) then consists of four parts, pertaining to the four straight portions of C. On the part that includes the point Q we have $F_t(Q) = -F_x(Q)$, so that, to the first order,

$$F_t(Q) = -\left(F_x(P) + \frac{1}{2}\frac{\partial F_x}{\partial y}\,dy\right). \tag{2}$$

Consequently, if dx and dy are small enough, the integral over this portion of C is

$$-\left(F_x(P) + \frac{1}{2}\frac{\partial F_x}{\partial y}\,dy\right) dx. \tag{3}$$

When the three remaining portions of the integral are evaluated in a similar way, the value of the entire integral proves to be

$$\left(\frac{\partial}{\partial x}F_y - \frac{\partial}{\partial y}F_x\right)a, \tag{4}$$

where $a = dx\,dy$.

Now, for the orientations shown in Fig. 191, a normal component relative to S is the z component. Therefore, when we write $\text{curl}_z\,\mathbf{F}$ for $\text{curl}_n\,\mathbf{F}$ and substitute (4) for the integral in (1), we get

$$\operatorname{curl}_z \mathbf{F} = \frac{\partial}{\partial x} F_y - \frac{\partial}{\partial y} F_x. \tag{5}$$

Since (5) agrees with (8^{78}), we have verified (1) for the special case when the positive normal to S points along the z axis. The agreement holds also,

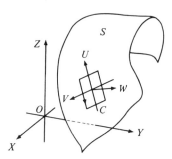

Fig. 192. The origin of the cartesian frame UVW lies on the surface S. The U and V axes are tangent to S.

however, for any other orientation of the portion of S enclosed by C. For example, if C is located as in Fig. 192, we would simply adopt the UVW frame as the XYZ frame, and proceed as before.

EXERCISE

1. Outline in detail the steps from (3) to (4).

81. STOKES'S THEOREM

Let C be a signed contour and S a surface spanning C. Given a field $\mathbf{F}$, we then have

$$(4^{22}) \qquad \text{flux of curl } \mathbf{F} \text{ across } S = \iint_S (\operatorname{curl} \mathbf{F})_n \, da, \tag{1}$$

where $(\operatorname{curl} \mathbf{F})_n$ is the normal component of curl $\mathbf{F}$, which we write in simpler formulas as $\operatorname{curl}_n \mathbf{F}$. Now,

$$(1^{79}) \qquad \qquad \operatorname{div} \operatorname{curl} \mathbf{F} = 0, \tag{2}$$

and therefore, according to §32, the numerical value of the flux of the curl of $\mathbf{F}$ is the same for every properly signed surface spanning C. Therefore, it should be possible to rewrite the right-hand side of (1) in terms of $\mathbf{F}$ and C rather than $\mathbf{F}$ and S. This was done by Stokes,[1] who showed that

---
[1] Sir George Gabriel Stokes (1819-1903).

(78^A)
$$\iint_S (\operatorname{curl} \mathbf{F})_n \, da = \oint_C F_t \, dl, \tag{3}$$

where S is any properly signed simple surface spanning C. With the aid of the terms "flux through a contour" and "circulation," mentioned in §§32 and 51, Equation (3) may be stated as follows: *The flux of curl* $\mathbf{F}$ *through a contour* C *is equal to the circulation of* $\mathbf{F}$ *around* C. We will now proceed to make this theorem plausible.

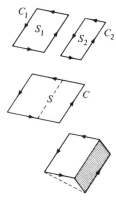

Figure 193

Write a_1 for the area of a flat surface S_1 bounded by a rectangular contour C_1, as at the top of Fig. 193. If a_1 is small enough, the exact equation (1^{80}) can be replaced by the approximate equation

$$a_1(\operatorname{curl} \mathbf{F})_n = \oint_{C_1} F_t \, dl \qquad \text{(nearly)}, \tag{4}$$

and then by

$$\iint_{S_1} (\operatorname{curl} \mathbf{F})_n \, da = \oint_{C_1} F_t \, dl \qquad \text{(nearly)}. \tag{5}$$

If a surface S_2 and its boundary C_2 satisfy similar conditions, we have

$$\iint_{S_2} (\operatorname{curl} \mathbf{F})_n \, da = \oint_{C_2} F_t \, dl \qquad \text{(nearly)}. \tag{6}$$

Now suppose that S_1 and S_2 adjoin each other as in the middle of Fig. 193—or at the bottom of this figure, where S_1 and S_2 do not lie in the same plane. Write S for the surface obtained by combining S_1 and S_2, and C for the contour bounding S. When we add the left-hand sides of (5) and (6), we then get the left-hand side of (3). Also, adding the right-hand sides of (5) and (6), we get the right-hand side of (3), because the portions of the line integrals pertaining to the common portion of the edges of S_1 and S_2 cancel each other.

Thus, Stokes's theorem holds in the limit of vanishingly small flat surface elements; if it holds for each of two adjoining elements, it holds also for their combination. Our arguments can be extended to the infinitely many elements into which a finite surface can be imagined to be subdivided, and a rigorous proof of Stokes's theorem can be formulated along these lines.

If the field **F** is the electric intensity **E**, Stokes's theorem reads

$$\iint_S (\operatorname{curl} \mathbf{E})_n \, da = \oint_C E_t \, dl. \tag{7}$$

Consequently, the definition

$$(9^{52}) \qquad \operatorname{emf}_C = \oint_C E_t \, dl \tag{8}$$

leads to the formula

$$\operatorname{emf}_C = \iint_S (\operatorname{curl} \mathbf{E})_n \, da, \tag{9}$$

which plays an important part in the formulation of Faraday's law of induction.

Direction lines and flux lines of curls. In §79 we inferred from the equation div curl **F** = 0 that the *direction lines* of a curl have neither beginnings nor ends. The flux integrals of the present section lead to a related question: Can the *flux lines* of a curl have beginnings or ends? The answer is that if the divergence of a vector field is zero, then the flux lines of this field *may* or *may not* have beginnings or ends. These possibilities were brought out in §32, where we discussed the equation div **B** = 0 and found that the flux lines of **B** may or may not have beginnings and ends; flux lines with beginnings and ends are shown in parts (d) and (e) of Fig. 91[32], which should be contrasted with part (a) of that figure.

A direction line of the field curl **F** passing through a point P gives the direction of curl **F** at P; but the density of *direction lines* that one might picture near P does not mean anything in particular. By contrast, the conditions imposed on the *flux lines* of curl **F** are much more stringent; they must have not only the correct directions but also the correct densities, because, by definition, they describe not only the direction but also the magnitude of the field curl **F**. In short, the direction lines of curls have no beginnings and no ends, but the flux lines of curls may or may not have beginnings and ends.

EXERCISES

1. Note that curl $(y\mathbf{1}_x) = -\mathbf{1}_z$ and use Stokes's theorem to solve Exercise 1^{51}. Also, solve Exercise 3^{51} in a similar way, without referring to (10^{51}).

2. Show that if curl $\mathbf{E} \neq 0$, then the electric field in question is certainly not conservative.

3. Let $\mathbf{F} = -\frac{1}{2}\rho^2 \mathbf{1}_z$ and verify Stokes's theorem, using the surface S whose back is shown in Fig. 103^{35}. [The surface integral can be checked by using (5^{35}).]

4. Show that

$$\frac{1}{\rho}\mathbf{1}_\phi = -\text{curl}\left(\ln\frac{\rho}{\rho_0}\mathbf{1}_z\right), \tag{10}$$

where $\rho > 0$, and ρ_0 is an arbitrarily chosen nonzero length.

5. Use Stokes's theorem and a convenient contour to show that the equation

$$\text{curl}\left(\frac{1}{\rho}\mathbf{1}_\phi\right) = 0 \quad \text{for } \rho > 0 \tag{11}$$

can be generalized to read

$$\text{curl}\left(\frac{1}{\rho}\mathbf{1}_\phi\right) = 2\pi\,\delta(x)\,\delta(y)\mathbf{1}_z. \tag{12}$$

6. Use cartesian coordinates and components to derive (73^A).

82. CONSERVATIVE FIELDS

Comparison of (8^{78}) with equations (12^{55}), (13^{55}), and (14^{55}) show that we used the term "curl" in §55 correctly; but we have yet to show that the condition

$$\text{curl }\mathbf{E} = 0 \tag{1}$$

implies that the electric field is conservative—a conclusion made plausible in §55 by a discussion of electric pinwheels. To make this conclusion still more plausible, we will first derive the identity

(76^A) $$\text{curl grad } f = 0. \tag{2}$$

We have

$$\text{grad } f = \mathbf{1}_x\frac{\partial f}{\partial x} + \mathbf{1}_y\frac{\partial f}{\partial y} + \mathbf{1}_z\frac{\partial f}{\partial z}, \tag{3}$$

and therefore, according to (8^{78}), the x component of the field curl grad f is

$$\frac{\partial}{\partial y}\frac{\partial}{\partial z}f - \frac{\partial}{\partial z}\frac{\partial}{\partial y}f. \tag{4}$$

Assuming that the order of differentiation can be reversed, we conclude that the x component of curl grad f is zero. The y and z components vanish for similar reasons, and therefore (2) has been verified.

Equation (2) has an important corollary: If the curl of a vector field is zero, then this field is the gradient of a scalar field. That is, if

$$\operatorname{curl} \mathbf{F} = 0, \tag{5}$$

then there are scalar fields f such that

$$\mathbf{F} = \operatorname{grad} f. \tag{6}$$

This theorem, which we take for granted, is listed at the end of Appendix A; but see Exercises 1 and 2.

Now, if an electric field is conservative, it is derivable from a scalar potential V, so

$$\mathbf{E} = -\operatorname{grad} V \tag{7}$$

and curl $\mathbf{E} = -$curl grad V. The equation curl $\mathbf{E} = 0$ then follows from (2). The fact that curl $\mathbf{E}$ vanishes if the electric field is conservative can, of course, be deduced directly from Stokes's theorem.

Conversely, if curl $\mathbf{E} = 0$, we conclude from (5) and (6) that $\mathbf{E} = \operatorname{grad} f$, where f is some scalar field. If we denote this field by $-V$, we get (7). Consequently, $\mathbf{E}$ is derivable from a scalar potential, and the electric field in question is conservative. Thus, the curl of $\mathbf{E}$ is identically zero if and only if the electric field is conservative.

What is the physical significance of $\mathbf{E}$? The answer is straightforward: The field $\mathbf{E}$ describes an electromechanical property of an electric field, namely, its force-exerting property. What is the physical significance of the curl of $\mathbf{E}$? The answer is again straightforward: The field curl $\mathbf{E}$ describes another

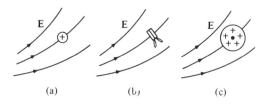

Fig. 194. Conceptual devices for exploring electric fields: a test charge for measuring $\mathbf{E}$, a two-coin probe for measuring $\mathbf{D}$, and an electric pinwheel for measuring curl $\mathbf{E}$.

electromechanical property of an electric field, namely, its ability to exert torques on symmetric charge distributions. A torque is a more complicated entity than a force, and therefore the concept of the field curl E is more complicated than the concept of the field E.

The properties of electric fields of greatest interest to us are described by the fields E, D, and curl E. The imagined devices for exploring these properties—a test charge, a two-coin probe, and an electric pinwheel—are pictured in Fig. 194, a sequel to Fig. 163[61].

EXERCISES

1. Let M and N be functions of x and y. A differential equation of the form $M\,dx + N\,dy = 0$ is then said to be *exact* if M and N satisfy the condition $\partial N/\partial x - \partial M/\partial y = 0$. Changing the symbols M and N to F_y and F_x, we may write this condition in the form $\partial F_y/\partial x - \partial F_x/\partial y = 0$, which states that the z component of the curl of a certain vector field is zero.

    If you are familiar with the theory of exact differential equations, derive (6) from (5) for the case when $F_z = 0$.

2. If curl $\mathbf{F} = 0$, the field f in (6) is not determined completely. Describe the arbitrary feature of f.

CHAPTER 20

Static Magnetic Fields in Nonmagnetic Mediums

Our attention will now shift for a time from electric to magnetic fields. Only vacuum is truly nonmagnetic, so the term "nonmagnetic mediums" in the title of this chapter stands for mediums, such as air or copper, whose magnetic properties can usually be ignored. We have already discussed almost all the necessary mathematics and can, therefore, proceed with the physics faster than before.

The existence of *isolated electric charges*, say electronic charges, is easy to demonstrate—all we need is a glass rod and a piece of silk. By contrast, although *isolated magnetic charges*, called *magnetic monopoles*, may exist, their production would require extremely high energies, and consequently they would not be involved in the applications of the theory discussed in this book, such as the propagation of radio waves from a short antenna. As we will see in §84, Maxwell's field equations do not provide for monopoles; the possibility that they may exist was, however, pointed out in 1931 by Dirac on quantum-mechanical grounds. An experimental identification of monopoles would, of course, be of great scientific significance and would, in particular, call for a generalization of Maxwell's equations.

83. PRELIMINARIES

The simplest readily available magnetic entity is a *magnetic dipole*, say a compass needle. One way of magnetizing a short iron rod is analogous to making an electret: The rod is placed in a strong magnetic field and is then tapped or stroked in order to stimulate some of the atomic magnetic moments to line up with the external magnetic field, so that the rod becomes *polarized* in a magnetic sense. Near its ends, such a rod acquires reasonably well-defined magnetic poles—a positive (north-seeking) pole near one end and a negative pole near the other. Experiment shows that the interaction between bar magnets can be interpreted in terms of interactions between their poles, which prove to exert on one another forces of the inverse-square kind, similar to the Coulomb forces between electric charges. Therefore, it is often useful to think of the poles of a magnet as accumulations of *magnetic charges*; we will sometimes call them magnetic "polarization" charges, since they are in some ways analogous to the polarization charges near the ends of an electret.

We will use the term "magnetic field" in a general way, but when precision is required we will speak of two specific vector fields, **B** and **H**, defined for nonmagnetic mediums later in this chapter. The top row of Fig. 195 pertains to the electret of Fig. 168[62]; the bottom row depicts a stubby permanent magnet, whose "poles" or "magnetic polarization charges" are assumed to be confined to the flat faces. The source density of **B** is zero everywhere; by contrast, the

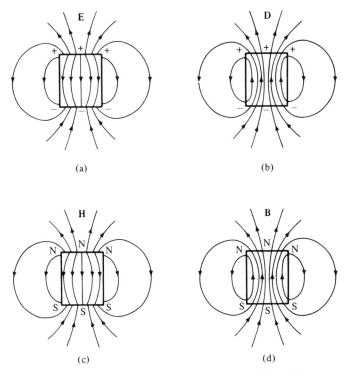

Fig. 195. Lines of **E** and **D** for an electret, and lines of **H** and **B** for a bar magnet.

lines of **H** begin on north-seeking and end on south-seeking magnetic polarization charges. In the figure, **B** is taken as equal to **H** in the vacuum or air outside the magnet; that is, the factor μ_0 in (1^{85}), given in (2^{85}), is replaced by unity.

The unit of measurement for magnetic charge and the "strength" of a pole of a magnetic dipole, used in this and in some other books, is the *weber*.[1] This choice is suggested, but not dictated, by the table on the inside front cover, which reveals a symmetry between the weber and the coulomb—see the exercise below.

The concept of magnetic charge must, of course, be used with caution. It can even be avoided altogether: Magnetic fields can be caused by electric currents, and it proves to be possible to ascribe the magnetic field of a bar magnet to currents—the "Ampèrian currents"—circulating near the surface of the bar.

Since magnetic monopoles have not been available in the laboratory, the concepts of the fields **B** and **H** have emerged from experiments very different from those that led to the fields **E** and **D**.

EXERCISE

1. Interchange the words "coulomb" and "weber" everywhere in the table on the inside front cover, whether they appear explicitly or implicitly; verify that the content of the table remains unchanged. (Note, for example, that since

    1 ampere = 1 joule/weber and 1 volt = 1 joule/coulomb,

    interchanging the coulomb and the weber implies interchanging also the volt and the ampere.)

84. MAGNETIC FLUX DENSITY

When charged particles move in regions pervaded by static magnetic fields but free from electric fields, the speeds of the particles are not affected, but their paths are, in general, curved. Experimental studies of these paths show that the force of magnetic origin acting on a charged particle at a point P can be described by the following equation involving a cross product:

$$\mathbf{F}_{\text{magnetic}} = q\mathbf{v} \times \mathbf{B}. \tag{1}$$

Here q is the electric charge on the particle, **v** is its velocity at P, and **B** is a vector associated with the magnetic field at P. We will call **B** the *magnetic flux density*; other names are listed on the inside back cover. The magnitude and direction of **B** at P can be found experimentally by investigating with the help of (1) the paths of two charged particles that pass through P, one at a time,

---

[1] Wilhelm Eduard Weber (1804–1891).

with the respective velocities v_1 and v_2, which make a suitable angle with each other. (Recall Exercise 18[77].) Equation (1) can, therefore, be taken as the *definition* of **B**, as we do in this book.

Since the force (1) is perpendicular to the velocity of the particle, we called it in §43 a "charge-deflecting" rather than a "charge-driving" force. The MKS-Giorgi unit for measuring **B** that follows from (1) is the (newton·second)/(meter·coulomb), which is equivalent to the *weber per square meter* and is often called the *tesla*, after Nikola Tesla (1857–1943). In middle latitudes, the magnitude of the horizontal component of **B** is about 1.5×10^{-5} weber/meter^2; near the equator, it is about twice as large. A conversion formula reads:

$$1 \text{ weber/meter}^2 = 1 \text{ tesla} = 10^4 \text{ gauss}.$$

Let us apply (1) to a positive ion (mass m, charge q) injected into a uniform magnetic field with a velocity **v** perpendicular to **B**, and let us ignore relativistic effects. The component of the force (1) parallel to **B** is *always* zero; in this example the component of the initial velocity of the ion parallel to **B** is also zero, and therefore the ion will remain in a plane perpendicular to **B**. The force (1) is *always* perpendicular to **v**; therefore, the speed v of the ion will remain constant; the force (1) will have the constant magnitude qvB, and the ion will move with a constant speed in a circle. The magnitude of the centripetal force required for uniform motion in a circle of radius r is mv^2/r, the magnitude of the available centripetal force is qvB, and therefore the radius of the circular path of the ion will be

$$r = \frac{mv}{qB}. \tag{2}$$

The formula for the force acting on a charged particle moving in a region pervaded by both an electric and a magnetic field is the sum of (1) and (2[43]):

$$\mathbf{F}_{\text{total}} = \mathbf{F}_{\text{electric}} + \mathbf{F}_{\text{magnetic}} = q(\mathbf{E} + \mathbf{v} \times \mathbf{B}). \tag{3}$$

This formula is named after Hendrik Antoon Lorentz (1853–1928).

Magnetic fields can be caused by permanent magnets, by electric currents (§87), and by time-varying electric fields (§95). For the moment, let us write $\mathbf{B}_i$ for the magnetic flux density produced by an electric circuit carrying a current i. In §87 we will introduce formulas for computing $\mathbf{B}_i$ and will show that

$$\text{div } \mathbf{B}_i = 0. \tag{4}$$

In Maxwell's theory, (4) is generalized to read

$$\text{div } \mathbf{B} = 0, \tag{5}$$

where **B** is the flux density of *any* magnetic field, whatever its cause. The step from (4) to (5) implies that Maxwell's theory precludes the existence of magnetic monopoles. Indeed, (5) implies that, although the *flux lines* of **B** may have breaks, as in Fig. 91[32], the *direction lines* of **B** have no beginnings or ends—which they would have on magnetic monopoles.

The flux of **B** across a surface S is called the *magnetic flux* across S; we denote it by Φ_S or simply Φ:

(5[22]) $$\Phi = \Phi_S = \iint_S B_n \, da. \qquad (6)$$

Since we express da in square meters and B in webers per square meter, our unit for measuring magnetic flux is the *weber*.

Let several surfaces, all properly signed, span the same signed contour C. In the light of §32, Equation (5) implies that the flux of **B** across each of these surfaces has the same numerical value. Therefore, we may speak of the flux of **B** across any one of these surfaces as the magnetic flux *through* the contour C or the magnetic flux *linking* the contour C, and may denote it by Φ_C.

EXERCISES

1. A proton is injected, at right angles to **B**, into a uniform magnetic field. Show that, as long as the nonrelativistic formula (2) is applicable, the period of the revolution of the proton in its circular path will not depend on its speed. (The operation of rudimentary cyclotrons is based on this fact.)

2. If a positive ion (mass m, charge q, speed v) is injected into a uniform magnetic field at the angle γ to **B**, it will move in a helix. Describe the orientation of this helix and compute its radius and pitch.

3. A positive ion moves in a *straight line* in a region pervaded by uniform electric and magnetic fields **E** and **B**. If $\mathbf{B} = B_0 \mathbf{1}_y$, and the velocity of the ion is $v \mathbf{1}_x$, what are the magnitude and direction of **E**?

85. MAGNETIC FIELD INTENSITY

The vector field called *magnetic field intensity* or *magnetic field strength* is denoted by **H**. In this book, we need **H** only for nonmagnetic mediums, but remarks on bar magnets, and diagrams such as Fig. 195 help to explain why, in addition to **B**, a second field is required for a complete description of magnetic fields in general. The sources and the sinks of **H** are taken to be the magnetic polarization charges, as in Fig. 195(c). The continuity of the lines of **B** in Fig. 196(d) emphasizes the fact that, in Maxwell's theory, div **B** = 0. Outside the magnet, the lines of **B** and **H** agree both in shape and in direction, suggesting that these fields differ only by a constant factor. In fact, as long as we restrict ourselves to vacuum, **H** can be defined by the equation

$$\mathbf{H} = \frac{1}{\mu_0} \mathbf{B}, \tag{1}$$

where **B** is the magnetic flux density and μ_0 is the *inductivity* or *magnetic permeability* of free space, given by the formula

$$\mu_0 = 4\pi \times 10^{-7} \; \frac{\text{henry}}{\text{meter}}. \tag{2}$$

We will show in §89 that (2) follows from the definition of the ampere. In general,

$$\mathbf{B} = \mu \mathbf{H}, \tag{3}$$

where μ is a complicated function of **H**; in some materials it depends on the previous history of the sample (hysteresis). The condition

$$\mu = \mu_0 \tag{4}$$

is exact for vacuum, but it also holds well enough for most practical purposes in such mediums as air and copper; in fact,

$$\mu_{\text{air}} = 1.0000004\mu_0, \qquad \mu_{\text{copper}} = .99999\mu_0. \tag{5}$$

Our unit for measuring H is the *ampere/meter*, or its equivalent, the *newton/weber* (Exercise 1).

Inside the magnet of Fig. 195, the lines of **H** and **B** run in nearly opposite directions, and hence (1) must be modified. When the microscopic viewpoint is emphasized, the field **H** at a point inside a magnet is defined by Equation (2^{104}), namely $\mathbf{H} = (\mathbf{B} - \mathbf{P}_m)/\mu_0$, where the field $\mathbf{P}_m$—the "intensity of magnetization"—is the magnetic moment of the magnetized material per unit volume at that point. This formula for **H** reduces to (1) outside the magnet; it is given in different units in Exercise 2^{104}.

An important quantity involving **H** and discussed in detail in Chapter 21 is the line integral, $\int H_t \, dl$, of the tangential component of **H** along a path. As shown on the inside back cover, this integral, called the *magnetomotive force* in the path, resembles the integral for the electromotive force in a path.

The magnetic properties of matter can be accounted for by the orbital motions of electrons and the intrinsic "spin" magnetic moments of electrons and atomic nuclei. The concept of magnetic charges has nevertheless a long history of usefulness.

EXERCISE

1. Verify that the units ampere/meter and newton/weber for **H** are consistent with (1) and (2). The unit for μ_0 in (2) will be verified in Exercise 1[89].

86. FORCES ON CURRENT-CARRYING WIRES

Straight wire. A straight portion of a thin current-carrying wire is located in vacuum in a uniform magnetic field, as in Fig. 196. The conduction electrons drift toward the left with the average speed v_d; the charge on each is

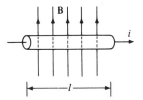

Fig. 196. A current-carrying wire located in a uniform magnetic field.

$q = -e$. According to (1[84]), the average force of magnetic origin acting on a conduction electron is directed toward the reader and has the magnitude $ev_d B$. The length and the cross-sectional area of this portion of the wire are l and A. If it contains n conduction electrons per unit volume, the magnitude of the average total force acting on all of them—and transmitted to the wire—is $nev_d lAB$. Now,

$$(11^1) \qquad v_d = \frac{i}{neA}, \qquad (1)$$

and hence the magnitude of the force is simply ilB.

If the straight segment of the wire is not perpendicular to **B**, the equation for the force of magnetic origin acting on the wire takes the form

$$\mathbf{F} = i\mathbf{l} \times \mathbf{B}; \qquad (2)$$

here the vector $\mathbf{l}$ has a magnitude equal to l and has the direction of the current i (Exercise 1).

To compute the force of magnetic origin acting on a thin *curved* current-carrying wire, we subdivide the wire into short straight segments of length dl, associate with each segment a vector $d\mathbf{l}$, write (2) as

$$d\mathbf{F} = i\,d\mathbf{l} \times \mathbf{B}, \qquad (3)$$

let $dl \to 0$, and integrate along the curved wire. The same procedure applies if the wire—whether straight or curved—lies in a nonuniform magnetic field, so that **B** changes from point to point.

Force and torque on a current loop. Next we consider the force acting on a rigid circular current-carrying loop of slender wire—a "current loop"—lying in the xy plane in a uniform field **B**, which is parallel to the zx

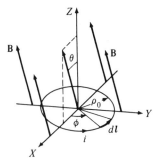

Fig. 197. A circular current loop located in the xy plane in a uniform magnetic field that makes the angle θ with the z axis.

plane and makes the angle θ with the axis of the loop, as in Fig. 197, where the battery driving the current in the loop is omitted. (The reader may have to refer to Appendix A for relations among the coordinates and the base-vectors used below.)

We have
$$\mathbf{B} = \mathbf{1}_x B \sin\theta + \mathbf{1}_z B \cos\theta, \tag{4}$$

and, from the figure,
$$d\mathbf{l} = \mathbf{1}_\phi \rho_0 \, d\phi = -\mathbf{1}_x \rho_0 \sin\phi \, d\phi + \mathbf{1}_y \rho_0 \cos\phi \, d\phi. \tag{5}$$

Therefore, (3) becomes
$$d\mathbf{F} = i\rho_0 B(\mathbf{1}_x \cos\theta \cos\phi + \mathbf{1}_y \cos\theta \sin\phi - \mathbf{1}_z \sin\theta \cos\phi) \, d\phi. \tag{6}$$

To compute the total force **F** acting on the loop, we integrate (6) with respect to ϕ once around the loop, and find that
$$\mathbf{F} = 0. \tag{7}$$

To compute the torque about the center of the loop acting on the loop as a whole, we begin with a segment of the loop of length dl, and find from (18[77]) that the torque $d\mathbf{T}$ acting on this segment about the origin is $\rho_0 \mathbf{1}_\rho \times d\mathbf{F}$; that is, in view of (6),
$$d\mathbf{T} = i\rho_0^2 B(-\mathbf{1}_x \sin\theta \cos\phi \sin\phi + \mathbf{1}_y \sin\theta \cos^2\phi) \, d\phi. \tag{8}$$

Integrating once around the loop, we get
$$\mathbf{T} = \mathbf{1}_y iaB \sin\theta, \tag{9}$$

where a is the area of an imagined flat surface spanning the loop. Thus, the torque points in the $+y$ direction and tends to tilt the loop so as to align its axis with **B**; that is, should the z axis tilt with the loop, the angle θ in Fig. 197 would decrease.

Consider the contour formed by the loop and take its positive sense to be that of the current i. According to Fig. 197, the flux of **B** through the contour is then positive and equal to $aB \cos \theta$. Therefore, we conclude from the direction of **T** given by (9) that *the torque tends to turn the loop so as to increase the flux of* **B** *through it.*

Coordinate-free formulas. The torque acting on an *electric dipole* placed in a uniform electric field is given in Exercise 16[77] as

$$\mathbf{T} = \mathbf{p}_e \times \mathbf{E}, \tag{10}$$

where $\mathbf{p}_e$ is the *electric moment* of the dipole; that is, in the notation of Exercise 4[42],

$$\mathbf{p}_e = q\mathbf{l}. \tag{11}$$

Our unit for p_e is the *coulomb · meter*.

The reader should verify (Exercise 3) that (9) can be generalized to read

$$\mathbf{T} = (i a \mathbf{1}) \times \mathbf{B} \tag{12}$$

or, which is the same thing for a current loop located in vacuum,

$$\mathbf{T} = (\mu_0 i a \mathbf{1}) \times \mathbf{H}. \tag{13}$$

In these coordinate-free formulas the symbol **1** stands for a unit vector normal to the plane of the loop and related by the right-hand rule to the current i. The similarity between these formulas and (10) suggests introducing the concept of the *magnetic moment*, say $\mathbf{p}_m$, of the current loop. Some writers define $\mathbf{p}_m$ as $ia\mathbf{1}$, others as $\mu_0 ia\mathbf{1}$. In this book we let

$$\mathbf{p}_m = \mu_0 i a \mathbf{1} \tag{14}$$

and write (13) as

$$\mathbf{T} = \mathbf{p}_m \times \mathbf{H}. \tag{15}$$

Our unit for p_m is the *weber · meter*. When $\mathbf{p}_m$ is defined by (14), the vector $ia\mathbf{1}$ is called the *area moment* of the loop.

EXERCISES

1. Derive (2).
2. Set $\theta = 0$ in (6) and express $d\mathbf{F}$ in cylindrical coordinates. What does the magnetic field tend to do to the area of the loop? Consider the cases $\theta = 90$

deg and $\theta = 180$ deg along similar lines. Correlate the results with the statement made in italics beneath Equation (9).
3. Transform (9) to the coordinate-free forms (12) and (13).
4. Show in detail that Equations (7), (9), and (15) hold for a *rectangular* current loop, and extend them to flat loops of any shape.

87. THE dB FORMULA

From the force exerted on a current-carrying wire by a given magnetic field, we now turn to a very different subject: the magnetic field produced by a given current.

Many people had experimented with magnets and electrified bodies before 1819, suspecting—but not finding—a relation between electricity and magnetism. In 1819, however, Hans Christian Oersted (1777-1851) noticed that a magnetized needle tends to turn at right angles to a current-carrying wire, and he thus became the founder of the science of electromagnetism.

Oersted's discovery led at once to several detailed studies of the new effect, and to attempts to describe it mathematically. Within a few months, Jean Baptiste Biot (1774-1862) and Félix Savart (1791-1841) reported that the magnetic field near a long straight current-carrying wire is circular, and that its strength is inversely proportional to the distance from the axis of the wire, as in Equation (7) below. At the same time Ampère began a series of brilliant experiments on the force of interaction between two circuits carrying steady currents, and inferred a formula for this force. We will not need Ampère's two-circuit formula, called *Ampère's law*; instead, we give Equation (1), which we call for short the "dB formula."[1] This formula can be used to compute, by

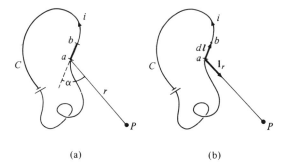

Fig. 198. The dB formula gives the contribution to **B** made by the segment ab of the current-carrying circuit C.

[1] Some authors credit a formula equivalent to (1) to Biot and Savart, some to Ampère, and some to Laplace.

integration, the field **B** produced at any point *in vacuum* by any steady "filamentary" current—say a current in a thin wire; and the results can be extended by further integration to thick current-carrying conductors. In this book we regard the $d\mathbf{B}$ formula as an assumption, justified by the fact that the results it gives after integration agree with the results of innumerable experiments.

The thin-wire circuit in Fig. 198 carries a *steady* current i; it is located in vacuum and is not necessarily flat. The short segment ab of the circuit has the length dl, which at the end of a computation is made to tend to zero. The point P is chosen arbitrarily; its distance from ab is r. We write **B** for the magnetic flux density produced at P by the entire circuit, and $d\mathbf{B}$ for the contribution to **B** made by the current in ab. The vector $d\mathbf{B}$ can be described most concisely by the following formula, which is one of the basic physical assumptions made in this book:

$$d\mathbf{B} = \frac{\mu_0}{4\pi} \frac{i dl \times \mathbf{1}_r}{r^2}; \qquad (1)$$

here dl and $\mathbf{1}_r$ are the vectors defined in Fig. 198(b), and $\mu_0 = 4\pi \times 10^{-7}$ henry/meter, as in (2^{85}). Note that the vector $d\mathbf{B}$ is perpendicular to the plane containing P and the segment ab; its direction is related by the right-hand rule to the direction of the current in ab (in Fig. 198 it points into the page); its magnitude is

$$dB = \frac{\mu_0}{4\pi} \frac{i \sin \alpha \, dl}{r^2}, \qquad (2)$$

where α is the angle defined in the figure. Note also that (2) is an *inverse-square* law so far as the distance r is concerned.

If i should change with time, the effect of a change will propagate with the speed of light and will not be "felt" at P immediately. Consequently, the validity of (1) depends not only on the rate of change of i but also on the size of the apparatus, which may range from an a. c. bridge on a laboratory table to a transcontinental telephone line. We will illustrate the effect of time-varying currents when we come in Chapter 24 to the radiation from an antenna.

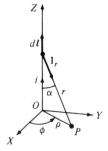

Fig. 199. Computation of the magnetic field of a straight-line current.

Steady straight-line current. A steady current i flows upward along the z axis, as in Fig. 199; we ignore the effects of the return portion of the circuit, located far away. Our task is to compute **B** at an arbitrary point P, which we put without loss of generality in the xy plane, at the distance ρ from the z axis. We then have

$$d\mathbf{l} = \mathbf{1}_z\, dz \tag{3}$$

and, in contrast to (22^A),

$$\mathbf{1}_r = \frac{x}{r}\mathbf{1}_x + \frac{y}{r}\mathbf{1}_y - \frac{z}{r}\mathbf{1}_z. \tag{4}$$

Consequently,

$$d\mathbf{l} \times \mathbf{1}_r = \left(\frac{x}{r}\mathbf{1}_y - \frac{y}{r}\mathbf{1}_x\right) dz = \frac{\rho}{r}\, dz\, \mathbf{1}_\phi, \tag{5}$$

and

$$d\mathbf{B} = \frac{\mu_0 i \rho}{4\pi} \frac{dz}{(z^2 + \rho^2)^{3/2}}\, \mathbf{1}_\phi. \tag{6}$$

Finally, we integrate (6) from $z = -\infty$ to $z = \infty$ and get

$$\mathbf{B} = \frac{\mu_0}{2\pi} \frac{i}{\rho}\, \mathbf{1}_\phi \qquad (\rho > 0). \tag{7}$$

This is the inverse first-power *law of Biot and Savart*. According to it, the lines of **B** around a long straight current-carrying wire are circles, whose sense is related by the right-hand rule to the direction of the current in the wire.

The divergence of B. Equation (1) pertains to the magnetic field produced by a filamentary circuit carrying a current i. To stress this fact, let us replace for a moment the symbol $d\mathbf{B}$ by $d\mathbf{B}_i$. Taking the divergence of both sides of (1), we then get

$$\operatorname{div}(d\mathbf{B}_i) = \frac{\mu_0 i}{4\pi} \operatorname{div}\left(d\mathbf{l} \times \frac{\mathbf{1}_r}{r^2}\right). \tag{8}$$

Now, in view of (73^A),

$$\operatorname{div}\left(d\mathbf{l} \times \frac{\mathbf{1}_r}{r^2}\right) = \frac{\mathbf{1}_r}{r^2} \cdot \operatorname{curl}(d\mathbf{l}) - d\mathbf{l} \cdot \operatorname{curl}\left(\frac{\mathbf{1}_r}{r^2}\right). \tag{9}$$

But $\operatorname{curl}(d\mathbf{l}) = 0$ because $d\mathbf{l}$ is a fixed constant vector (or a uniform vector field); also, $\operatorname{curl}(\mathbf{1}_r/r^2) = 0$, because the curl of any field of the form $f(r)\mathbf{1}_r$ is zero. It then follows from (8) that $\operatorname{div}(d\mathbf{B}_i) = 0$. Since the total field $\mathbf{B}_i$ is a superposition of the various $d\mathbf{B}_i$'s, we conclude that

$$\operatorname{div} \mathbf{B}_i = 0. \tag{10}$$

As we remarked in §84, this result is generalized in Maxwell's theory to

$$\text{div } \mathbf{B} = 0. \tag{11}$$

One can, of course, assume outright that there are no magnetic monopoles and can postulate (11) without first showing with the help of the $d\mathbf{B}$ formula that (10) holds for the magnetic fields of filamentary currents. Our derivation of (10) would then amount to proving that the $d\mathbf{B}$ formula is consistent with Maxwell's equation (11). One can derive the $d\mathbf{B}$ formula from Maxwell's equations, but we will not do that in this book.

88. THE $d\mathbf{H}$ FORMULA

The appearance of some formulas becomes simpler if we focus our attention on $\mathbf{H}$ rather than $\mathbf{B}$. Therefore, we introduce the "$d\mathbf{H}$ formula" that follows from (1^{85}) and (1^{87}):

$$d\mathbf{H} = \frac{i}{4\pi} \frac{d\mathbf{l} \times \mathbf{1}_r}{r^2}. \tag{1}$$

The Biot-Savart law (7^{87}) then reads

$$\mathbf{H} = \frac{i}{2\pi} \frac{1}{\rho} \mathbf{1}_\phi \quad (\rho > 0). \tag{2}$$

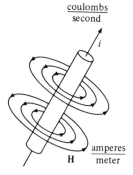

Fig. 200. The right-hand relation between a current i and its field $\mathbf{H}$.

The right-hand relation between $\mathbf{H}$ and i in (2) is pictured in Fig. 200. In the general case, let C be the contour formed by a filamentary current, as in Fig. 198^{87}. The total field $\mathbf{H}$ at the point P in the figure is then given by the line integral of (1) taken around C; that is,

$$\mathbf{H} = \int_C d\mathbf{H} = -\frac{i}{4\pi} \int_C \left(\frac{\mathbf{1}_r}{r^2}\right) \times d\mathbf{l}. \tag{3}$$

We will now illustrate how the formulas given above for filamentary currents can be used for currents in extended conductors.

Current in a thin sheet. A thin copper sheet (the dark line in Fig. 201) is parallel to the zx plane and carries a steady and uniformly distributed

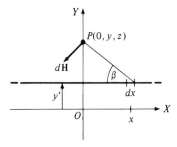

Fig. 201. A current sheet (thick line) parallel to the zx plane. Its y coordinate is y'. The current flows toward the reader.

current toward the reader; the lateral dimensions of this sheet are infinite. The *linear density* of the current is J^* amperes per meter of width of the sheet. The intensity of the magnetic field produced by this current at a point $P(x, y, z)$ does not depend on x and z; therefore, we may place P in the yz plane.

The current in the strip of width dx in the figure is $J^* dx$ amperes. For a sufficiently small dx, it is a straight-line current flowing at the distance $[x^2 + (y - y')^2]^{1/2}$ from P. Therefore, according to (2), the magnitude of the contribution made by this current to **H** at P is

$$dH = \frac{J^*}{2\pi} \frac{dx}{\sqrt{x^2 + (y - y')^2}}; \tag{4}$$

the direction of $d\mathbf{H}$ is shown in the figure. The $d\mathbf{H}$ contributed by another strip of width dx, whose coordinate in the figure is $-x$ rather than x, will cancel the y component of the $d\mathbf{H}$ shown in the figure, and will double its x component. Consequently, the resultant of these two $d\mathbf{H}$'s will point leftward in the figure, and its x component will be $-2 \sin \beta$ times the expression (4). Since $\sin \beta = (y - y')/[x^2 + (y - y')^2]^{1/2}$, integration over the entire current-carrying plane yields

$$H_x = -2\frac{J^*}{2\pi} \int_0^\infty \frac{(y - y') dx}{x^2 + (y - y')^2} = -\frac{J^*}{\pi} \left[\tan^{-1} \frac{x}{(y - y')} \right]_0^\infty. \tag{5}$$

Using the principal values of the arctangents, we finally get

$$\mathbf{H} = \begin{cases} -\tfrac{1}{2} J^* \mathbf{1}_x & \text{if } y > y' \\ +\tfrac{1}{2} J^* \mathbf{1}_x & \text{if } y < y'. \end{cases} \tag{6}$$

Consequently, as long as P does not lie *in* the current-carrying plane, the value of H does not depend on the distance from P to the plane, but the direction of **H** reverses if we cross this plane. The field **H** is pictured in perspective in Fig. 241[111].

Current in a thick plate. The copper plate in Fig. 202 is parallel to the zx plane; its thickness is $2s$; its lateral dimensions are infinite. The density of the current flowing in it is

$$\mathbf{J} = J_0 \mathbf{1}_z, \qquad (7)$$

where J_0 is a constant (amperes/meter^2). The thin slice of this plate of thickness dy', shown in the figure, is similar to the thin sheet of current in Fig. 201; its

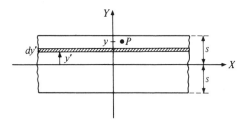

Fig. 202. A thick current-carrying plate.

contribution to the magnetic flux density at P is given by (6), provided that we replace J^* by $J_0\, dy'$.

Next, imagine that the entire plate is subdivided into similar slices, each of thickness dy'. According to (6), each slice located down-the-page from P will make the same contribution to **H** at P, directed leftward; since the total thickness of these slices is $s + y$, their total contribution will be $-\tfrac{1}{2}J_0(s + y)\mathbf{1}_x$. The total thickness of the layers located up-the-page from P is $s - y$, and hence their total contribution is $\tfrac{1}{2}J_0(s - y)\mathbf{1}_x$. As long as P lies *in* the plate, the field **H** at P is the sum of these contributions; that is

$$\mathbf{H}(x, y, z) = -J_0 y \mathbf{1}_x, \qquad -s < y < s. \qquad (8)$$

TABLE 7. Current Density and Magnetic Field Intensity at a Point P in Fig. 202

Location of $P(x,y,z)$	**J**	**H**
$\infty > y \geq s$	0	$-sJ_0\mathbf{1}_x$
$s \geq y \geq 0$	$J_0\mathbf{1}_z$	$-\lvert y\rvert J_0\mathbf{1}_x$
$0 \geq y \geq -s$	$J_0\mathbf{1}_z$	$+\lvert y\rvert J_0\mathbf{1}_x$
$-s \geq y > -\infty$	0	$+sJ_0\mathbf{1}_x$

A line map of a field of type (8) is shown in Fig. 76[21] with a reversed sign. A more detailed description of H appears in Table 7.

EXERCISES

1. Verify Table 7 and show that, at least in this example,

$$\operatorname{curl} \mathbf{H} = \mathbf{J}, \qquad -\infty < y < \infty. \tag{9}$$

2. The thin circular current loop in Fig. 203 lies in the xy plane and is centered on the origin; its radius is ρ_0. Show that at a point $P(0, 0, z)$ on the z axis,

$$\mathbf{H} = \frac{i\rho_0^2}{2(\rho_0^2 + z^2)^{3/2}} \mathbf{1}_z. \tag{10}$$

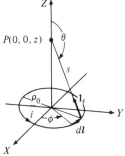
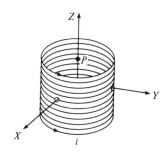

Fig. 203. A circular current loop in the xy plane.

Fig. 204. A solenoid, presumably tightly wound.

3. Circular loops of wire are stacked as in Fig. 204 into a tightly wound solenoid extending from $z = -\infty$ to $z = \infty$ and having n turns per meter of length; each turn has the radius ρ_0 and carries i amperes. A slice of this solenoid extending vertically from z to $z + dz$ meters includes $n\,dz$ turns and is similar to the current loop of Fig. 203, provided that the i in that figure is replaced by $ni\,dz$. Show with the help of (10) that, at any point on the z axis,

$$\mathbf{H} = ni\,\mathbf{1}_z. \tag{11}$$

Note that this H does not depend on ρ_0. (In fact, H vanishes everywhere *outside* the solenoid and is given by (11) at every point *inside* it; see Exercise 6[93].)

4. A steady current i flows in a thin wire along the z axis. Consider a portion of this wire extending from $z = -\frac{1}{2}dl$ to $z = \frac{1}{2}dl$ and show that its contribution to H at a point $P(r, \theta, \phi)$ is

320 Static Magnetic Fields in Nonmagnetic Mediums

$$dH = \frac{i \, dl \sin \theta}{4\pi} \frac{1}{r^2} \mathbf{1}_\phi. \tag{12}$$

5. A constant current i flows along the y axis from $y = -\infty$ to the origin and then along the z axis from the origin to $z = +\infty$. The (x, y, z) coordinates of a point P are $(x_0, 0, 0)$. Integrate $d\mathbf{H}$ to find $\mathbf{H}$ at P (magnitude and direction) and show that

$$H = \frac{\sqrt{2}}{4\pi} \frac{i}{x_0}. \tag{13}$$

6. Use an equation displayed in this section to solve Exercise 5 without explicit integration.

89. THE NUMERICAL VALUE OF μ_0

The evaluation of μ_0 involves the definition of the ampere, given in Appendix B as follows:

"The ampere is the constant current which, if maintained in each of two straight parallel conductors of infinite length, of negligible circular sections, and placed 1 meter apart in a vacuum, will produce between these conductors a force equal to 2×10^{-7} newton per meter of length."

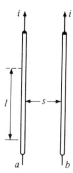

Fig. 205. Parallel current-carrying wires involved in the definition of the ampere.

The wires a and b in Fig. 205 each carry a current of i amperes; the distance between them is s. The magnitude of the field **B** produced at the position of the wire a by the current flowing in b is

(7[87]) $$B = \frac{\mu_0}{2\pi} \frac{i}{s}. \tag{1}$$

Consequently, the force acting on the length l of wire a has the magnitude

(2[86]) $$F = \frac{\mu_0}{2\pi} \frac{i^2 l}{s} \tag{2}$$

and is directed toward the wire b. If we now set $i = 1$ ampere and $l = s = 1$ meter in the right-hand side of (2), the definition of the ampere reduces the left-hand side to 2×10^{-7} newton. Taking into account the units of measurement involved in (2), we then find that

$$\mu_0 = 4\pi \times 10^{-7} \frac{\text{henry}}{\text{meter}}, \tag{3}$$

as in (2^{85}). Thus, we have at last used the definition of the ampere.

EXERCISES

1. Verify that the unit for μ_0 in (3) and (2^{85}) follows from (2).

2. Show that two parallel wires carrying currents in *opposite* directions repel each other. Suggest an analogy between this conclusion and the results of Exercise 2^{86}. Why does not the analogy apply to wires carrying currents in the *same* direction? Or does it?

CHAPTER 21

Magnetomotive Force

By now we have defined the field **B**, postulated the d**B** formula (1^{87}) for computing the field **B** caused by electric currents, and postulated Maxwell's equation div **B** = 0. We have also defined the field **H**, at least for vacuum and for nearly nonmagnetic mediums; and we have mentioned magnetomotive forces and introduced the d**H** formula (1^{88}) for computing the field **H** associated with electric currents.

In this chapter we consider magnetomotive forces in detail and derive Ampère's law (2^{93}). This law proved to be inadequate for time-varying fields, and Maxwell corrected it by introducing the concept of a displacement current and replacing (2^{93}) by (4^{95})—a master stroke that led to the electromagnetic theory of light.

Our method of arriving at (2^{93}) is essentially Ampère's own method. It is not the shortest or mathematically the most attractive method, but it will invite the reader to visualize spatial relationships—an opportunity that a student of the electromagnetic theory should grasp whenever it appears.

The magnetostatic potential, described in §91, is of little interest in practice, but it does make the passage from the d**H** formula to Ampère's law more graphic.

90. MAGNETIC FIELDS AND SOLID ANGLES

Hints for a relation between magnetostatic fields and solid angles come from simple special cases.

Circular current loop. The field **H** at a point P on the axis of the circular loop of radius ρ_0, shown in Fig. 203^{87}, is

Magnetic Fields and Solid Angles § 90

(10⁸⁸)
$$\mathbf{H} = \frac{i\rho_0^2}{2(\rho_0^2 + z^2)^{3/2}} \mathbf{1}_z, \tag{1}$$

where i is the current in the loop. A derivative of the solid angle subtended at P by a surface spanning the loop is

(14⁹)
$$\frac{\partial \Omega}{\partial z} = \frac{2\pi \rho_0^2}{(\rho_0^2 + z^2)^{3/2}}. \tag{2}$$

Consequently, at points on the z axis,

$$H_z = \frac{i}{4\pi} \frac{\partial \Omega}{\partial z}. \tag{3}$$

Another way of writing (3) is

$$H_z = \mathrm{grad}_z \left(\frac{i\Omega}{4\pi} \right). \tag{4}$$

Equation (4) resembles the formula $E_z = -\mathrm{grad}_z V$, which holds in electrostatics. To make the resemblance superficially more complete, we define the symbol ω by the equation

$$\omega = -\Omega \tag{5}$$

and write (4) as

$$H_z = -\mathrm{grad}_z \left(\frac{i\omega}{4\pi} \right). \tag{6}$$

Straight-line current. A rectangular current loop is located in the zx plane, as in Fig. 206(a). If the loop is made larger, as in Fig. 206(b), the

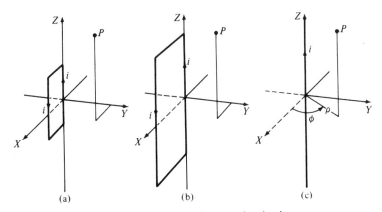

Fig. 206. A sequence of rectangular circuits.

solid angle that it subtends at P will change, and so will the field **H** at P. If the dimensions of the loop in the x direction and the $\pm z$ directions are increased without limit, we get Fig. 206(c). In this figure the current i flows upward along the entire z axis, and the return part of the loop lies infinitely far from P and does not affect the magnetic field at P. In this limiting case the following equations hold at P:

$$(2^{88}) \qquad \mathbf{H} = \frac{i}{2\pi}\frac{1}{\rho}\mathbf{1}_\phi \qquad (7)$$

and, in view of (5) and (20⁹),

$$\omega = -2\phi + 2\pi + 4\pi n, \qquad n = 0, \pm 1, \pm 2, \ldots, \qquad (8)$$

where ϕ is the angle shown in Fig. 206(c). The terms 2π and $4\pi n$ in (8) do not depend on the coordinates of P. Therefore, grad $\omega = -2$ grad ϕ and, according to (56^A),

$$\text{grad } \omega = -2\frac{1}{\rho}\mathbf{1}_\phi. \qquad (9)$$

Comparison with (7) now shows that, in this example,

$$\mathbf{H} = -\text{grad}\left(\frac{i\omega}{4\pi}\right). \qquad (10)$$

Equations (8) and (10) show that although ω is multivalued, its gradient is single-valued. Note that (6) is a special case of (10).

The general case. We will now show that (10) holds for filamentary currents in general. Let Ω be the solid angle subtended at the point P in Fig. 207 by a surface spanning a thin wire that carries the current i and forms a contour C, which is not necessarily flat. The angle Ω depends on the position of P relative to the contour; for example, if the coordinates of P are (x, y, z), then $\Omega = \Omega(x, y, z)$. If P should be moved through a small vector step $-d\mathbf{s}$ to the position P', the angle Ω will change to $\Omega + d\Omega$, where

$$(6^{57}) \qquad \begin{aligned} d\Omega &= (\text{grad } \Omega) \cdot (-d\mathbf{s}) \\ &= -(\text{grad } \Omega) \cdot d\mathbf{s}. \end{aligned} \qquad (11)$$

The angle Ω will change by the same amount if we translate the entire circuit C through the step $+d\mathbf{s}$ into the position C', instead of displacing P by $-d\mathbf{s}$. Thus, in view of (21⁹), $d\Omega$ is *minus* the solid angle subtended at P by the ribbon-like surface connecting C and C'. That is, $d\Omega = -\Omega_{\text{ribbon}}$.

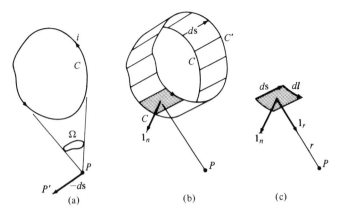

Fig. 207. Diagrams for deriving the formula $\mathbf{H} = \text{grad}\,(i\Omega/4\pi)$.

Let $\Delta\Omega$ be the solid angle subtended at P by the small flat *shaded* surface shown twice in Fig. 207. In view of (1^{88}),

$$\Delta\Omega = -\left(\frac{\mathbf{1}_r}{r^2} \times d\mathbf{l}\right) \cdot d\mathbf{s} = \frac{4\pi}{i} d\mathbf{H} \cdot d\mathbf{s}, \tag{12}$$

a formula that applies to every sufficiently narrow strip of the kind marked on the ribbon in Fig. 207(b). (Recall Exercise 19^{77}.) The line integral of $\Delta\Omega$ around C is Ω_{ribbon}; since $d\mathbf{s}$ is a constant, the line integral of the last term in (12) is $(4\pi/i)\mathbf{H} \cdot d\mathbf{s}$. Therefore, $\Omega_{\text{ribbon}} = (4\pi/i)\mathbf{H} \cdot d\mathbf{s}$, and

$$d\Omega = -\Omega_{\text{ribbon}} = -\frac{4\pi}{i}\mathbf{H} \cdot d\mathbf{s}. \tag{13}$$

Comparing (13) with (11), we find that $\text{grad}\,\Omega = (4\pi/i)\mathbf{H}$. The formula

$$\mathbf{H} = -\text{grad}\left(\frac{i\omega}{4\pi}\right) \tag{14}$$

now follows in view of (5). If, for example, we use cartesian coordinates, then $\mathbf{H} = \mathbf{H}(x, y, z)$ and $\omega = \omega(x, y, z)$ in (14).

91. MAGNETOSTATIC POTENTIAL

Equation (14^{90}) is formally so similar to the equation $\mathbf{E} = -\text{grad}\,V$ that the field $i\omega/4\pi$ is called the *magnetostatic potential* or, more precisely, the *magnetic scalar potential*. (We will later discuss a magnetic *vector* potential.) If we use (5^{90}) and replace ω by $-\Omega$, the definition reads:

$$\left.\begin{array}{l}\text{magnetic scalar potential produced}\\\text{at a point } P \text{ by a filamentary}\\\text{current } i \text{ flowing in a contour } C\end{array}\right\} = -\frac{i\Omega}{4\pi}; \tag{1}$$

here Ω is the solid angle subtended by C at P. According to (1), our unit for measuring magnetostatic potentials is the *ampere*, which is equivalent to the *joule per weber*.

If C is a *flat* contour, it is natural to compute Ω by evaluating the integral (12⁹) over the flat surface S spanning C. But if C is not flat, we run into a complication, because then there is no satisfactory general rule for choosing a particular surface S spanning C. In fact, *by definition*, the solid angle Ω subtended at a point P by a signed contour C is the solid angle subtended at P by *any* properly signed surface S spanning C. This definition, which holds whether C is flat or not, implies that, if C and P are given, the numerical value of Ω is *not unique*.

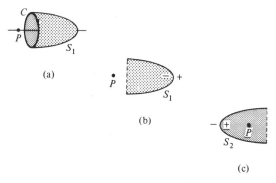

Fig. 208. The solid angle subtended at P by the contour C is positive if we use the surface S_1 but negative if we use S_2.

To illustrate, consider Fig. 208, where the signed contour C is the rim of an imagined butterfly net; in parts (b) and (c) of the figure, C is shown edge-on. The surface S_1 is the net; the surface S_2 is also the net, but pulled to the left through the rim. The point P lies in the same position relative to C in all parts of the figure. It sees the back of S_1, so $\Omega > 0$; in part (c), however, it sees the front of S_2, so $\Omega < 0$. Now, both S_1 and S_2 are spanning C and have the proper signatures, and therefore the definition given above implies that the value of the solid angle subtended by C at P is not single-valued.

Since Ω is multivalued, the magnetostatic potential at any point P is also multivalued, in contrast to the electrostatic potential V. As we show in the next section, this makes it necessary to impose a restriction on the paths of integration when we are evaluating the magnetic analogue of the emf.

EXERCISES

1. Write Ω for the solid angle subtended at P in Fig. 208 by the contour C, and let $\Omega = \Omega^*$ if the surface S_1 is used for computing Ω. Then show that $\Omega = \Omega^* - 4\pi$ if S_2 is used instead.
2. Let the point P in Fig. 208(b) move clockwise once around a circle linking C, and suppose that the net is impenetrable but flexible and stretchable, and that it keeps getting out of the way of P. When P returns to its starting point, the net will be distorted and will form a new surface spanning C, say S_3. Sketch S_3 in a diagram in which C is viewed edge-on, use S_3 to compute Ω and compare with the Ω's of Exercise 1. Repeat for the case when P moves counterclockwise around a circle linking C. Consider Fig. 208(c) along similar lines.

92. MAGNETOMOTIVE FORCE

The definition of the electromotive force in a fixed path ayb reads

$$(4^{52}) \qquad \text{emf}_{ayb} = \int_{ayb} E_t \, dl. \qquad (1)$$

In *electrostatics*, the choice of the path leading from a to b does not affect the value of the integral in (1), and we have the formula

$$(3^{56}) \qquad \text{emf}_{ab} = \int_{ab} E_t \, dl = V_a - V_b. \qquad (2)$$

The definition of the magnetomotive force in a path ayb is patterned after (1) and reads

$$\text{mmf}_{ayb} = \int_{ayb} H_t \, dl. \qquad (3)$$

In the case of a filamentary current, the magnetostatic potential (1^{91}) at a point P is $-(i/4\pi)\Omega_P$, where Ω_P is the solid angle subtended at P by the circuit. As we have seen, this potential is analogous to V so far as gradients are concerned; Equation (2) then leads to the question of whether or not the formula

$$\text{mmf}_{ab} = \int_{ab} H_t \, dl = -\frac{i}{4\pi}(\Omega_a - \Omega_b) \qquad (4)$$

can be made to play a rôle similar to that played by (2) in electrostatics, despite the fact that the right-hand side of (4) is multivalued. The answer is yes, provided that, when evaluating the integral in (4), we put a restriction on the paths leading from a to b.

Suppose that a surface S spanning the contour C in Fig. 208 is flexible and stretchable, that initially it coincides with the surface S_1, and that it is then made to coincide with S_2. The value of Ω will drop abruptly by 4π when S moves across P. If we let S stay fixed in any position and move the point P, then Ω will in general change—but it will change smoothly as long as P does not cross S. However, if P should cross S, the value of Ω will jump abruptly by $\pm 4\pi$. (Recall Exercise 8^9.) In the light of these remarks, we will accept the following statement without a rigorous proof: If P moves from place to place and returns to its original position *without crossing S*, then the solid angle Ω will return to its original value.

We conclude that the factor $\Omega_a - \Omega_b$ in Equation (4)—and hence also this equation itself—will be free from ambiguities, provided that

- we associate with a current-carrying contour C an arbitrarily chosen— and then fixed—imagined surface S that spans C; and
- we restrict ourselves to paths that lead from a to b without crossing S.

Closed paths are discussed in the next section.

The emf (1) has a straightforward physical interpretation: It is the work per unit charge (joules/coulomb, or simply volts) done by the electric field when a small charge is moved from a to b along the path γ. The mmf (3) may be interpreted in a similar way, even though magnetic monopoles are not available. For example, let a long slender magnet with two well-defined poles—say a carefully magnetized knitting needle—be moved from one position in a magnetic field to another. Equation (3) can then be used to compute, for one pole at a time, the work per unit pole (joules/weber, or simply amperes) that the magnetic field will do during this process.

93. AMPÈRE'S mmf LAW IN INTEGRAL FORM

We now come to an important theorem, usually called *Ampère's circuital theorem* or *law*, or simply *Ampère's law*; we will call it *Ampère's mmf law*, as in the chart in the preface. It has to do with the mmf in a closed path, say (γ),

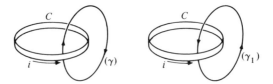

Fig. 209. Examples of mathematical paths linking a current-carrying circuit.

that links a current-carrying contour C. Two examples of linkages are shown in Fig. 209, where C is pictured as a band to help the perspective, and where the paths (γ) and (γ_1) each lie in a plane and have the same shape but opposite signatures. Since the magnetic fields in the two cases are the same, the magnetomotive forces in (γ) and (γ_1) have opposite signs; therefore, to avoid ambiguities, we need a sign convention.

Imagine a properly signed flat surface spanning the path (γ) in Fig. 209. The contour C crosses this surface only once, and crosses it *from back to front*; we then say that *the path (γ) links the current i once in the positive sense*. By contrast, the path (γ_1) links the current i once in the negative sense. An equivalent statement is that, in Fig. 209, (γ) links the current i and (γ') links the current $-i$. This sign convention is used for either flat or not flat paths and current contours.

Ampère's mmf law applies in magnetostatics; that is, when currents are constant, and electric and magnetic fields are also constant. It states that the mmf in any closed path (γ) that links a filamentary current of magnitude i once in the positive sense satisfies the equation

$$\mathrm{mmf}_{(\gamma)} = i \tag{1}$$

or, more explicitly,

$$\oint_{(\gamma)} H_t \, dl = i. \tag{2}$$

The current i in (1) and (2) is the "true" current linking (γ); it includes conduction and convection currents, but does not include the "Ampèrian" currents of electrons moving in their atomic or molecular orbits. We begin with examples.

Straight-line current. A steady current i flows upward along the z axis, as in Fig. 199[87]. At points ρ meters from the axis we then have

(2^{88})
$$\mathbf{H} = \frac{i}{2\pi\rho}\mathbf{1}_\phi \qquad (\rho > 0). \tag{3}$$

First, let (γ) be a circular path in the xy plane, shown in Fig. 210(a) (radius ρ_0, center at O). For this path, $H_t = i/2\pi\rho_0$ and $dl = \rho_0 \, d\phi$, so

$$\int_{(\gamma)} H_t \, dl = \frac{i}{2\pi\rho_0}\int_0^{2\pi} \rho_0 \, d\phi = i, \tag{4}$$

as required by (1).

Next let (γ) be the noncircular path lying in the xy plane and shown in Fig. 210(b). We distort it by replacing the smooth arc ab by the line $aa'b'b$,

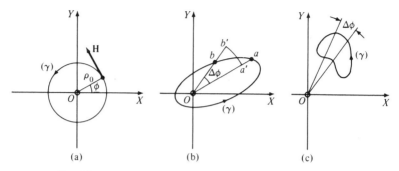

Fig. 210. Examples of paths in the magnetic field of a current flowing along the z axis.

where $a'b'$ is a circular arc of radius ρ' centered on O, and the straight segments aa' and $b'b$ lie on radial lines passing through O. The value of H_t for the straight segments is zero, and hence the portion $aa'b'b$ of the distorted contour contributes to the mmf the amount $(i/2\pi\rho') \cdot (\rho'\Delta\phi)$, which is equal to $(i/2\pi)\Delta\phi$, and does not depend on the radius of the arc $a'b'$. Therefore, it is perhaps apparent that if we approximate the entire contour by circular arcs similar to $a'b'$, each subtending the angle $\Delta\phi$ at O, and if we then let $\Delta\phi \to 0$ and increase the number of arcs without limit, then the jagged contour will merge into the original smooth contour, and (2) will be verified once again.

Similar arguments show that for the path pictured in Fig. 210(c) the left-hand side of (2) is zero; the right-hand side is also zero, because no current flows through this path. These arguments are readily extended to contours that do not lie in the xy plane or are not even flat (Exercise 1), and we conclude that Ampère's mmf law does hold for a straight-line current.

Current in a rod. Figure 211 shows the circular cross section of a long copper rod carrying toward the reader a uniformly distributed current i; the axis of the rod is the z axis. Symmetry suggests that H should depend only on the distance ρ from the z axis and that $\mathbf{H}$ should be circular both inside and outside the rod; the direction of $\mathbf{H}$ then follows from the right-hand rule.

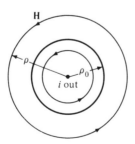

Fig. 211. The magnetic field inside and outside a current-carrying copper rod.

Let us compute the mmf in the circular path shown in the figure *outside* the rod. For this path, $H_t = H$ and

$$\text{mmf} = \int_0^{2\pi} H_t \rho \, d\phi = \rho H \int_0^{2\pi} d\phi = 2\pi\rho H. \tag{5}$$

To express **H** in terms of i, imagine that the total current in the rod is subdivided into infinitely many filamentary straight-line currents. Since (2) holds for each of them, it holds for all of them taken together. The sum of all the filamentary currents is simply i, and therefore, in view of (5), Equation (2) gives $2\pi\rho H = i$; that is, $H = i/2\pi\rho$, and $\mathbf{H} = (i/2\pi\rho)\mathbf{1}_\phi$, as though the current i were confined to the axis of the rod.

Inside the rod, H is smaller, because the smaller circular path in Fig. 211 encircles less current (Exercise 3). The complete formula, which we have given in §35 without proof, is

(2^{35})
$$\mathbf{H} = \begin{cases} \dfrac{\rho}{\rho_0} H_0 \mathbf{1}_\phi & \text{if } \rho \leq \rho_0 \\[2pt] \dfrac{\rho_0}{\rho} H_0 \mathbf{1}_\phi & \text{if } \rho \geq \rho_0; \end{cases} \tag{6}$$

here H_0 is the value of H at the surface of the rod:

$$H_0 = \frac{i}{2\pi\rho_0}. \tag{7}$$

The general magnetostatic case. After the two special cases, we will now consider Equation (2) from a more general standpoint, and will first compute the mmf in the closed path (γ) that links, as in Fig. 212(a), the flat current-carrying contour C, shown edge-on in Fig. 212(b). As required by the rules stated near the end of §92, we choose arbitrarily a particular surface

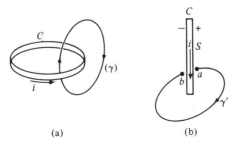

Fig. 212. Derivation of Ampère's mmf law. The contour C, shown as a band, is a line. The surface S, shown as thick, has zero thickness.

spanning C, say the flat surface S, whose front and back are marked in Fig. 212(b) by plus and minus signs. Next, we consider the portion, say γ', of (γ) that begins at the point a near the front of S and ends at a point b near the back of S. The solid angles subtended by S at a and at b are nearly -2π and 2π; that is, $\Omega_a = -2\pi(1 - \delta_1)$ and $\Omega_b = 2\pi(1 - \delta_2)$, where the δ's are positive and small compared to unity. Substitution into (4^{92}) yields

$$\int_{ab} H_t\, dl = i - \tfrac{1}{2}(\delta_1 + \delta_2)i. \tag{8}$$

Finally, we let a and b approach a common point on S in such a way that, in the limit, the path $a\gamma'b$ becomes the path (γ). In this process, $\delta_1 \to 0$ and $\delta_2 \to 0$, and (8) reduces to (2). To complete the derivation of Ampère's mmf law, one should show that the proof given above for a flat current contour can be extended to a contour that is not flat—but this is perhaps obvious.

EXERCISES

1. Verify in detail that Ampère's mmf law (2) holds for the path pictured in Fig. 210(c). Also, verify (2) for the case of a straight-line current and paths that are not flat.

2. Lines of **H** near a current-carrying rod are pictured in Fig. 200[88]. To what extent can one determine their general character, using only symmetry arguments? (Note that one can look at the field lines while facing either in the direction of the current or in the opposite direction.) Consider also the symmetry arguments that apply to the lines of **D** from a charged rod, pictured in Fig. 96[34].

3. Verify the upper part of (6) and review Figs. 73[20] and 74[20].

4. A path (γ) runs along the z axis from $z = -n\rho_0$ to $z = n\rho_0$ and returns to its starting point in a semicircle of radius $n\rho_0$. Consider the circular current loop of Fig. 203[88] and verify (2) for that path in the special case when $n \to \infty$.

5. Use Ampère's mmf law (2) and symmetry arguments to compute **H** for the thick current-carrying plate described in Fig. 202[87] and Table 7[88]. (Consider several representative rectangular paths.)

6. Use (2) and the result of Exercise 3[88] to show that, for the solenoid described in that exercise and in Fig. 204[88], we have $\mathbf{H} = ni\mathbf{1}_z$ everywhere inside the solenoid and $\mathbf{H} = 0$ everywhere outside. (Use rectangular paths, with two sides parallel to the z axis.) Then use (2) to check these conclusions without using the result of Exercise 3[88].

7. The circular cross section of a nonhomogeneous current-carrying rod has the radius ρ_0. The rod is coaxial with the z axis. The current density in it is $\mathbf{J} = (\rho/\rho_0)J_0\mathbf{1}_z$, where J_0 is a constant. Show that

$$\mathbf{H} = \begin{cases} \frac{1}{3}\rho_0 J_0 \left(\frac{\rho}{\rho_0}\right)^2 \mathbf{1}_\phi & \text{for } \rho \leq \rho_0 \\ \frac{1}{3}\rho_0 J_0 \left(\frac{\rho_0}{\rho}\right) \mathbf{1}_\phi & \text{for } \rho \geq \rho_0. \end{cases} \quad (9)$$

8. The inner and outer radii of a long straight current-carrying copper pipe are ρ_1 and ρ_2. The current i is distributed uniformly across the copper portion of the cross section of the pipe. Show that $H = 0$ for $\rho \leq \rho_1$, $H = i/2\pi\rho$ for $\rho \geq \rho_2$, and

$$H = \frac{i}{2\pi} \frac{\rho_1}{(\rho_2^2 - \rho_1^2)} \left(\frac{\rho}{\rho_1} - \frac{\rho_1}{\rho}\right) \quad \text{for } \rho_1 \leq \rho \leq \rho_2. \quad (10)$$

94. AMPÈRE'S LAW IN DIFFERENTIAL FORM

Let us find the curl of the field $\mathbf{H}$ given in (6^{93}). According to (60^A), curl ($\rho \mathbf{1}_\phi$) = $2\mathbf{1}_z$, so the top line of (6^{93}) yields curl $\mathbf{H} = (i/\pi\rho_0^2)\mathbf{1}_z$, or simply curl $\mathbf{H} = \mathbf{J}$, where $\mathbf{J}$ is the current density in the rod. The bottom line, which yields curl $\mathbf{H} = 0$, pertains to the space outside the rod, where $\mathbf{J} = 0$; therefore, in this case too, curl $\mathbf{H} = \mathbf{J}$. Hence the equation

$$\text{curl } \mathbf{H} = \mathbf{J} \quad (1)$$

holds both inside and outside the rod. We have already seen this equation in Exercise 1^{88}. As we will now show, it holds whenever currents, electric fields, magnetic fields, charge densities, and so on, do not vary with time. Under these conditions

(9^{38}) $$\text{div } \mathbf{J} = 0, \quad (2)$$

so that the direction lines of $\mathbf{J}$ have neither beginnings nor ends.

The current flowing *across a surface*, say S, is

(2^{15}) $$i = \iint_S J_n \, da. \quad (3)$$

In view of (2), the current flowing *through a closed path* (γ) is also given by (3), provided that S is *any* properly signed surface spanning (γ). Consequently, Ampère's law (2^{93}) can be written as

$$\oint H_t \, dl = \iint J_n \, da. \quad (4)$$

Here the integral on the right is taken over any properly signed surface spanning the path used on the left. Now, according to Stokes's theorem,

334   Magnetomotive Force

$$(3^{81}) \qquad \oint H_t \, dl = \iint (\operatorname{curl} \mathbf{H})_n \, da, \qquad (5)$$

so that the right-hand sides of (4) and (5) are equal. Furthermore, since S is chosen arbitrarily, they are equal for *every* surface. Hence a step similar to that from (45^5) to (46^5) leads to the equation

$$\operatorname{curl} \mathbf{H} = \mathbf{J}, \qquad (6)$$

which is the *differential form* of Ampère's mmf law.

EXERCISES

1. Verify that the fields $\mathbf{H}$ and $\mathbf{J}$ of Exercise 7^{93} satisfy (6).
2. Identify the formula for $\mathbf{J}$ in the idealized case of a current i confined to the z axis. (Recall Exercise 5^{81}.)

95. MAXWELL'S mmf LAW

The identity $\operatorname{div} \operatorname{curl} \mathbf{F} = 0$ implies pictorially that the direction lines of a curl have neither beginnings nor ends. Consequently, the equation

$$(6^{94}) \qquad \operatorname{curl} \mathbf{H} = \mathbf{J} \qquad (1)$$

implies that the direction lines of $\mathbf{J}$ are continuous. In Fig. 80^{24}, however, these lines begin on one plate of the capacitor, pass through the generator, and end on the other plate without bridging the gap between the plates. This figure illustrates the fact that the direction lines of $\mathbf{J}$ are not necessarily continuous even when $\mathbf{J}$ does not vary with time, and that consequently Equation (1) is not completely general. In fact, in our derivation of (1), we specified that we are dealing with the case when not only the currents, but also the electric fields, magnetic fields, charge densities, and so on are all constant. These conditions are not met in Fig. 80^{24}, because, even though the current is constant, the charge densities on the plates and the electric field between the plates change with time.

Equation (1) was generalized by Maxwell, who introduced for this purpose the concept of displacement currents. In our discussion of these currents in §76, we began with the equations $\operatorname{div} \mathbf{J} = -\partial \rho / \partial t$ and $\operatorname{div} \mathbf{D} = \rho$, which are, respectively, the law of conservation of charge and Maxwell's form of Coulomb's law, and we found that

$$(4^{76}) \qquad \operatorname{div} \left(\mathbf{J} + \frac{\partial \mathbf{D}}{\partial t} \right) = 0. \qquad (2)$$

The vector field $\partial \mathbf{D}/\partial t$ in (2) is the *displacement current density*. Maxwell's generalization consists in replacing the field $\mathbf{J}$ in (1) by $\mathbf{J} + \partial \mathbf{D}/\partial t$. The resulting equation,

$$\operatorname{curl} \mathbf{H} = \mathbf{J} + \frac{\partial \mathbf{D}}{\partial t}, \tag{3}$$

is one of Maxwell's four basic field equations, listed in the chart in the preface. Equation (3) reduces to (1) when $\mathbf{D}$ does not depend on t. Also, in view of (2), it does not violate the equation div curl $\mathbf{F} = 0$; note, for example, that the lines of the field $\mathbf{J} + \partial \mathbf{D}/\partial t$ in Fig. 185[76] *are* continuous.

We will call (3) Maxwell's mmf law in differential form. Applying to it Stokes's theorem, we get the integral form of this law for a closed path (γ):

$$\operatorname{mmf}_{(\gamma)} = \int_{(\gamma)} H_t\, dl = \iint_S \left(\mathbf{J} + \frac{\partial \mathbf{D}}{\partial t} \right)_n da; \tag{4}$$

here S is any properly signed surface spanning the path (γ). The integral of J_n in (4) is the conduction (or conduction-plus-convection) current linking (γ), and the integral of $(\partial \mathbf{D}/\partial t)_n$ is the displacement current linking (γ). As in §76, we sometimes write $\mathbf{J}_d$ and $\mathbf{J}_c$, respectively, for the densities of the displacement and the conduction-plus-convection currents, and write i_d and i_c for the corresponding currents.

When Maxwell added the extra term $\partial \mathbf{D}/\partial t$ to Ampère's mmf law, he not only extended a law of magnetostatics but also, as we will see, provided for the possibility of existence of electromagnetic waves in empty space.

Sometimes it is preferable to rewrite (3) in terms of $\mathbf{B}$ and $\mathbf{E}$. Since $\mathbf{B} = \mu\mathbf{H}$, $\mathbf{D} = \epsilon\mathbf{E}$, and $\mathbf{J} = g\mathbf{E}$, we have

$$\operatorname{curl} \left(\frac{1}{\mu} \mathbf{B} \right) = g\mathbf{E} + \frac{\partial}{\partial t}(\epsilon\mathbf{E}). \tag{5}$$

If the medium is homogeneous, (5) becomes

$$\frac{1}{\mu} \operatorname{curl} \mathbf{B} = g\mathbf{E} + \epsilon \frac{\partial}{\partial t} \mathbf{E}. \tag{6}$$

In vacuum, $g = 0$ (as in any insulator) and (6) reduces to

$$\operatorname{curl} \mathbf{B} = \epsilon_0 \mu_0 \frac{\partial}{\partial t} \mathbf{E}. \tag{7}$$

We will see in §108 that, according to Maxwell's theory, the speed of light in vacuum, denoted by c, is equal to $(\epsilon_0 \mu_0)^{-1/2}$. In anticipation of this conclusion, we recall the values of ϵ_0 and μ_0 in (4^{59}) and (2^{85}), write and rewrite (7) as

$$c = \frac{1}{\sqrt{\epsilon_0 \mu_0}} = 3.00 \times 10^8 \; \frac{\text{meters}}{\text{second}}, \qquad (8)$$

$$\text{curl } \mathbf{B} = \frac{1}{c^2} \frac{\partial}{\partial t} \mathbf{E}. \qquad (9)$$

Sinusoidal currents. Let g and ϵ be the conductivity and capacitivity at a point P in a circuit and let the electric intensity at P be

$$\mathbf{E} = E_0 \sin \omega t \, \mathbf{1}_x, \qquad (10)$$

where E_0 is the amplitude of $\mathbf{E}$ and ω is 2π times the frequency of the alternating electric field. Since $D = \epsilon E$ and $d(\sin \omega t)/dt = \omega \cos \omega t$, the amplitude of $\mathbf{J}_d$ is

$$J_d = \omega \epsilon E_0. \qquad (11)$$

Since convection currents are not involved in this example, the amplitude of $\mathbf{J}_c$ is given by Ohm's law:

$$J_c = g E_0, \qquad (12)$$

so that

$$\frac{J_d}{J_c} = \frac{\epsilon}{g} \omega. \qquad (13)$$

The right-hand side of (13) is called the "Q" of the conducting medium.

The factor ϵ/g in (13) is the "relaxation time" introduced in §72. This time is long for good insulators but very short for even moderately good conductors. Therefore, in the sinusoidal case, except at extremely high frequencies,

$$J_c \ll J_d \quad \text{in good insulators} \qquad (14)$$

and

$$J_d \ll J_c \quad \text{in good conductors.} \qquad (15)$$

EXERCISES

1. In the text we used pictorial considerations to show that the equation curl $\mathbf{H} = \mathbf{J}$ cannot hold in general. Draw the same conclusion without referring to field lines, but by showing that Ampère's law (4^{94}) is unmanageable in the case of a capacitor discharging through a wire, as in Fig. 213. Remember the sentence that follows (4^{94}).

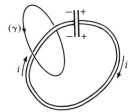

Fig. 213. A capacitor discharging through a wire.

2. Describe, as functions of t, the directions of $\mathbf{J}_d$ and $\mathbf{J}_c$ when $\mathbf{E}$ in a copper block is given by (10).

3. The dielectric separating the plates of a parallel-plate capacitor is slightly conducting; the edge effects can be ignored. When $t = 0$, the charge stored in the capacitor is q_0 (that is, the charges on the plates are q_0 and $-q_0$). Show that the formula for the charge stored at a later time is $q_0 \exp(-gt/\epsilon)$, where g and ϵ are the conductivity and the capacitivity of the dielectric. Compute the densities of the conduction and the displacement currents in the dielectric. Finally, compute the magnetic field produced in the dielectric during the discharge.

CHAPTER 22

Faraday's Law of Induction

By now we have set up three of Maxwell's four field equations. In this chapter we turn to the remaining equation, concerned with Faraday's law of induction. As we do this, we will speak of the time rate of change of magnetic flux linking *stationary* contours—contours that have fixed shapes and are at rest relative to the observer. Nonstationary contours will not be needed in our work, but leaving them out altogether does mean, of course, that we are limiting ourselves to only a portion of the electromagnetic theory.

96. FARADAY'S DISCOVERY

After Oersted noticed in 1819 that a current-carrying wire exerts a torque on a magnetized needle, several people began to look for an inverse phenomenon—an electric effect of a magnetic field; on August 29, 1831, Faraday wrote the following entry in his celebrated diary:[1]

> "Have had an iron ring made (soft-iron), iron round and 7/8 inches thick and ring 6 inches in external diameter. Wound many coils of copper wire round one half, the coils being separated by twine and calico—there were 3 lengths of wire each about 24 feet long, and they could be connected as one length or used as separate lengths. By trial with a trough each was insulated from the other. Will call this side of the ring A. On the other side but separated by an interval was wound wire in two pieces together amounting to about 60 feet in length, the direction being as with the former coils; this side call B.
>
> "Charged a battery of 10 pr. plates 4 inches square. Made the coil on B side one coil and connected its extremities by a copper wire passing to a distance and just over

---

[1] *Faraday's Diary*, Vol. 1, p. 367 (1932); quoted with permission of The Royal Institution of Great Britain and of the publishers, G. Bell and Sons, Ltd. (London). See also W. H. Bragg, "Faraday's Diary," *Reviews of Modern Physics*, Vol. 3, p. 449 (1931).

a magnetic needle (3 feet from iron ring). Then connected the ends of one of the pieces on *A* side with battery: immediately a sensible effect on needle. It oscillated and settled at last in original position. On breaking connection of *A* side with battery again a disturbance of the needle."

What Faraday found that day is that a change in the magnetic flux linking a closed metallic circuit induces in this circuit an electromotive force, which reveals itself by driving an electric current in the circuit—the current that Faraday detected by the torque it exerted on a magnetized needle. The law underlying the phenomenon discovered by Faraday is called *Faraday's law of induction*, but we will often call it *Faraday's emf law*, as in the chart in the preface.

After Faraday's discovery, he and others made various related experiments, such as studies of the effects of the relative motion of a loop of wire and a magnet. Presently it became apparent that a current was produced in a closed wire loop whenever the total magnetic flux linking it was changing with time. The dependence of this current on the resistance of the circuit was investigated, and it was eventually found that the *electromotive force*—the "voltage"—induced in a circuit of a given shape by a magnetic flux changing in a given way does not depend on this resistance. The rule of signs for the induced emf was formulated in 1834 by Lenz,[2] but it was only in 1845 that Franz Ernst Neumann (1798–1895) succeeded in developing an adequate mathematical formula for the induced emf, the formula that we will call the *integral form* of Faraday's law. Sometimes, for emphasis, we have been writing $\mathbf{E}^{es}$ and $\mathbf{E}^{f}$ for the respective intensities of electrostatic fields and electric fields of the Faraday type. In this notation, Neumann's equation for Faraday's law takes the form (5^{98}). If both types of fields are present, their total intensity is $\mathbf{E} = \mathbf{E}^{es} + \mathbf{E}^{f}$;

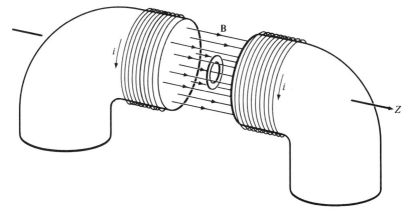

Fig. 214. A ring held in a time-varying magnetic field.

[2] Heinrich Friedrich Emil (Emilij Christianovich) Lenz (1804–1865).

but the substitution of $E_t^{es} + E_t^f$ for E_t^f in (5^{98}) does not affect any physical implications of that equation because the integral of E_t^{es} around a closed path is zero, as in (6^{51}).

Faraday's law implies, for example, that if the field **B** in the region between the jaws of the electromagnet in Fig. 214 is changing with time, a Faraday field will be induced in that region, and hence an emf will be induced in any contour located there in such a position that the net magnetic flux through it is changing. If a slender wooden ring is placed there, as in the figure, a Faraday field—and hence an emf—will be induced in the ring. If the ring is made of copper, the same Faraday field will be induced in *this* ring, but in the copper ring it will produce a current, as indicated in Fig. 215. In either case, the changing magnetic field will induce a Faraday field not only in the ring but also in the space outside it.

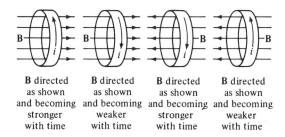

Fig. 215. The current induced in a copper ring held in a time-varying magnetic field.

Conventional introductions to Faraday's law begin with discussions of Neumann's formula, written in one of several equivalent notations. This procedure follows historical precedent, but it implies introducing emf's—which are line integrals of $\mathbf{E}^f$—before discussing the simpler concept of the field $\mathbf{E}^f$ itself, which is the "charge-driving" field. This circumstance led Sears[3] to suggest that a better starting point for an introductory discussion of Faraday's law is a formula appearing in a paper by Ramanathan.[4] We will adopt this procedure, but before giving Ramanathan's formula we should note that the order of presentation suggested by Sears also helps to clarify, among other things, the rôle played in electromagnetic induction by surface charges. The effects of these charges are not involved in Neumann's formula; therefore they are sometimes given only scant attention, even though they are essential

---

[3] Francis W. Sears, "Faraday's Law and Ampere's Law," *American Journal of Physics*, Vol. 31 (1963), p. 439.

[4] K. G. Ramanathan, "Faraday's Law of Induction and the Force on a Body due to Change in Its Magnetization in an Electric Field," *Contemporary Physics*, Vol. 3 (1962), p. 286.

EXERCISES

1. Let us call the respective windings on sides *A* and *B* of Faraday's iron ring the "primary" and the "secondary." One might think offhand that Faraday's magnetic needle responded to the magnetic field produced by the current flowing in the *primary*. What part of our quotation from his diary establishes the fact that the needle responded to the magnetic field produced by a current induced in the *secondary*?

2. If the ring in Fig. 214 is a conductor and **B** changes with time, a current, say i, will be induced in the ring and will produce a magnetic field of its own, say $\mathbf{B}_i$. Lenz's law is based on the conservation of energy and can be stated as follows: The field $\mathbf{B}_i$ will tend to oppose the changes of **B** that cause the current i. Use this law to check all four parts of Fig. 215.

97. THE dE FORMULA

The filamentary circuit of Fig. 216(a) carries a steady current of i amperes (coulombs/second). If a magnetic pole (webers)—say the north-seeking end of a compass needle—is placed at the point P, it will experience a force (newtons).

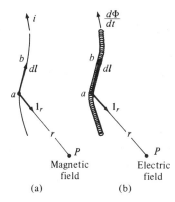

Fig. 216. Diagrams pertaining to the dH formula and the dE formula.

In particular, the current in the segment ab of the circuit will make the following contribution to the force per weber of the magnetic pole located at P:

$$(1^{88}) \qquad d\mathbf{H} = \frac{i}{4\pi} \frac{d\mathbf{l} \times \mathbf{1}_r}{r^2} \quad \frac{\text{newtons}}{\text{weber}}. \qquad (1)$$

In the present context, the newton/weber is a more suggestive unit for $d\mathbf{H}$ than its equivalent, the ampere/meter.

Next we turn from (1) to a phenomenon in which the rôles of electricity and magnetism are in essence interchanged. As to notation, we write Φ for the flux of $\mathbf{B}$ inside the very slender "filamentary" solenoid, a part of which is shown in Fig. 216(b); we assume that the solenoid forms a closed loop and is wound so tightly that the magnetic field vanishes outside it, as in Exercise 6^{93}. We also write

$$i_m = \frac{d\Phi}{dt} \quad \frac{\text{webers}}{\text{second}} \tag{2}$$

and call i_m the *magnetic displacement current*, a name used by some but not by all writers. In the rest of this section we assume that i_m is constant during the time under consideration.

Briefly put, the phenomenon in question is this: If an electric charge of q coulombs—say an electron or a positive ion—is placed at P in Fig. 216(b), it will experience a force. In particular, we assert with Ramanathan that the changing magnetic flux in the segment ab of the solenoid will make the following contribution to the force per coulomb of charge located at P:

$$d\mathbf{E}^f = -\frac{i_m}{4\pi} \frac{d\mathbf{l} \times \mathbf{1}_r}{r^2} \quad \frac{\text{newtons}}{\text{coulomb}}. \tag{3}$$

The corresponding contribution to the force acting on the entire charge q is

$$q\, d\mathbf{E}^f \quad \text{newtons}. \tag{4}$$

Let us write $\mathbf{B}$, as before, for the magnetic flux density inside the slender solenoid. The value of i_m in (3) is then *positive*

- if $\mathbf{B}$ has the direction of $d\mathbf{l}$ and is becoming stronger, or
- if $\mathbf{B}$ has the direction of $-d\mathbf{l}$ and is becoming weaker.

Otherwise i_m is either negative or zero (Exercise 1).

Except for the minus sign, the structures of (1) and (3) are very similar. Roughly speaking, the $d\mathbf{H}$ formula tells us that a magnetic pole located near (but not necessarily *in*) a region pervaded by an electric current will experience a driving force, and a magnetic dipole will experience a torque. Similarly, the $d\mathbf{E}^f$ formula—or simply the $d\mathbf{E}$ formula—tells us that an electric charge located near (but not necessarily *in*) a region pervaded by a time-varying magnetic field will experience a driving force, and an electric dipole will experience a torque.

Example. A long portion of the slender solenoid of Fig. 216(b) lies along the z axis; its return portion lies far enough from a point $P(\rho, \phi, z)$ to be ignored. To get a diagram for this setup, we need only to replace i by i_m in Fig. 199[87].

Our problem is to find the intensity $\mathbf{E}^f$ of the Faraday field induced at P by the entire solenoid. To do this, we use (5[87]) and get for $d\mathbf{E}^f$ a formula very similar to (6[87]). Integration from $z = -\infty$ to $z = \infty$ then gives

$$\mathbf{E}^f = -\frac{i_m}{2\pi} \frac{1}{\rho} \mathbf{1}_\phi \qquad (\rho > 0). \tag{5}$$

This formula can be obtained, of course, directly from the equation

(2[88]) $$\mathbf{H} = \frac{i}{2\pi} \frac{1}{\rho} \mathbf{1}_\phi \qquad (\rho > 0), \tag{6}$$

by writing $\mathbf{E}^f$ for $\mathbf{H}$ and $-i_m$ for i, which are the substitutions that lead from (1) to (3). Figure 217 is a sketch of $\mathbf{H}$ outside a copper rod carrying a steady current and of $\mathbf{E}^f$ outside a tightly wound solenoid that has a finite cross section and carries a current that, for a time, is changing at a constant rate.

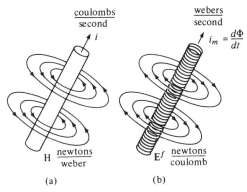

Fig. 217. (a) Direction lines of the field $\mathbf{H}$ associated with an electric current. (b) Direction lines of the electric field of the Faraday type associated with a time-varying magnetic flux in a solenoid.

Let us ignore gravity and assume that the space outside the solenoid is free from electrostatic and magnetic fields.[5] The field (5) is then the only field in that space. A free electron placed there would be accelerated and would spiral outward from the solenoid. The details of this motion are not simple, however,

---

[5] The assumption that $\mathbf{E}$ vanishes outside a solenoid implies that the conductivity of the wire forming the solenoid is infinite. Otherwise an electric field is needed to maintain the current inside the wire and, according to the continuity condition (4[70]), this field extends outside the wire. We ignore this field in our remarks on solenoids.

344 Faraday's Law of Induction § 97

because the moving electron would generate a magnetic field, this field would affect the current in the winding of the solenoid, and $d\Phi/dt$ would not stay constant unless extra energy is provided.

Next, suppose that the space outside the solenoid is pervaded by a magnetic field, produced, say, by a bar magnet located nearby. Then a charge q would experience not only the "Faraday force" $q\mathbf{E}^f$, but also the force (1^{84}), and the total force would be

$$q(\mathbf{E}^f + \mathbf{v} \times \mathbf{B}); \tag{7}$$

here $\mathbf{v}$ is the velocity of the charge at the instant in question, and $\mathbf{B}$ is the magnetic flux density at the point where the charge is located at that instant.

EXERCISES

1. Verify that the rule of signs for i_m stated below (4) is consistent with Figs. 215 and 217.

2. Generalize (7) to the case when an electrostatic field of intensity $\mathbf{E}^{es}$ is also present outside the solenoid.

98. FARADAY'S LAW IN INTEGRAL FORM

The small circle in Fig. 218 outlines the cross section of a slender solenoid lying along the z axis. Let us suppose that the Faraday field (5^{97}) is the only field outside the solenoid, and let us compute the emf in the signed circular contour C pictured in the figure. In other words, let us compute the work per coulomb done by the Faraday field when we take a small positive electric charge once around C. This problem is very similar to that of the mmf in the contour (γ) in the first part of Fig. 210^{93}; we can use the simpler symbol C instead of (γ), because we will not need to deal with pairs of contours, as we did in Fig. 209^{93}.

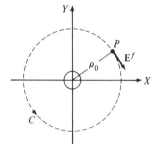

Fig. 218. Computation of the Faraday field of a solenoid whose axis is the z axis and whose flux varies with the time.

The vector $\mathbf{E}^f$ in Fig. 218 is drawn for a positive value of i_m, so the coefficient of $\mathbf{1}_\phi$ in (5^{97}) is negative; in this case the Faraday field will oppose us all the way and will do negative work. To compute the emf in detail, we write

$$\text{emf}_C = \int_C E_t\, dl, \tag{1}$$

and, using (5^{97}), find that

$$\int_C E_t^f\, dl = -i_m, \tag{2}$$

as one can infer directly from (4^{93}). Now,

(6^{84})
$$\Phi = \iint_S B_n\, da, \tag{3}$$

where S is a surface—say the flat surface—spanning C. Since $\mathbf{B} = 0$ outside the solenoid, the flux Φ given by (3) is the flux inside, which appears in the formula $i_m = d\Phi/dt$. Therefore,

$$i_m = \frac{d}{dt} \iint_S B_n\, da, \tag{4}$$

and (2) becomes

$$\int_C E_t^f\, dl = -\frac{d}{dt} \iint_S B_n\, da. \tag{5}$$

A generalization. Equation (5) is the integral form of Faraday's law, set forth by Neumann. We have derived it from the $d\mathbf{E}$ formula (3^{97}) only for the rather artificial case of a time-varying magnetic flux confined to a slender tube along the z axis. We now assert, however, that it holds in general. To make this assertion plausible, we list the main steps that led us to Ampère's mmf law.

In Chapter 20 we took for granted the $d\mathbf{B}$ formula, and with it the $d\mathbf{H}$ formula (1^{88}). Next, we considered an electric current confined to the z axis, derived the Biot-Savart law (2^{88}), and worked out a relation among magnetic fields, current loops, and solid angles; and we showed that the equation

(4^{93})
$$\int_C H_t\, dl = i \tag{6}$$

holds for a filamentary current of any shape. Finally, assuming that currents in extended regions can be treated as bundles of filamentary currents, we converted (6) into the equation

(4[94])
$$\int_C H_t\, dl = \iint_S J_n\, da, \tag{7}$$

which resembles (5).

If we should begin with the $d\mathbf{E}$ formula and go through similar steps, we will be led to Faraday's law in the form (5), provided that we treat a magnetic field pervading an extended region as though it is produced by bundles of tightly wound slender solenoids. We will not discuss these steps in detail, however, because they are so very similar to the steps that led us to (7).

Further equations. Let us suppose that an electrostatic field of intensity $\mathbf{E}^{es}$ may be present in addition to the Faraday field, so that

$$\mathbf{E} = \mathbf{E}^{es} + \mathbf{E}^{f}, \tag{8}$$

and let us restrict ourselves to *stationary contours*, that is, contours that have fixed shapes and do not move relative to the observer. Since electrostatic fields are conservative, the integral of $\mathbf{E}^{es}$ around C is zero, and we can write E_t for E_t^f in (5). If C is stationary, we can take S to be stationary, and rewrite the right-hand side of (5) with the help of (42[5]) and (6[75]), The final result then reads

$$\int_C E_t\, dl = -\iint_S \left(\frac{\partial \mathbf{B}}{\partial t}\right)_n da. \tag{9}$$

The left-hand side of (5) is the *induced voltage* or *Faraday emf* in the contour C. This contour may be an imagined mathematical contour and can be chosen arbitrarily; S is any properly signed surface spanning it. The whole of C or any part of it may lie in a vacuum, in other insulators, or in conductors. The time rate of change of the flux of $\mathbf{B}$ through C depends on the changes of $\mathbf{B}$ and also on the motion and changes of shape of C itself. Equation (5) provides for all these possibilities, while (9) is restricted to stationary contours.

In view of (1) and (3), Equation (5) can be written as

$$\text{emf}_C = -\frac{d}{dt}\Phi_C. \tag{10}$$

Note that (1) does *not* describe a law of nature; it merely defines the term "electromotive force" and the symbol "emf." Similarly, (3) does *not* describe a law of nature; it merely defines the symbol Φ_C as the flux of $\mathbf{B}$. But (10) *does* describe a law of nature—the law discovered by Faraday.

The left-hand side of (5) is the work per unit charge done by the electric field when a small electric charge is taken around the contour C. Consequently, the field $\mathbf{E}^f$ is *not conservative*. As shown in §56, a scalar potential V can be defined uniquely only for conservative fields. Therefore, the field $\mathbf{E}$ cannot be derived from a scalar potential in the presence of a time-varying magnetic field; that is, in general,

$$\mathbf{E} = -\operatorname{grad} V + \mathbf{E}^f, \tag{11}$$

where the Faraday field $\mathbf{E}^f$ is not a gradient of any scalar field.

Nonstationary contours. We remarked at the end of §52 that, if a curve, say a contour C, is moving relative to the observer or is changing shape, and if (from the observer's standpoint) a magnetic field is present in addition to an electric field, then the emf in this curve must be redefined, because the force $q\mathbf{E}$ becomes replaced by the Lorentz force (7^{97}), namely, $q(\mathbf{E} + \mathbf{v} \times \mathbf{B})$. At the same time, because of the motion of C, the formula for the time rate of change of the flux of $\mathbf{B}$ through C acquires a term that involves $\mathbf{v} \times \mathbf{B}$ and proves to be equal to the extra term in emf_C. As a result, *Faraday's law in its integral form* (10) *remains valid for nonstationary contours.*[6]

Kirchhoff's loop laws. In circuit theory the magnetically induced emf's are treated in the same way as the emf's provided by batteries. To illustrate, we consider the circuit shown in Fig. 219 and located in a time-varying

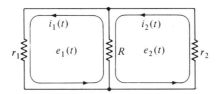

Fig. 219. A circuit driven by Faraday emf's.

magnetic field, which induces the respective emf's of $e_1(t)$ and $e_2(t)$ volts in the first and the second loop. As to signs, we take the current $i_1(t)$ in the figure to be positive if it flows counterclockwise, take the emf $e_1(t)$ to be positive if it tends to make $i_1(t)$ positive, and use the same convention for $i_2(t)$ and $e_2(t)$. According to Kirchhoff's law, $e_1(t)$ is then equal to the total *ir* drop in the first loop; that is, writing e for $e(t)$ and i for $i(t)$, we have $e_1 = r_1 i_1 + R(i_1 - i_2)$. The second loop gives $e_2 = r_2 i_2 + R(i_2 - i_1)$. Consequently,

$$(r_1 + R)i_1 - R i_2 = e_1, \tag{12}$$

and

---

[6] For a detailed discussion of these matters, see P. J. Scanlon, R. N. Henriksen, and J. R. Allen, *Americal Journal of Physics*, Vol. 37 (1969), p. 698; note especially part IX on p. 708.

$$-Ri_1 + (r_2 + R)i_2 = e_2, \tag{13}$$

so that

$$i_1 = \frac{(r_2 + R)e_1 + Re_2}{r_1r_2 + (r_1 + r_2)R}, \tag{14}$$

and

$$i_2 = \frac{Re_1 + (r_1 + R)e_2}{r_1r_2 + (r_1 + r_2)R}. \tag{15}$$

Axially symmetric magnetic fields. If

$$\mathbf{B} = B_\rho(\rho, z, t)\mathbf{1}_\rho + B_z(\rho, z, t)\mathbf{1}_z, \tag{16}$$

we say that the magnetic field is symmetric about the z axis. The uniform time-varying field

$$\mathbf{B} = B_z(t)\mathbf{1}_z \tag{17}$$

is a special case of (16). In the laboratory one cannot attain either perfect axial symmetry or perfect uniformity. However, if a small ring is placed between the jaws of a large electromagnet, as in Fig. 214, the magnetic field pervading the ring and its vicinity can be assumed to be practically uniform and practically symmetric about the axis of the ring. The assumption of uniformity implies that any currents induced in the ring are so feeble that the extra magnetic field that they produce can be ignored.

EXERCISES

1. Figure 220 shows lines of the field $\partial \mathbf{B}/\partial t$ between the jaws of an electromagnet. Verify that this figure provides as much information as Fig. 215.

Fig. 220. An abbreviated version of Fig. 215.

2. Assume that $r_2 = r_1 = r$ and $e_2 = e_1 = e$ in Fig. 219 and show on intuitive grounds that then the values of i_1 and i_2 should not depend on R. Use (14) and (15) to check.

3. Assume that **B** in Fig. 214 has the axially symmetric form (16) and show that then the Faraday E is a circular field of the form $\mathbf{E} = E_\phi(\rho, z, t)\mathbf{1}_\phi$.

4. The ring in Fig. 214 is replaced by a copper washer (thick lines in Fig. 221).

Show that, if **B** is given by (17), then $\mathbf{E} = f(t)\rho \mathbf{1}_\phi$. Express E_ϕ in terms of ρ and $B_z(t)$.

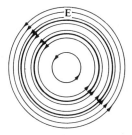

Fig. 221. The Faraday field induced in and near a copper washer held symmetrically in an axially symmetric time-varying magnetic field.

5. Most students find the $d\mathbf{E}$ formula (3^{97}) more mystifying than the $d\mathbf{H}$ formula (1^{97}). Explain why.

6. Equation (5^{97}) was derived in §97 from the $d\mathbf{E}$ formula. In this section we used it in the step from (1) to (2). Now derive it from the integral form (5) of Faraday's law.

99. A. C. VOLTMETERS

In the presence of time-varying magnetic fields, the current in the wire of Fig. 132^{53} will depend, in general, not only on the time t but also on the part of the wire where it is measured; for example, the current entering the wire at a need not be equal to the current leaving it at the same instant at b. Also, the current density near the axis of the wire need not be equal to that near the surface (skin effect). When these effects cannot be ignored, there is no longer any simple way of interpreting the reading of a voltmeter. Therefore, we will restrict ourselves to fields that do not change too rapidly, so that, although i is a function of t, it is at any instant practically constant along the wire, and there is no appreciable skin effect. (One criterion for sufficiently slow variation of sinusoidal currents is that $2\pi f \ll R/L$, where f is the frequency, R the resistance of the wire, and L its self-inductance.)

We consider an "instantaneous" a. c. voltmeter that has a negligible mechanical inertia and a very rapid response (as a cathode-ray oscillograph), a high resistance, and a compact coil; its zero mark lies in the middle of the scale. The arguments of §53 then lead us once again from the manufacturer's calibration rule, namely,

(1^{53}) reading of voltmeter $= iR$, (1)

to the equation,

(9^{53}) reading of voltmeter $=$ emf in the leads and coil, (2) but now both sides of (2) are functions of the time t. The electric fields of §53 are conservative, and consequently we could have said in §53 that the voltmeter reads the potential difference between the ends of its leads. In this section, however, time-varying magnetic fields are involved, the electric fields are not conservative, and (2) cannot be restated in terms of the electrostatic potential.

We will discuss only *instantaneous* a. c. voltmeters because they illustrate most clearly the basic function and purpose of voltmeters. The various averaging meters, such as the usual root-mean-square a. c. meters, will not concern us.

Our examples involve a slender copper ring held between the jaws of the electromagnet of Fig. 214[96]. We assume that the fields **B** and $\partial \mathbf{B}/\partial t$ are at all times nearly uniform in and near the ring, and that the whole setup is axially symmetric. The Faraday field $\mathbf{E}^f$ induced in and near the ring is then circular,

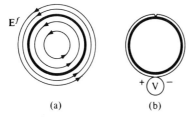

Fig. 222. (a) A *slender* copper ring (dark line) replacing the ring of Fig. 214, and the field $\mathbf{E}^f$ induced in and near the ring. (b) A voltmeter connected to measure the emf induced in the ring.

as in Fig. 222(a). If the leads of a voltmeter are made into a circular loop hugging the ring, as in Fig. 222(b), the emf induced in the leads and coil of the meter is practically equal to that induced in the ring. For definiteness, we assume that the meter then reads 12 volts at an instant t_0, so that, at least roughly,

$$\text{emf}_{\text{ring}} = 12 \text{ volts} \quad \text{when } t = t_0. \tag{3}$$

At this instant, the magnitude of the Faraday field in the ring is

$$E_0^f = \frac{12}{l} \frac{\text{volts}}{\text{meter}}, \tag{4}$$

where l is the circumferential length of the ring. Another way of writing (3) is

$$\text{emf}_{\text{ring}} = \oint_{\text{ring}} E_t \, dl = 12 \text{ volts}, \tag{5}$$

where E_t is the E_0^f of (4). If the ring is not sufficiently slender, the emf's induced in its outer and inner circumferences may be different, and (3) becomes unprecise.

Example. What will the meter read when $t = t_0$ if it is connected to the slender ring of Fig. 222 as shown in Fig. 223(a)? We will find the answer in two ways.

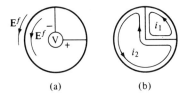

Fig. 223. (a) A voltmeter connected in a different way to the slender copper ring of Fig. 222. (b) A circuit diagram for computing the reading of the meter.

The field $\mathbf{E}^f$ is circular, but the leads of the meter are radial. Therefore, $\mathbf{E}^f$ will tend to drive the conduction electrons *across* the leads, not along them. Consequently (if the coil of the meter is sufficiently compact), no current will be driven through the meter, and the meter will read zero. Under these circumstances, the resistance of the meter should not matter.

To solve this problem in detail by circuit theory, we consider the one-quadrant loop and the three-quadrant loop in Fig. 223(b). When $t = t_0$, the time rate of change of magnetic flux through the ring is 12 volts; since the field **B** is presumably uniform at all times, the respective emf's induced in the smaller and in the larger loop are proportional to their areas, and hence are equal to 3 volts and 9 volts. If R' is the resistance of the ring and R the resistance of the leads and coil of the meter, Kirchhoff's laws yield the equations $(R'/4)i_1 + R(i_1 - i_2) = 3$ and $(3R'/4)i_2 + R(i_2 - i_1) = 9$; that is,

$$(\tfrac{1}{4}R' + R)i_1 - Ri_2 = 3 \tag{6}$$

$$-Ri_1 + (\tfrac{3}{4}R' + R)i_2 = 9. \tag{7}$$

Therefore (Exercise 1),

$$i_1 = i_2. \tag{8}$$

These currents flow in the meter in opposite directions, and hence the meter reads zero, as we found in the preceding paragraph by inspection.

One might say offhand that the leads of the meter are connected to the ring "at points 3 volts apart," and that consequently the meter should read 3 volts. However, in this sense, the leads are also connected "at points -9 volts apart," so the meter should read -9 volts. But the meter cannot read both 3 and -9 volts; in fact, as we have shown by inspection and also by circuit theory, it will read zero.

Second example. Next, suppose that the voltmeter, whose leads now have unequal lengths, is connected to the ring as in Fig. 224(a). Its coil and the radial portions of its leads then make no contribution to the emf, as before,

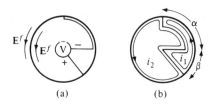

Fig. 224. A voltmeter connected to a slender copper ring in a still different way.

(a) (b)

but the circular portion of the lead marked minus does, because in it E_t is positive. This portion lies close to the ring, so the value of E_t in it is about equal to E_0^f, namely, $12/l$ volts per meter. Its length is about $l/4$ meters, and conseqently the integral in (11^{53}) is about 3 volts. Therefore, if its resistance is high enough, the meter will read about 3 volts when $t = t_0$.

EXERCISES

1. Consider the ring of Fig. 223, verify (8), compute the current in the ring, and note that the resistance of the meter does not matter in this case. What field cancels the tendency of $\mathbf{E}^f$ to drive charges *across* the leads?

2. In the text, we estimated that, when $t = t_0$, the meter in Fig. 224(a) reads about 3 volts. Use circuit theory to verify this estimate. [Figure 224(b) may help to systematize the computations.] Note that the resistance of the meter *does* matter in this case. What field drives the current in the radial portions of the leads?

3. Assume (3) and show by inspection that in Fig. 225 the readings of the voltmeters at the instant t_0 are about (a) 1.5 volts, (b) 3 volts, (c) zero, and (d)

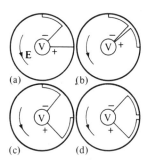

Fig. 225. Further examples of voltmeters connected to slender copper rings.

−3 volts. Use some of these examples to illustrate the fact that, in the presence of time-dependent magnetic fields, voltmeters connected in parallel need not agree in their readings.

4. A voltmeter located at the center of the ring of Fig. 225 is provided with leads of various lengths and negligible resistances. Assume (3) and show by diagrams how the meter can be connected to the ring at points one-quarter

of the circumference apart, so that its reading at the instant t_0 will be about (a) 6 volts, (b) 9 volts, (c) −4.5 volts, (d) 15 volts. Check the case (d) by circuit theory. (Recall the type of surfaces illustrated in Fig. 64(b)[17].)

5. Suppose that $\mathbf{E} = ky\mathbf{1}_x$, place the points a and b on the x axis, and show that the path γ leading from a to b can be so chosen that the emf$_{a\gamma b}$ will have any value you please, positive or negative. Illustrate by diagrams.

100. MAGNETIC ENERGY

In §74 we derived the formula

$$u_e = \tfrac{1}{2}\mathbf{E} \cdot \mathbf{D} \tag{1}$$

for the density of electric energy (joules/meter³). Now we will use Faraday's emf law to derive the formula

$$u_m = \tfrac{1}{2}\mathbf{H} \cdot \mathbf{B} \tag{2}$$

for the density of magnetic energy.

We began the proof of (1) by considering an idealized parallel-plate capacitor, because the electric field inside it is uniform. For a similar reason we begin the proof of (2) by speaking of an idealized, infinite cylindrical solenoid of the type pictured in Figs. 204[88] and 226. We write a for its cross-sectional

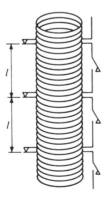

Fig. 226. Computation of the density of magnetic energy in a solenoid.

area and n for the number of turns per meter of length, and assume that the solenoid is located in vacuum. If the winding carries a constant current, say i_0, we have the uniform fields

$$\mathbf{H} = ni_0\mathbf{1}_z, \qquad \mathbf{B} = \mu_0 ni_0\mathbf{1}_z \tag{3}$$

inside the solenoid, and $\mathbf{H} = 0$ and $\mathbf{B} = 0$ outside, as in Exercise 6[93]. This solenoid presumably consists of infinitely many segments, each l meters long, connected in series by the switches shown in Fig. 226 at the left.

Let us now imagine that when $t = 0$ we open *all* the switches at the left of the figure and, at the same instant, close *all* the switches at the right, so that the segments of the solenoid become disconnected from one another, and each becomes shorted. If the current in the winding will subside slowly enough, the magnetic field inside each segment will remain uniform, but its strength will tend to zero; that is, Equations (3) will become

$$\mathbf{H} = ni\mathbf{1}_z, \qquad \mathbf{B} = \mu_0 ni\mathbf{1}_z, \tag{4}$$

where i tends to zero; the magnetic field will still be zero outside. Our program is to compute the energy converted into heat after the shorting switches are closed, and to claim that this energy is equal to the magnetic energy originally "stored" in the solenoid.

The magnitude of the magnetic flux across a cross section of the solenoid is aB, and hence the flux linkage for a whole segment is $nlaB$. Therefore, by Faraday's law and in view of the second equation in (4),

$$\text{emf}_{\text{segment}} = -nal\frac{dB}{dt} = -\mu_0 n^2 al\frac{di}{dt}. \tag{5}$$

To find the current i, we write R for the resistance of a shorted segment (including the shorting switch) and set the iR drop equal to the emf given in (5):

$$iR = -\mu_0 n^2 al \frac{di}{dt}. \tag{6}$$

The solution of (6) that reduces to i_0 when $t = 0$ is

$$i = i_0 e^{-t/\tau}, \tag{7}$$

where

$$\tau = \mu_0 n^2 \frac{al}{R}. \tag{8}$$

Therefore, the instantaneous rate of the generation of heat is

$$i^2 R = i_0^2 R e^{-2t/\tau}. \tag{9}$$

Assuming that all the magnetic energy, say U_m, originally stored in a segment is converted into heat, we write

$$\begin{aligned} U_m &= \int_0^\infty i^2 R \, dt = i_0^2 R \int_0^\infty e^{-2t/\tau} \, dt \\ &= \tfrac{1}{2}\tau i_0^2 R = \tfrac{1}{2}\mu_0 (ni_0)^2 al. \end{aligned} \tag{10}$$

Dividing U_m by the volume al of a segment, we get $u_m = \tfrac{1}{2}\mu_0(ni_0)^2$. In view of (3), this result reduces to (2).

101. FARADAY'S LAW IN DIFFERENTIAL FORM

The equation

$$\oint_C E_t\, dl = -\frac{d}{dt} \iint_S B_n\, da \tag{1}$$

(5^{98})

presumably holds for any signed contour C and any fixed (simple, two-sided) and properly signed surface S spanning C. Now, by Stokes's theorem (3^{81}),

$$\oint_C E_t\, dl = \iint_S (\operatorname{curl} \mathbf{E})_n\, da, \tag{2}$$

and, for a stationary contour,

$$-\frac{d}{dt}\iint_S B_n\, da = -\iint_S \left(\frac{\partial \mathbf{B}}{\partial t}\right)_n da, \tag{3}$$

(9^{98})

so that the right-hand sides of (2) and (3) are equal. The surface S in (1) is an arbitrarily chosen surface spanning an arbitrarily chosen stationary contour. Consequently, a step similar to that leading from (4^{94}) and (5^{94}) to (6^{94}) will now lead to the equation

$$\operatorname{curl} \mathbf{E} = -\frac{\partial \mathbf{B}}{\partial t}, \tag{4}$$

which is the *differential form* of Faraday's emf law.

In cartesian coordinates, for example, Equation (4) turns into a set of three equations:

$$\frac{\partial}{\partial y}E_z - \frac{\partial}{\partial z}E_y = -\frac{\partial}{\partial t}B_x, \tag{5}$$

$$\frac{\partial}{\partial z}E_x - \frac{\partial}{\partial x}E_z = -\frac{\partial}{\partial t}B_y, \tag{6}$$

$$\frac{\partial}{\partial x}E_y - \frac{\partial}{\partial y}E_x = -\frac{\partial}{\partial t}B_z. \tag{7}$$

Even when the field $\mathbf{B}$ has a simple form, this set of partial differential equations has a variety of solutions, as illustrated in a small way in the next section.

EXERCISE

1. Derive (5^{97}) from (4).

102. INDUCED EMF'S, CURRENTS, AND SURFACE CHARGES

The differential forms of Maxwell's mmf law and Faraday's emf law are curl $\mathbf{H} = \mathbf{J} + \partial \mathbf{D}/\partial t$ and curl $\mathbf{E} = -\partial \mathbf{B}/\partial t$. In this section we consider the curl $\mathbf{E}$ equation and assume that the speed of light is *infinite*. (See Exercise 4.)

A mathematical circuit. The imagined mathematical "circuit" shown by a dashed line in Fig. 227 is a square of side $2a$. It is located in a region pervaded by a nearly uniform magnetic field, which is normal to the page and

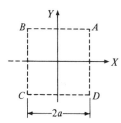

Fig. 227. A square mathematical contour. (Positive sense: $ABCDA$.)

varies sinusoidally with the time.[7] The positive normal to the xy plane in the figure points toward the reader, so the positive sense of the contour is the counterclockwise sense. To avoid some minus signs later, we assume that

$$\mathbf{B} = -\mathbf{1}_z B_0 \sin \omega t. \tag{1}$$

To simplify the arithmetic, we write

$$a^2 \omega B_0 = 1 \text{ volt}. \tag{2}$$

The area of the circuit is $4a^2$; the magnetic flux linking it is

$$\Phi = B_n(4a^2) = -4a^2 B_0 \sin \omega t, \tag{3}$$

and consequently the emf induced in it is

$$\text{emf}_{\text{circuit}} = -\frac{d}{dt} \Phi = 4a^2 \omega B_0 \cos \omega t. \tag{4}$$

---

[7] The examples given in this section were suggested to the author by the discussion of rectangular coils in L. V. Bewley's *Flux Linkages and Electromagnetic Induction* (New York: Dover Publications, Inc., 1964), p. 85. That discussion, however, does not illustrate the variety of $\mathbf{E}$'s allowed by the curl $\mathbf{E}$ equation, or the effects of surface charges. But see D. R. Moorcroft, *American Journal of Physics*, Vol. 38 (1970), p. 376, and Sherwood Parker, *ibid.*, p. 720.

In particular, when $t = 0$,

$$\text{emf}_{\text{circuit}} = 4a^2\omega B_0 = 4 \text{ volts}. \tag{5}$$

Next we will try to compute the field $\mathbf{E}^f$ induced at points on the square contour. In view of (1), the final equations of § 101 are

$$\frac{\partial}{\partial y} E_z^f - \frac{\partial}{\partial z} E_y^f = 0, \tag{6}$$

$$\frac{\partial}{\partial z} E_x^f - \frac{\partial}{\partial x} E_z^f = 0, \tag{7}$$

$$\frac{\partial}{\partial x} E_y^f - \frac{\partial}{\partial y} E_x^f = \omega B_0 \cos \omega t. \tag{8}$$

This system of equations has an infinite variety of solutions. For example, one family of solutions is

$$E_x^f = [-\tfrac{1}{2}(1 - k)\omega B_0]y \cos \omega t, \tag{9}$$

$$E_y^f = [\tfrac{1}{2}(1 + k)\omega B_0]x \cos \omega t, \tag{10}$$

$$E_z^f = 0, \tag{11}$$

where k is an arbitrary constant.

If, for example, we set $k = -1$ and $t = 0$, we get

$$E_x^f = -\omega B_0 y, \quad E_y^f = 0, \quad E_z^f = 0. \tag{12}$$

The only nonvanishing component is E_x^f, so that, when $t = 0$, the field $\mathbf{E}^f$ is directed to the right or to the left in Fig. 227, depending on the value of y. On the other hand, if we set $k = 1$ and $t = 0$, we get

$$E_x^f = 0, \quad E_y^f = \omega B_0 x, \quad E_z^f = 0, \tag{13}$$

and hence $\mathbf{E}^f$ is directed up or down the page. The discrepancy between (12) and (13) illustrates the fact that the curl $\mathbf{E}$ equation does not determine the Faraday field completely unless it is supplemented by adequate boundary or symmetry conditions. The same arbitrariness persists in the case of actual wire circuits. We will illustrate this after verifying that the field (12) produces a counterclockwise 4-volt emf in the square contour, as required by (5); the case of (13) is similar.

On the upper side of the square we have $y = a$ and $E_x^f = -a\omega B_0$. To compute the Faraday emf along this side, we should integrate E_x^f from $x = a$ to $x = -a$, but since in this example E_x^f is a constant, we need only to multiply it by $-2a$. Now, $2a^2\omega B_0 = 2$ volts, the result for the lower side is also 2 volts, and (12) gives zero for the vertical sides. Therefore, when $t = 0$,

$$\text{emf}^f_{AB} = 2\text{v}, \quad \text{emf}^f_{BC} = 0\text{v}, \quad \text{emf}^f_{CD} = 2\text{v}, \tag{14}$$

and $\text{emf}^f_{DA} = 0$ volts. The total emf around all four sides amounts to 4 volts and has the counterclockwise sense.

A wire circuit. The circuit shown in Fig. 228 is made of thin and uniform copper wire; it has the same size as the mathematical circuit in Fig. 227, it is oriented in the same way, and it is pervaded by the same magnetic field.

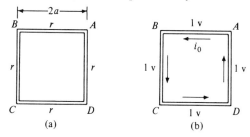

Fig. 228. A square circuit made of uniform copper wire and held in a time-varying magnetic field. (Positive sense: $ABCDA$.)

The resistance of each of its four sides is r ohms. We assume that the current induced in this circuit is so feeble that its magnetic effects can be ignored. Since the emf around the entire circuit is then 4 volts, there is a 1-volt ir drop in each side, as in Fig. 228(b). This drop is the sum of two emf's: one caused by the time-varying magnetic field, the other caused by the surface charges that have settled on the wire. So, when $t = 0$,

$$\text{emf}^{total}_{AB} = 1\text{v}, \quad \text{emf}^{total}_{BC} = 1\text{v}, \quad \text{emf}^{total}_{CD} = 1\text{v}, \tag{15}$$

and $\text{emf}^{total}_{DA} = 1$ volt. The total emf along the side AB, say, can be read on a voltmeter whose leads are connected to the points A and B, and lie very close to that side.

Since the wire is uniform, the magnitude of the electric intensity $\mathbf{E}$ has, at any instant, the same value at all points inside the wire; in particular, when $t = 0$, we have $E_0 = a^2\omega B_0/2a = \tfrac{1}{2}a\omega B_0$. Writing $\mathbf{E}^f_0$ and $\mathbf{E}^{es}_0$ for the respective intensities of the Faraday field and the field of the surface charges, we then get

$$E_0 = |\mathbf{E}^f_0 + \mathbf{E}^{es}_0| = \frac{1}{2}a\omega B_0. \tag{16}$$

All the terms in (16) pertain to the instant when $t = 0$. However, while E_0 has *the same value* at all points inside the wire, *Equation* (16) *permits the fields* $\mathbf{E}_0^f$ *and* $\mathbf{E}_0^{es}$ *to vary from point to point in the wire, even on the same side of the square.*

The copper wire is nonmagnetic, so its presence does not affect the magnetic field in which it may be placed, and the Faraday electric field caused by a time-varying magnetic field can be investigated theoretically without regard for the wire. That is, we can proceed as we did in the case of a mathematical contour.

Example. One of the solutions of the curl **E** equation that we found above is the $\mathbf{E}^f$ described in (12). The sizes and orientations of the squares in Figs. 227 and 228 are the same. Consequently the evaluation of the emf's caused by that $\mathbf{E}^f$ will lead once again to (14), as pictured in the first part of Fig. 229. We should stress the fact that the 2-volt and zero-volt emf's in this

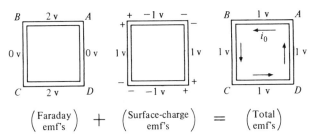

Fig. 229. The emf's induced in the circuit of Fig. 228 by a particular solution of the curl E equation. (Positive sense: $ABCDA$.)

figure are the *Faraday* emf's. The total emf's—"Faraday plus surface charges" —are shown in Fig. 228(b) and amount to 1 volt along each of the four sides; they are also shown in the third part of Fig. 229.

Now about surface charges. As the field $\mathbf{E}^f$ in the side AB builds up to its maximum value, it drives a current toward the left, so that (in conventional terms) positive charges stream toward the corner B and negative charges become uncovered at the corner A. The charges arriving at B cannot go any further to the left; they keep settling on the surface of the wire and exert increasing repulsive forces on the further charges that are moving toward B. These forces cause the oncoming charges to veer away from the surface at B and to proceed downward toward C—much as the surface charges in Fig. (53¹⁴) cause oncoming charges to veer away from the hole in the conductor. In the side CD the situation is the same, except for a reversal of signs.

The resulting distribution of surface charges is indicated roughly by the plus and minus signs in the middle part of Fig. 229. Their exact distribution is not easy to compute, but one can expect from the results of Exercises 1[42] and 4[60] that their magnitudes are very small.[8]

To find the emf's caused by the surface charges, we compare (14) and (15). The results, shown in Fig. 229, are

$$\text{emf}^{es}_{AB} = -1 \text{ v}, \quad \text{emf}^{es}_{BC} = 1 \text{ v}, \quad \text{emf}^{es}_{CD} = -1 \text{ v}, \quad (17)$$

and $\text{emf}^{es}_{DA} = 1$ volt. Since the origin of these emf's is electrostatic, they are also called "potential drops," and the equations given just above can be written in the notation of §56 as

$$V_{AB} = -1 \text{ v}, \quad V_{BC} = 1 \text{ v}, \quad V_{CD} = -1 \text{ v}, \quad (18)$$

and $V_{DA} = 1$ volt.

The interplay of the Faraday field and the electrostatic field of the surface charges can be illustrated further as follows. Suppose that a tube, made of glass of negligible thickness, has the same external diameter and the same length as the wire in Fig. 229. Suppose that we bend it into a square and, using an electrostatic machine, charge its outer surface in exactly the same way as the surface of the wire in Fig. 229 is charged when $t = 0$. The potential differences between adjacent corners of the tube will then be equal to ± 1 volt, as in the figure. Finally, suppose that, instead of the wire circuit, this *tube* is placed in the magnetic field and the Faraday emf's shown in Fig. 229 are induced in it. The total emf's induced in it when $t = 0$ will then be given by the third part of the figure, although no current will flow in the glass tube.

Second example. When we set $k = 0$ and $t = 0$ in (9), (10), and (11), we get

$$E^f_x = -\tfrac{1}{2}\omega B_0 y, \quad E^f_y = \tfrac{1}{2}\omega B_0 x, \quad E^f_z = 0. \quad (19)$$

It is perhaps obvious that now the Faraday emf equals 1 volt along each side. We also note that, in the upper and lower sides of the circuit, the y component of $\mathbf{E}^f$ is directed across the wire and varies from point to point in the wire. This component is canceled by the field of the charges which this very component causes to settle on the surface of the wire. The situation is similar in the other sides of the square circuit.

Next we recall the formula $-y\mathbf{1}_x + x\mathbf{1}_y = \rho\mathbf{1}_\phi$, reintroduce the factor $\cos \omega t$, and find that

$$\mathbf{E}^f = (\tfrac{1}{2}\omega B_0 \cos \omega t)\rho\mathbf{1}_\phi. \quad (20)$$

---

[8] See W. G. V. Rosser, "Magnitudes of Surface Charge Distributions Associated with Electric Current Flow," *American Journal of Physics*, Vol. 38 (1970), p. 265.

Therefore, *in this example*, the Faraday field is circular and symmetric about the z axis.

Comments. We have displayed above three particular solutions of the curl equations (6), (7), and (8) pertaining to the instant when $t = 0$. They are (12), (13), and (19), and they are all different. What is more, they differ not only in mathematical form, but also in physical content. In particular, although they all lead to the same total emf in the circuit—and therefore to the same current—they imply different distributions of surface charges. Hence this question: *Which of these solutions, if any, applies to the circuit that we have described, located in the magnetic field that we have described?* The fact is that our description of the magnetic field is not complete enough to make it possible to answer this question.

The correct solution for a given problem is determined by the boundary conditions and other information supplied with the problem. But in our example we have *already* used all the information given to us about the physical setup; therefore, we must conclude that we do not have enough information to determine either the intensity E^f of the Faraday field or the intesity E^{es} of the electrostatic field produced by surface charges. All that we can do is to compute the magnitude of the sum of the these fields and to note that, when $t = 0$, the result is given by (16).

The additional information can be provided by a more detailed description of the physical setup. Suppose, for example, that the square circuit, as the ring in Fig. 214[96], is placed with its sides at right angles to **B** and its center on the axis of the poles. Suppose also that the field **B** produced by the electromagnet is at every instant practically uniform in the region of the circuit and that, although it gets weaker farther out from the z axis, it remains for all practical purposes axially symmetric. The extra condition provided by *this* description of the setup is *axial symmetry*. Under this condition, the applicable solution of the curl **E** equations is the axially symmetric solution (20).

Suppose that a given magnetic field can be easily interpreted as produced by an assembly of slender solenoids carrying time-varying currents, and that no magnetic materials are involved. The simplest way of computing the Faraday field, E^f, produced by this magnetic field may then be to work directly from the dE formula (3[97]), rather than from the integral or differential forms of Faraday's law. This would give E^f without ambiguity.

Four square circuits. The periphery $ABCDA$ of the four identical wire circuits in Fig. 230 has the same size as the square in Fig. 228. The field **B** in the vicinity of these circuits is again given by (1), and we assume axial symmetry throughout the magnetic field. When $t = 0$, we then have the following emf's: 4 volts around $ABCDA$, 1 volt along each side of $ABCDA$, 1 volt around each small square, and $\frac{1}{4}$ volt along each side of a small square.

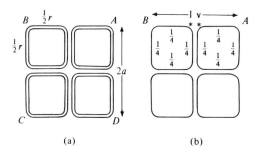

Fig. 230. Four adjacent square circuits replacing the circuit of Fig. 228. The "missing" half a volt appears between the starred points.

Figure 230(b) now suggests that the emf in the line AB is *half a volt*, in contradiction to our earlier conclusion that it is *one volt*. To resolve this paradox, the reader should show that the missing emf of half a volt is localized on the short portion of the line AB lying between the starred points in the figure (Exercise 2).

EXERCISES

1. Fig. 229 pertains to the choice $k = -1$ in (9) and (10). Sketch similar figures for the following choices of solutions of the curl equations:
   (a) if $k = +1$ in (9) and (10);
   (b) if the term $-\omega B_0 a \mathbf{1}_x$ is added to the $\mathbf{E}^f$ in (12);
   (c) if
   $$E_x^f = -(\tfrac{1}{2}\omega B_0 \cos \omega t)(x+y), \qquad E_x^f = (\tfrac{1}{2}\omega B_0 \cos \omega t)(x+y), \tag{21}$$
   and $E_x^f = 0$. Also, verify that the fields described in (b) and (c) satisfy the curl **E** equations.

2. Each of the four circuits in Fig. 230 has a corner near the center of the figure. Imagine that each of these four corners is grounded; consider a path leading from one starred point to the other *via* the center, and evaluate the emf in the portion of the straight line AB lying between the starred points. Does the same conclusion hold when the central corners are not grounded?

3. Replace the four wire circuits in Fig. 230 by four square mathematical contours, ignore the spaces between the adjacent sides (so that these sides are radial), set $t = 0$, and compute the emf's in each of the 16 sides for the cases (12) and (20).

4. In this section we began with the formula (1) for **B** and found various Faraday **E**'s that satisfy Faraday's emf equation. Show, however, that the given

B and our Faraday **E**'s *do not* satisfy Maxwell's mmf equation for free space, unless the speed of light can be taken as infinite. [Recall (9^{95}).]

5. A fixed point P lies on the z axis between the jaws of the electromagnet of Fig. 214; the field **B** is symmetric about the z axis. A copper ring (full lines in Fig. 231) can be held in either of the two positions shown in the figure; the point P lies in the copper in either case. When the ring is held as in Fig. 231(a), a *clockwise* current flows in it at a certain instant, hence the

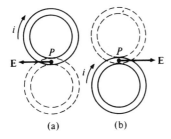

Fig. 231. In this diagram, the z axis (not shown) is normal to the page and points toward the reader.

vector **E** at P points leftward at that instant. If the ring were held as in Fig. 231(b) at that instant, the current would *also* flow clockwise, and the vector **E** at P would point rightward. What causes the directions of **E** at P to be different? Describe **E** at P when the ring is removed.

CHAPTER 23

Maxwell's Equations, Wave Equations, and the Flow of Energy

In this chapter we assemble together the four field equations of Maxwell that we discussed earlier one by one; and we recapitulate the "constitutive relations" interconnecting pairs of vector fields, such as **E** and **D**. Next, we take up fields in uncharged vacuum and illustrate various ways of writing Maxwell's equations in terms of *two* of the field vectors.

When Maxwell's equations are displayed together, three kinds of general questions are likely to arise: Which of the fields **B**, **D**, **E**, and **H** are more *important* than others? Which are most *convenient* to use in equations? Which electric quantities are *analogous* to which magnetic quantities? These questions are touched upon in §104.

The rest of this chapter is a preparation for the study of electromagnetic waves. Wave equations are introduced in §105 and the propagation of energy —Poynting's vector—in §106. Most of the applications will wait until the later chapters.

103. MAXWELL'S EQUATIONS

We took many pages to lead up to Maxwell's equations and to make them plausible. But now that this is done, we must stress the fact that they are the *assumptions* of Maxwell's theory. In a less elementary presentation of this theory, they might well appear on page one. In vector notation, they are:

Maxwell's Equations   § 103

(4^{101})
$$\text{curl } \mathbf{E} = -\frac{\partial \mathbf{B}}{\partial t}, \tag{1}$$

(3^{95})
$$\text{curl } \mathbf{H} = \mathbf{J} + \frac{\partial \mathbf{D}}{\partial t}, \tag{2}$$

(1^{67})
$$\text{div } \mathbf{D} = \rho, \tag{3}$$

(11^{87})
$$\text{div } \mathbf{B} = 0. \tag{4}$$

The "constitutive relations" interconnecting $\mathbf{D}$ with $\mathbf{E}$ and $\mathbf{B}$ with $\mathbf{H}$ are (4^{62}) and (3^{85}):

$$\mathbf{D} = \epsilon \mathbf{E}, \quad \mathbf{B} = \mu \mathbf{H}. \tag{5}$$

The letter $\mathbf{J}$ in (2) stands for the sum of the densities of conduction and convection currents. If the latter are ignored, as in this book, we have the further relation:

(6^{44})
$$\mathbf{J} = g\mathbf{E}. \tag{6}$$

Note that Maxwell's equations do not involve the electrostatic potential V, which we have discussed at some length. Nor do they involve the magnetic scalar potential of Chapter 21 or the magnetic vector potential to be introduced in Chapter 27. Potentials are very useful, however, for solving a great variety of physical problems.

If the medium is an uncharged insulator, we have $\rho = 0$ and $\mathbf{J} = 0$. If it is a charge-free vacuum, then ϵ and μ have the particular constant values ϵ_0 and μ_0, so the divergence equations (3) and (4) become

$$\text{div } \mathbf{E} = 0, \quad \text{div } \mathbf{H} = 0. \tag{7}$$

The curl equations (1) and (2) can then be written in various ways, such as

$$\text{curl } \mathbf{E} = -\mu_0 \frac{\partial \mathbf{H}}{\partial t}, \quad \text{curl } \mathbf{H} = \epsilon_0 \frac{\partial \mathbf{E}}{\partial t}, \tag{8}$$

or

$$\text{curl } \mathbf{D} = -\epsilon_0 \frac{\partial \mathbf{B}}{\partial t}, \quad \text{curl } \mathbf{B} = \mu_0 \frac{\partial \mathbf{D}}{\partial t}, \tag{9}$$

or, in view of (7^{95}),

$$\text{curl } \mathbf{E} = -\frac{\partial \mathbf{B}}{\partial t}, \quad \text{curl } \mathbf{B} = \frac{1}{c^2} \frac{\partial \mathbf{E}}{\partial t}. \tag{10}$$

Here

$$c^2 = \frac{1}{\epsilon_0 \mu_0};\qquad(11)$$

as we remarked in §95, c proves to be the speed of light in vacuum.

EXERCISE

1. Show that **D** and **H** satisfy equations similar to (10), namely

$$\operatorname{curl}\mathbf{H} = \frac{\partial \mathbf{D}}{\partial t},\qquad \operatorname{curl}\mathbf{D} = -\frac{1}{c^2}\frac{\partial \mathbf{H}}{\partial t}.\qquad(12)$$

104. THE FIELDS B, D, E, AND H

Is **E** in electricity analogous to **B** in magnetism, or is it analogous to **H**? Here are some reasons for saying that **E** is similar to **H**, and **D** is similar to **B**:

(a) Maxwell's equations contain the curls of **E** and **H**, but the divergences of **D** and **B**.

(b) A comparison of an electret with a bar magnet, as in Fig. 195[83], suggests pairing **E** with **H**, and **D** with **B**.

(c) The table on the inside front cover and Exercise 1[83] suggest that the weber is the "natural" magnetic counterpart of the coulomb and hence the "natural" choice of measure of magnetic pole strength; if we make this choice, and arrange various quantities according to the structure of their units, we get the table on the inside back cover, which correlates **E** with **H**, and **D** with **B**.

The remarks (a) and (b) are true, of course, but the situation is nevertheless not clear-cut, because one can certainly challenge the implications drawn above in (c). In fact, as in Exercise 1, we can choose the ampere · meter instead of the weber as a measure of magnetic pole strength; if we make this choice, **E** becomes correlated with **B**, and **D** with **H**, so far as the structure of their units is concerned. The standard argument for supporting this correlation and emphasizing the fields **E** and **B** has to do with microscopic considerations and runs as follows:

The definition of **E**, given in §42, makes no reference to atoms or molecules and holds equally well both macro- and microscopically. By contrast, after defining **D** in §61 in a macroscopic fashion, we remarked in §62 that a microscopic definition of **D** at a point in a dielectric reads

$$\mathbf{D} = \epsilon_0 \mathbf{E} + \mathbf{P},\qquad(1)$$

where **P** is the intensity of polarization of the dielectric at that point. Similarly, the definition of **B**, given in §84, holds both macro- and microscopically, but this is not so for **H**. In fact, after defining **H** in §85 in a macroscopic way, we mentioned the microscopic definition of **H** at a point in a magnetic material, namely,

$$\mathbf{H} = \frac{1}{\mu_0}(\mathbf{B} - \mathbf{P}_m), \tag{2}$$

where $\mathbf{P}_m$ is the intensity of magnetization of the material at that point, expressed in webers per square meter.

Thus, the fields **E** and **B** can be defined without referring to electric or magnetic properties of atoms and molecules. By contrast, the microscopic definitions (1) and (2) of **D** and **H** involve **P** and $\mathbf{P}_m$, and hence depend on atomic or molecular properties; the step from the micro- to the macroscopic definitions involves taking averages over multitudes of atoms. From this standpoint, **E** and **B** are similar, **D** is similar to **H**, and the pair to be emphasized is **E** and **B**.

So there are arguments for correlating **E** with **B**, and there are also arguments for correlating **E** with **H**. If and when magnetic monopoles are observed, this subject will have to be reviewed in a new setting. In the meantime, the question is not so much which correlation is correct, but which pair of fields—**E** and **B**, or **E** and **H**—is more meaningful, more descriptive, and more convenient in specific computations.

In particle dynamics (the study of motion of charged particles in electric and magnetic fields) the Lorentz force, $\mathbf{F} = q(\mathbf{E} + \mathbf{v} \times \mathbf{B})$, comes into play, with its emphasis on **E** and **B**. Also, the two equations (10^{103}), which interrelate **E** and **B**, contain only one parameter, the speed of light c. For this and other reasons, this pair of equations is particularly convenient for relativistic transformations.

In contrast to particle dynamics, in studies of electromagnetic waves the "intrinsic impedance" of the medium comes into play, a parameter that has to do with ratios of amplitudes of the electric and magnetic fields comprising a wave (§110). In our units, impedances are expressed in terms of the ohm, which is also our unit for the ratio of E to H (the volt/meter divided by the ampere/meter). As a result, the pair **E** and **H** is somewhat more convenient in a study of electromagnetic waves than the pair **E** and **B**, if the waves are discussed in terms of the MKS-Giorgi units. This is one of the main reasons why we lean in this book toward the correlations shown on cover three.

EXERCISES

1. For the purposes of this exercise, give the name "mag" to a unit of magnetic pole strength and set 1 mag = 1 ampere · meter. Express in terms of the mag

368 Maxwell's Equations

the entries in the fourth column on cover three (the weber, the newton/weber, and so on). Note that, so far as the structure of the units is concerned, **H** becomes correlated with **D**, and **B** with **E**. Identify any other correlations.

2. In books using the ampere · meter as the unit of magnetic pole strength, intensity of magnetization is denoted by **M**, and (2) is replaced by $\mathbf{H} = \mathbf{B}/\mu_0 - \mathbf{M}$. Check the units in this equation.

105. WAVE EQUATIONS

Let us find an equation satisfied by **E** when the medium is uncharged vacuum. To do this, we eliminate **B** from equations (10^{103}), as follows:

$$\text{curl curl } \mathbf{E} = -\text{curl}\frac{\partial \mathbf{B}}{\partial t} = -\frac{\partial}{\partial t}\text{curl } \mathbf{B} = -\frac{1}{c^2}\frac{\partial^2 \mathbf{E}}{\partial t^2}. \tag{1}$$

According to (47^A), curl curl $\mathbf{E} = \text{grad div } \mathbf{E} - \nabla^2 \mathbf{E}$. Also, in an uncharged medium div $\mathbf{E} = 0$, and consequently (1) reduces to

$$\nabla^2 \mathbf{E} = \frac{1}{c^2}\frac{\partial^2 \mathbf{E}}{\partial t^2}. \tag{2}$$

An equation of this kind is called a *wave equation*; it is a generalization of (48^5).

In cartesian coordinates we have the formula (51^A), and (2) reduces to the following scalar equations:

$$\nabla^2 E_x = \frac{1}{c^2}\frac{\partial^2}{\partial t^2}E_x, \quad \nabla^2 E_y = \frac{1}{c^2}\frac{\partial^2}{\partial t^2}E_y, \quad \nabla^2 E_z = \frac{1}{c^2}\frac{\partial^2}{\partial t^2}E_z. \tag{3}$$

More explicitly,

$$\frac{\partial^2}{\partial x^2}E_x + \frac{\partial^2}{\partial y^2}E_x + \frac{\partial^2}{\partial z^2}E_x - \frac{1}{c^2}\frac{\partial^2}{\partial t^2}E_x = 0, \tag{4}$$

with similar equations for E_y and E_z.

Example. Let

$$\mathbf{E} = E_z(x, t)\mathbf{1}_z, \tag{5}$$

so that, at any point $P(x, y, z)$ the vector **E** does not depend on y or z, and is parallel to the z axis. Equation (4) then becomes

$$\frac{\partial^2}{\partial x^2}E_z = \frac{1}{c^2}\frac{\partial^2}{\partial t^2}E_z. \tag{6}$$

The following simple solution of (6) can be inferred from (48^5), (19^3), and the fact that, according to (5), the x and y components of **E** are zero:

$$E_x = 0, \qquad E_y = 0, \qquad E_z = E_0 \cos(kx - \omega t); \qquad (7)$$

here

$$k = \frac{\omega}{c}, \qquad (8)$$

and E_0 and ω are arbitrary constants. More briefly,

$$\mathbf{E} = E_0 \cos(kx - \omega t)\mathbf{1}_z. \qquad (9)$$

When **E** is given by (9), we have

$$\text{curl } \mathbf{E} = kE_0 \sin(kx - \omega t)\mathbf{1}_y. \qquad (10)$$

Consequently, the equation curl $\mathbf{E} = -\mu_0\, \partial \mathbf{H}/\partial t$ implies that an electric field of the form (9) must be accompanied by a magnetic field that satisfies the following equations:

$$\frac{\partial H_x}{\partial t} = 0, \qquad -\mu_0 \frac{\partial H_y}{\partial t} = kE_0 \sin(kx - \omega t), \qquad \frac{\partial H_z}{\partial t} = 0. \qquad (11)$$

A simple solution of the set (11) is

$$H_x = 0, \qquad H_y = -H_0 \cos(kx - \omega t), \qquad H_z = 0, \qquad (12)$$

where, in view of (8) and (11^{103}),

$$H_0 = \sqrt{\frac{\epsilon_0}{\mu_0}} E_0. \qquad (13)$$

More briefly,

$$\mathbf{H} = -H_0 \cos(kx - \omega t)\mathbf{1}_y. \qquad (14)$$

Taken together, the electric field (9) and the magnetic field (14) comprise a plane *electromagnetic wave* propagating in the $+x$ direction.

The last term in (7) might well be written as $E_0 \cos(\pm kx - \omega t)$. These matters will become systematized when we introduce in §112 a *propagation vector*.

Effect of conductivity. The step from Maxwell's curl equations to equations (10^{103}) holds only for nonconducting mediums. If we do not ignore the current density **J**, but otherwise proceed much as before, we obtain the generalized wave equation

$$\nabla^2 \mathbf{E} = \epsilon\mu \frac{\partial^2 \mathbf{E}}{\partial t^2} + g\mu \frac{\partial \mathbf{E}}{\partial t} \tag{15}$$

instead of (2). The term involving $\partial/\partial t$ in (15) implies that, unless the conductivity g is zero, the electromagnetic disturbance is attenuated—just as the term involving d/dt in the equation $m\, d^2x/dt^2 + b\, dx/dt + kx = 0$ implies that a mechanical oscillation is damped.

EXERCISES

1. Verify that the fields (9) and (14) satisfy *both* equations (8^{103}).
2. Derive (15) and state the conditions on g, ϵ, μ, and ρ under which it holds.
3. Show that, in (uncharged) vacuum, $\mathbf{H}$ satisfies a wave equation of the form (2).

106. THE FLOW OF ELECTROMAGNETIC ENERGY

In §40 we claimed that electric charge cannot be created or destroyed, but can only be moved from place to place. We then considered a region R bounded by a closed surface (S), wrote q for the total charge contained in R, and arrived at Equation (4^{40}), which can be written as

$$-\frac{dq}{dt} = \iint_{(S)} J_n \, da. \tag{1}$$

This equation implies that the rate at which the charge contained in a region is decreasing, is equal to the rate at which charge is crossing the surface of the region from the inner to the outer side.

In this section we consider along similar lines the behavior of electromagnetic energy, rather than the behavior of electric charge. This problem is more complicated because, although energy is conserved, *electromagnetic* energy as such is not necessarily conserved, for it may be converted into heat or into other forms of energy, say mechanical or chemical. We will begin with a stationary homogeneous, isotropic, nonconducting, and chemically inert medium, so that electromagnetic energy *is* conserved. Later in this section we will turn to a conducting medium and allow for the generation of heat. The Lorentz forces acting on conduction electrons will be ignored in the computations—we mention them briefly only at the end. We will need the formulas (1^{100}), (2^{100}), and (5^{45}), namely,

$$u_e = \tfrac{1}{2}\mathbf{E} \cdot \mathbf{D}, \quad u_m = \tfrac{1}{2}\mathbf{H} \cdot \mathbf{B}, \quad h_v = \mathbf{E} \cdot \mathbf{J}, \tag{2}$$

as well as the identity

$$\text{div } \mathbf{E} \times \mathbf{H} = \mathbf{H} \cdot \text{curl } \mathbf{E} - \mathbf{E} \cdot \text{curl } \mathbf{H}, \tag{3}$$

which follows from (73$^\text{A}$). In (2), the scalar fields u_e and u_m are the densities of electric and magnetic energies (joules/meter^3), and h_v is the rate of generation of heat per unit volume (watts/meter^3).

Nonconductors. When we set $\mathbf{J} = 0$ and use Maxwell's equations (1^{103}) and (2^{103}) to eliminate the curls from (3), we get

$$\text{div } \mathbf{E} \times \mathbf{H} = -\mathbf{E} \cdot \frac{\partial \mathbf{D}}{\partial t} - \mathbf{H} \cdot \frac{\partial \mathbf{B}}{\partial t}. \tag{4}$$

Now,

$$\begin{aligned}
\mathbf{E} \cdot \frac{\partial \mathbf{D}}{\partial t} &= \mathbf{E} \cdot \frac{\partial}{\partial t}(\epsilon \mathbf{E}) \\
&= \epsilon \mathbf{E} \cdot \frac{\partial \mathbf{E}}{\partial t} = \frac{1}{2} \epsilon \frac{\partial}{\partial t}(\mathbf{E} \cdot \mathbf{E}) \\
&= \frac{1}{2} \frac{\partial}{\partial t}(\mathbf{E} \cdot \mathbf{D}) = \frac{\partial}{\partial t} u_e.
\end{aligned} \tag{5}$$

Similarly,

$$\mathbf{H} \cdot \frac{\partial \mathbf{B}}{\partial t} = \frac{\partial}{\partial t} u_m, \tag{6}$$

and hence

$$-\frac{\partial}{\partial t}(u_e + u_m) = \text{div } \mathbf{E} \times \mathbf{H}. \tag{7}$$

As before, we write U_e and U_m for the electric and magnetic energy (joules) contained in a region R. Integration of (7) over this region then yields

$$-\frac{d}{dt}(U_e + U_m) = \iiint_R \text{div } \mathbf{E} \times \mathbf{H} \, dv. \tag{8}$$

Consequently, by the divergence theorem,

$$-\frac{d}{dt}(U_e + U_m) = \iint_{(S)} (\mathbf{E} \times \mathbf{H})_n \, da, \tag{9}$$

where (S) is the surface of R.

The field $\mathbf{E} \times \mathbf{H}$ is called *Poynting's vector*.[1] To find a physical interpretation for it, we note that the structure of (9) is identical with that of (1), and that the description of the left-hand side of (9) can be obtained from that of (1) by replacing the words "electric charge" by the words "electromagnetic energy." Accordingly, we interpret the right-hand side of (9) as the time rate

---

[1] John Henry Poynting (1852-1914).

at which electromagnetic energy is leaving a region R by flowing across the surface (S) of R. Furthermore, since **J** is the density of the flow of electric charge (coulombs per second per square meter, or simply amperes/meter²), we interpret the vector **E** × **H** as *the density of flow of electromagnetic energy* (joules per second per square meter, or simply watts/meter²).

Example. A long straight wire carries a current i, as in Fig. 232. The point P lies on the curved portion of the imagined closed surface (S), which is coaxial with the wire and has the height l and the radius ρ. The resistance of the part of the wire contained in (S) is r, so that, inside the wire,

$$E = \frac{ir}{l}. \tag{10}$$

We assume that P lies so close to the wire that, because of the continuity of the tangential component of **E**, the formula (10) applies also at P, and **E** at P

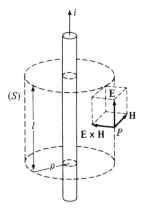

Fig. 232. At a point near a current-carrying wire, Poynting's vector, **E** × **H**, is directed toward the wire. This implies that, in the surrounding space, electromagnetic energy is streaming toward the wire, where it is converted into heat.

is parallel to the wire. The field **H** produced by the current is circular, and its magnitude at P is

$$(2^{88}) \qquad\qquad H = \frac{i}{2\pi\rho}. \tag{11}$$

The fields **E** and **H** in Fig. 232 are so oriented that Poynting's vector, **E** × **H**, is directed toward the wire. Consequently, according to the interpretation of this vector adopted above, *electromagnetic energy is streaming toward the wire from the surrounding space*. To find what happens to this energy, we will compute the integral of $(\mathbf{E} \times \mathbf{H})_n$ over (S) and evaluate the time rate at which this energy enters the region bounded by (S). Now, $(\mathbf{E} \times \mathbf{H})_n$ vanishes at points on the flat top and bottom of (S); on the curved part, the vector **E** × **H** points along the *negative* normal to (S), so that

$$(\mathbf{E} \times \mathbf{H})_n = -EH \sin 90° = -\frac{ir}{l} \cdot \frac{i}{2\pi\rho}. \tag{12}$$

To evaluate the integral, we multiply this result by the area of the curved part of (S), which is $2\pi\rho l$. Energy that flows *into* the region bounded by (S) crosses this surface *from front to back*. Therefore, multiplying the component $(\mathbf{E} \times \mathbf{H})_n$ in (12) by $2\pi\rho l$ and reversing the sign, we finally get the equation

$$\left.\begin{array}{l}\text{time rate at which electromagnetic} \\ \text{energy enters the region bounded} \\ \text{by } (S)\end{array}\right\} = i^2 r. \tag{13}$$

It follows that the rate at which electromagnetic energy streams from the surrounding space *toward* the wire is precisely the rate at which heat is generated *in* the wire. Thus, Maxwell's field theory accounts for the generation of heat without referring to such details as the transfer of kinetic energy from conduction electrons to the vibrations of the atomic lattice of the conductor—which we mentioned in §1.

Note that (13) does not involve the radius ρ of the surface of (S). This implies that electromagnetic energy is conserved until it reaches the wire, where it is converted into heat.

Conductors. Equation (9) does not hold *inside* the wire because in deriving it we ignored the term $\mathbf{J}$ in Maxwell's equation (2^{103}) for curl $\mathbf{H}$. If this term is kept, the right-hand side of (4) acquires the extra term $-\mathbf{E} \cdot \mathbf{J}$, which is $-h_v$ in the notation of (2); similarly, the right-hand side of (7) acquires the extra term $+h_v$, and (9) becomes

$$-\frac{d}{dt}(U_e + U_m) = h + \iint_{(S)} (\mathbf{E} \times \mathbf{H})_n \, da; \tag{14}$$

here, as in (6^{45}), h is the time rate (watts) of generation of heat in the region bounded by (S). Equation (14) is *Poynting's theorem*, which was also derived independently by Heaviside. It states that, if chemical and thermoelectric effects are ignored, then the rate at which electromagnetic energy is disappearing from a region R is equal to the rate at which it is converted into heat in R, plus the rate at which it is leaving across the surface of R.

Poynting's vector. We have interpreted an integral of $(\mathbf{E} \times \mathbf{H})_n$ over a *closed surface* as the time rate (watts) at which electromagnetic energy is crossing this surface. Equation (9) testifies to the soundness of this interpretation. In addition, we adopted the standard interpretation of the field $\mathbf{E} \times \mathbf{H}$ as the density (watts/meter^2) of the flow of electromagnetic energy. The added interpretation, however, does not rest on an equally straightforward basis, and arguments have been raised against it. Suppose, for example, that a small charged copper sphere is held in a uniform magnetic field, produced by a large permanent magnet. The field $\mathbf{E} \times \mathbf{H}$ is then circular near the sphere (Exercise 3). The standard interpretation of Poynting's vector then suggests that electromagnetic energy keeps circulating around the sphere, even though the fields $\mathbf{E}$ and $\mathbf{H}$ are not only static, but are also unrelated to each other. This may seem strange. A complete analysis involves the concepts of electromagnetic momentum and angular momentum. We will merely say that the standard interpretation of Poynting's vector has been very useful and has not led so far to any contradictions with experiment.

Lorentz forces. In such problems as the flow of charge past a cavity we ignored outright the magnetic fields caused by the currents. But since Poynting's formula includes $\mathbf{H}$, we should not dismiss these fields in this section without comment.

Let a current flow in a long copper rod of circular cross section. The average force acting on a conduction electron is given by (3^{84}) as $\mathbf{F} = -e(\mathbf{E} + \mathbf{v} \times \mathbf{B})$. Here $-e$ and $\mathbf{v}$ are, respectively, the charge and the average velocity of a conduction electron; $\mathbf{B}$ is the magnetic field caused at the position of this electron by the current flowing in the rod. Now, the vector $\mathbf{v} \times \mathbf{B}$ points inward, but $\mathbf{F}$ points along the rod. Hence $\mathbf{E}$ must have a transverse component. This component is provided by a nonuniform volume charge distribution that develops within the rod.[2] The term $\mathbf{v} \times \mathbf{B}$ is ordinarily very small compared to $\mathbf{E}$ (recall Exercise 2^1); but it does mean that our computations involving current densities call for a closer scrutiny, at least in principle.

EXERCISES

1. Write ρ_0 for the radius of the wire in Fig. 232, assume that the current density in the wire is uniform, and take the radius ρ of the surface (S) to be smaller than ρ_0. Then show that the time rate of generation of heat per unit volume (watts/meter^3) is constant throughout the wire. Use (3^{45}) as a check.

2. Assume that $\mathbf{E}$ and $\mathbf{H}$ are constant in time, and compute the divergence of Poynting's vector at a point in a conductor. Interpret the result in terms of energy flow, and use it to solve Exercise 1 without referring to surfaces.

3. A charged copper sphere, centered on the origin of a coordinate frame, is held in a uniform magnetic field of the form $H_0 \mathbf{1}_z$. Show that, according to the standard interpretation of Poynting's vector, electromagnetic energy keeps flowing outside the sphere in circles.

[2] M. A. Matzek and B. R. Russell, *American Journal of Physics*, Vol. 36 (1968), p. 905.

CHAPTER 24

Radiation from a Short Antenna

As our first example of the implications of Maxwell's four equations taken together, we will now consider the radiation emitted by an idealized "short" antenna, the limiting case of an antenna whose length tends to zero. We choose this problem primarily because this radiation has a readily visualizable source.

When Maxwell's equations are written in terms of the components of **E** and **H**, they turn into eight simultaneous scalar partial differential equations of the first order. Our problem is to find that solution of these equations which pertains to a short antenna, rather than to some other device that produces electromagnetic radiation. The standard procedure is to introduce an auxiliary vector field, called the magnetic vector potential and denoted by **A**. This field, however, does not appear in Maxwell's equations; to illustrate its usefulness, it is best to begin with Maxwell's equations as they stand, and to introduce **A** later. This order of presentation is likely to help the reader to appreciate the virtues of **A** more fully when he comes to Chapter 27.

In this chapter, we will find the proper solution by a series of reasonable guesses, followed by corrections. This time-honored process is a bit longer than one might wish, but it is straightforward at every step and illustrates the interdependence of **E** and **H** without the introduction of fields that are more abstract.

107. THE LOCAL FIELD

Our idealized antenna, which is far removed from the ground and other objects, consists of two straight wires with an a.e. generator connected between them, as in Fig. 233(a), and metal spheres at their free ends. The generator keeps sending electrons up and down the wires and reversing the charges on the spheres, so that the antenna has an alternating electric moment. We will presently let the length l of the antenna tend to zero.

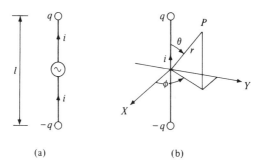

Fig. 233. An idealized antenna.

We assume that, at any instant, the current i in the antenna has the same value at every cross section of the antenna, given by the formula

$$i = I \cos \omega t. \tag{1}$$

We write c for the speed of propagation of electromagnetic effects in vacuum or air, and recall the definition of the propagation constant:

$$(21^3) \qquad k = \frac{\omega}{c}. \tag{2}$$

For definiteness, we let a positive i mean an *upward* current in Fig. 233. The charge q on the upper end of the antenna then satisfies the equation $dq/dt = i$. We choose the simplest solution, namely,

$$q = \frac{I}{\omega} \sin \omega t = \frac{I}{ck} \sin \omega t, \tag{3}$$

so that the electric moment of the antenna at the instant t is

$$p_e = ql = \frac{lI}{ck} \sin \omega t. \tag{4}$$

We assume that the antenna is so short that the distance from the point P in Fig. 233(b) to any part of the antenna is practically equal to r. The time required for an electromagnetic effect to travel from the antenna to P is then

r/c. Consequently, the electric field at P at an instant t will be determined by the electric moment of the antenna at an *earlier* instant, say τ, given by the formula

$$\tau = t - \frac{r}{c} \tag{5}$$

and called the *retarded time*. Thus, the effects "felt" at P at an instant t are caused not by the current and electric moment (1) and (4), but by their earlier values, namely,

$$i = I \cos \omega \tau \tag{6}$$

and

$$p_e = \frac{lI}{ck} \sin \omega \tau. \tag{7}$$

We write $\mathbf{E}^a$ and $\mathbf{H}^a$ for the first approximation to the complete fields $\mathbf{E}$ and $\mathbf{H}$, write $\mathbf{E}^b$ and $\mathbf{H}^b$ for the second approximation, and so on. To evaluate $\mathbf{E}^a$, we substitute (7) into (5^{60}) and find that

$$E_r^a = \frac{2M}{ck\epsilon_0} \frac{\cos\theta}{r^3} \sin\omega\tau, \quad E_\theta^a = \frac{M}{ck\epsilon_0} \frac{\sin\theta}{r^3} \sin\omega\tau, \quad E_\phi^a = 0, \tag{8}$$

where

$$M = \frac{lI}{4\pi}. \tag{9}$$

Remember that Equations (5^{60}) were derived for a "short" dipole.

To get our first approximation to $\mathbf{H}$ at P, we write l for dl and $\mathbf{H}^a$ for $d\mathbf{H}$ in (12^{88}), and find that

$$H_r^a = 0, \quad H_\theta^a = 0, \quad H_\phi^a = M \frac{\sin\theta}{r^2} \cos\omega\tau. \tag{10}$$

Note that $\mathbf{H}^a$ involves $\cos\omega\tau$ rather than $\sin\omega\tau$, and hence is 90 degrees out of phase with $\mathbf{E}^a$.

On the basis of our derivation of (8) and (10), one can say, at least roughly, that the field $\mathbf{E}^a$ is produced "directly" by the charges on the ends of the antenna and that $\mathbf{H}^a$ is produced "directly" by the current flowing in the antenna. These fields are, of course, not exact: to put things roughly again, the time-varying magnetic field $\mathbf{H}^a$ will contribute an extra time-varying electric field (Faraday's emf law); the field $\mathbf{E}^a$ will contribute an extra magnetic field (Maxwell's mmf law); and so on. The resultants of all these fields will satisfy Maxwell's equations. However, we would expect that, close enough to the antenna, the fields produced "directly" will be the dominant fields. Therefore, we impose the following conditions on the complete fields $\mathbf{E}$ and $\mathbf{H}$ at P:

$$\mathbf{E} \to \mathbf{E}^a, \quad \mathbf{H} \to \mathbf{H}^a, \quad \text{when } r \to 0. \tag{11}$$

In view of (11), the combination of the fields $\mathbf{E}^a$ and $\mathbf{H}^a$ is called the *local field* of the antenna; the region where this field is a sufficiently good approximation is called the *near zone*. Note that

$$\omega\tau = \omega t - kr, \tag{12}$$

and remember that $\tau \to t$ when $r \to 0$.

Point dipoles. The lines of $\mathbf{E}$ of an electric dipole, shown in Fig. 61[17], begin on the positive and end on the negative charge, located at the points P_1 and P_2, say. Therefore, the divergence of $\mathbf{E}$ is a delta function at P_1, a delta function of opposite sign at P_2, and zero elsewhere. Now, according to Exercise 8[60],

$$\text{div } \mathbf{E}^a = 0 \quad \text{for } r \neq 0. \tag{13}$$

Consequently, the lines of $\mathbf{E}^a$ can begin and end only at the origin, as though the length l of the antenna were zero. If so, the constant M in (9) would vanish, and with it the fields $\mathbf{E}^a$ and $\mathbf{H}^a$, unless we require that the amplitude I of the current increase without limit. Accordingly, our idealized antenna can be described as follows:

$$l \to 0, \quad I \to \infty, \quad lI = M = \text{constant}. \tag{14}$$

An antenna of this kind is called an oscillating *Hertzian dipole*,[1] a *differential antenna*, or, because in the limit it shrinks to a point, a *point-dipole* antenna. Besides our "electric" antenna, there are also "magnetic" or "loop" antennas. They radiate electromagnetic waves because of an alternating magnetic moment, produced by an alternating current maintained in a wire loop.

The theory of a Hertzian dipole serves as a stepping stone to the study of actual antennas. Note, however, that even in the case of the simplest actual antennas, two of our simplifying assumptions may fail to apply: The current in an antenna usually does not have the same value at every cross section of the antenna, and the proximity of the ground may cause the "image" of the antenna in the ground to play an essential role.

In the atomic domain, the theory of the point-dipole antenna is of interest in the study of the emission of light by individual atoms.

EXERCISE

1. Verify (13).

---

[1] Heinrich Rudolf Hertz (1857-1894).

108. THE COMPLETE FIELD

Maxwell's free-space equations can be written as

(7¹⁰³) $$\text{div } \mathbf{E} = 0, \quad \text{div } \mathbf{H} = 0, \tag{1}$$

and

(8¹⁰³) $$\text{curl } \mathbf{E} = -\mu_0 \frac{\partial \mathbf{H}}{\partial t}, \quad \text{curl } \mathbf{H} = \epsilon_0 \frac{\partial \mathbf{E}}{\partial t}. \tag{2}$$

The fields $\mathbf{E}^a$ and $\mathbf{H}^a$ do not satisfy these equations exactly. We will, therefore, proceed to use the simplest components to uncover discrepancies and will let these discrepancies guide us in a search for corrections. When several alternatives offer themselves, we will first try the simplest. In addition to the formulas for curls given in Appendix A, we will need two formulas that follow from (12¹⁰⁷):

$$\frac{\partial}{\partial r} \sin \omega\tau = -k \cos \omega\tau, \quad \frac{\partial}{\partial r} \cos \omega\tau = k \sin \omega\tau. \tag{3}$$

First correction. Maxwell's equations involve curl E and curl H. Since $\mathbf{H}^a$ is a simpler field than $\mathbf{E}^a$, we begin with curl H and see whether or not the tentative equation curl $\mathbf{H}^a = \epsilon_0 \, \partial \mathbf{E}^a/\partial t$ is satisfied. The ϕ components of both sides vanish, and the r components prove to be equal. But the θ components lead to a discrepancy: on the one hand

$$\text{curl}_\theta \mathbf{H}^a = M \frac{\sin \theta}{r^3} \cos \omega\tau - kM \frac{\sin \theta}{r^2} \sin \omega\tau, \tag{4}$$

while on the other hand, since $k = \omega/c$, we have

$$\epsilon_0 \frac{\partial}{\partial t} E_\theta^a = M \frac{\sin \theta}{r^3} \cos \omega\tau. \tag{5}$$

It is, perhaps, obvious that this discrepancy can be obviated by introducing a partly corrected field $\mathbf{E}^b$, such that

$$E_r^b = E_r^a, \quad E_\theta^b = E_\theta^a + \frac{kM}{\omega\epsilon_0} \frac{\sin \theta}{r^2} \cos \omega\tau, \quad E_\phi^b = 0. \tag{6}$$

When $r \to 0$, the added term in (6) becomes negligible compared to E_θ^a and does not violate (11¹⁰⁷). For uniformity of notation, we write

$$H_r^b = 0, \quad H_\theta^b = 0, \quad H_\phi^b = H_\phi^a. \tag{7}$$

Second correction. Next, we test the fields $\mathbf{E}^b$ and $\mathbf{H}^b$ by computing the ϕ components of the two sides of the equation curl $\mathbf{E} = -\mu_0 \partial \mathbf{H}/\partial t$. The results are

$$\text{curl}_\phi \, \mathbf{E}^b = \frac{\omega M}{c^2 \epsilon_0} \frac{\sin\theta}{r^2} \sin\omega\tau - \frac{2M}{c\epsilon_0} \frac{\sin\theta}{r^3} \cos\omega\tau \tag{8}$$

and

$$-\mu_0 \frac{\partial}{\partial t} H_\phi^b = \mu_0 \omega M \frac{\sin\theta}{r^2} \sin\omega\tau. \tag{9}$$

So far, the value of the speed c appearing in our equations has not been specified. But now we see that the terms in $\sin\omega\tau$ in (8) and (9) will agree if we set

$$c = \frac{1}{\sqrt{\epsilon_0 \mu_0}} = 3.00 \times 10^8 \, \frac{\text{meters}}{\text{second}}, \tag{10}$$

as we did without proof in (8^{95}) and again in (11^{103}).

The simplest way to remove the remaining discrepancy between the right-hand sides of (8) and (9) is to add to H_ϕ^b a term in $\sin\omega\tau$. Such a term, however, would have to include the factor $1/r^3$, would increase faster than H_ϕ^a when $r \to 0$, and hence would violate (11^{107}). The remaining possibility is to modify $\mathbf{E}^b$ in such a way as to make the term in $\cos\omega\tau$ disappear from (8). We have already corrected the θ component of the electric field, and, because of the axial symmetry of the antenna, we would expect that the ϕ component should be zero. Therefore, we turn to the r component and add to E_r^b an extra term, say f. According to (71^A), f would contribute to $\text{curl}_\phi \mathbf{E}$ the term $-r^{-1} \partial f/\partial\theta$. To make this term cancel the last term in (8), we let

$$-\frac{1}{r} \frac{\partial f}{\partial \theta} = \frac{2M}{c\epsilon_0} \frac{\sin\theta}{r^3} \cos\omega\tau. \tag{11}$$

Adding to E_r^b the simplest solution of (11), we then get the twice-corrected electric field, whose components are

$$E_r^c = E_r^b + \frac{2M}{c\epsilon_0} \frac{\cos\theta}{r^2} \cos\omega\tau, \qquad E_\theta^c = E_\theta^b, \qquad E_\phi^c = 0. \tag{12}$$

We also write $H_r^c = 0$, $H_\theta^c = 0$, and $H_\phi^c = H_\phi^b = H_\phi^a$.

Third correction. We have

$$\text{curl}_r \mathbf{H}^c = 2M \frac{\cos\theta}{r^3} \cos\omega\tau \tag{13}$$

and

$$\epsilon_0 \frac{\partial}{\partial t} E_r^c = 2M \frac{\cos\theta}{r^3} \cos\omega\tau - 2kM \frac{\cos\theta}{r^2} \sin\omega\tau. \tag{14}$$

The last term in (14), which causes the discrepancy, originates in the correction that led us from $\mathbf{E}^b$ to $\mathbf{E}^c$. Therefore, we will now modify $\mathbf{H}^c$ by adding to H_ϕ^c an extra term, say g. According to (69^A), g will contribute to curl$_r$ H^c the term $(r \sin \theta)^{-1} \partial(g \sin \theta)/\partial\theta$, which we want to be equal to the last term in (14). Therefore, we set

$$\frac{1}{r \sin \theta} \frac{\partial(g \sin \theta)}{\partial \theta} = -2kM \frac{\cos \theta}{r^2} \sin \omega\tau. \tag{15}$$

The simplest solution of (15) is $g = -kMr^{-1} \sin \theta \sin \omega\tau$. When we add it to H_ϕ^c and change superscripts from c to d, we get the field

$$H_r^d = 0, \qquad H_\theta^d = 0, \qquad H_\phi^d = H_\phi^c - kM \frac{\sin \theta}{r} \sin \omega\tau. \tag{16}$$

We also replace E_r^c by E_r^d, and so on.

Fourth and last correction. Our first correction involved H^a and E^a. Now that we have changed to H^d and E^d, we must return to the equation curl$_\theta$ H = $\epsilon_0 \partial E_\theta/\partial t$. The pertinent formulas are

$$\text{curl}_\theta \, \mathbf{H}^d = M \frac{\sin \theta}{r^3} \cos \omega\tau - kM \frac{\sin \theta}{r^2} \sin \omega\tau - k^2 M \frac{\sin \theta}{r} \cos \omega\tau \tag{17}$$

and

$$\epsilon_0 \frac{\partial}{\partial t} E_\theta^d = M \frac{\sin \theta}{r^3} \cos \omega\tau - kM \frac{\sin \theta}{r^2} \sin \omega\tau. \tag{18}$$

The discrepancy between (17) and (18) can be removed by adding to E_θ^d the term $-(kM/c\epsilon_0)r^{-1} \sin \theta \sin \omega\tau$. When this is done, *all* of Maxwell's field equations become satisfied (Exercise 1). This fact implies, in particular, that the speed of propagation of electromagnetic radiation in free space is, indeed, given by (10) and is equal to the experimental value of the speed of light in free space.

The complete field. When we correct the fields E^a and H^a as outlined above, we get the following complete electromagnetic field of a point-dipole antenna oriented as in Fig. 233(b):

$$E_r = \frac{2M}{\omega\epsilon_0} \frac{\cos \theta}{r^3} \sin \omega\tau + \frac{2M}{c\epsilon_0} \frac{\cos \theta}{r^2} \cos \omega\tau, \tag{19}$$

$$E_\theta = \frac{M}{\omega\epsilon_0} \frac{\sin \theta}{r^3} \sin \omega\tau + \frac{M}{c\epsilon_0} \frac{\sin \theta}{r^2} \cos \omega\tau - \frac{kM}{c\epsilon_0} \frac{\sin \theta}{r} \sin \omega\tau, \tag{20}$$

$$E_\phi = 0, \tag{21}$$

and

$$H_r = 0, \qquad H_\theta = 0, \qquad H_\phi = M \frac{\sin \theta}{r^2} \cos \omega\tau - kM \frac{\sin \theta}{r} \sin \omega\tau. \tag{22}$$

382 Radiation from a Short Antenna § 108

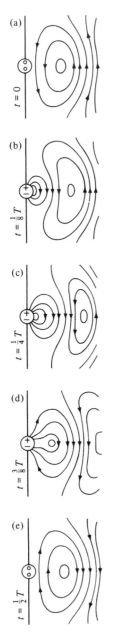

Fig. 234. Sucessive maps of the field **E** of a short antenna.

Line maps of **E**. The lines of the complete field **E** in the vicinity of the antenna are pictured in Fig. 234 for a half-plane containing the antenna and the case when oscillations of period T began long before time zero; this figure is patterned after that published in 1889 by Hertz.[2]

In part (a), where $t = 0$, the charge (3^{107}) on the antenna is zero, the current i in the antenna, given by (1^{107}), is at a maximum, and $di/dt = 0$. To lead up to (a), we will consider first the other parts of the figure.

In (b), where $t = \frac{1}{8}T$, the ends of the antenna have become charged. Very near to the antenna, lines of **E** run from positive to negative charges, and the electric field is essentially the local field $\mathbf{E}^a$.

In (c), the charges have reached their maximum magnitudes, and the number of lines of **E** running from one end of the antenna to the other is largest. At this time, however, the magnitude of di/dt is at a maximum, and hence the rate of change of **H** near the antenna is largest. Consequently, very close to the antenna we have the field $\mathbf{E}^a$ pertaining to $t = \frac{1}{4}T$, but farther out the changing magnetic field induces an additional electric field of considerable strength, and the lines of **E** are no longer shaped as the lines of a static dipole.

In (d), the magnitudes of the charges on the antenna have decreased, and so has the number of lines of **E** that begin and end on them. However, because of the electric field induced magnetically, the lines near the antenna no longer resemble the lines of a static dipole—they

[2]Heinrich Hertz, *Electric Waves* (New York: Dover Publications Inc., 1962), pp. 144–145.

even include *closed lines*, completely detached from the antenna. And the lines that still begin and end on the antenna are about to be pinched off and to close upon themselves.

In (e), a half-cycle is completed. The antenna is again uncharged, and no lines of E begin or end on it. The lines that were about to become detached in part (d) are now closed and are moving away.

Part (a) pertains to the end of the previous half-cycle, during which the upper end of the antenna was charged negatively. Consequently, parts (a) and (e) are just alike, except that the directions of the lines are reversed.

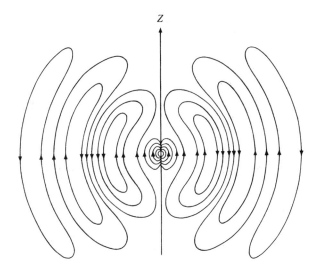

Fig. 235. An instantaneous map of E, extending from a short antenna beyond a wavelength of the radiation.

The general features of E farther out from the antenna are suggested for a particular instant of time in Fig. 235.[3]

EXERCISES

1. Verify that the fields (19) through (22) satisfy all four of Maxwell's free-space equations (1) and (2), which comprise eight scalar equations in the components of **E** and **H**.

---

[3] Patterned, by permission of the publishers, after Fig. 60 of *Fundamentals of Electric Waves* by H. H. Skilling (New York and London: John Wiley & Sons, Inc.), 2nd ed., 1948, p. 169.

384 Radiation from a Short Antenna

2. Use (47^A) to show in detail that, for the field **H** given in (22),

$$\nabla^2 \mathbf{H} = \left(-k^2 M \frac{\sin\theta}{r^2} \cos\omega\tau + k^3 M \frac{\sin\theta}{r} \sin\omega\tau\right) \mathbf{1}_\phi. \quad (23)$$

Also verify that this **H** satisfies a wave equation of type (2^{105}).

3. Describe the instantaneous direction of **H** at several points in Fig. 235.

109. THE DISTANT FIELD

If r is small enough, the complete field reduces to the local fields $\mathbf{E}^a$ and $\mathbf{H}^a$. In this section we consider the opposite extreme and let r grow large. The dominating terms in the complete field are then the terms in $1/r$, and this field reduces to the *distant field*, namely,

$$\mathbf{E} = \frac{kM}{c\epsilon_0} \frac{\sin\theta}{r} \sin(kr - \omega t) \mathbf{1}_\theta \quad (1)$$

and

$$\mathbf{H} = kM \frac{\sin\theta}{r} \sin(kr - \omega t) \mathbf{1}_\phi. \quad (2)$$

The region where this approximation is applicable is called the *far zone* of the antenna. Note that in this zone the fields **E** and **H** are mutually perpendicular and that, in contrast to the near zone, they are *in phase* with each other. These features are illustrated in Fig. 236 without regard for scales.

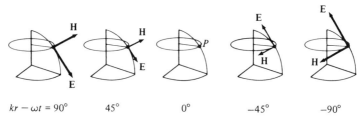

$kr - \omega t = 90°$ $45°$ $0°$ $-45°$ $-90°$

Fig. 236. Successive diagrams of **E** and **H** at a fixed point in the far zone of a short vertical antenna.

The lines of **E** in Fig. 235 *can* close upon themselves, because the complete field **E** has a radial component. The lines of the field (1) *cannot* do this, because now $E_r = 0$. Also, because of the factor $\sin\theta$, the density of these lines decreases as we move along a meridian toward the z axis. Consequently, they have beginnings and ends, and div $\mathbf{E} \neq 0$ (Exercise 3). The fields (1) and (2) are, nevertheless, very useful, despite this blemish.

The Distant Field §109

Poynting's vector in the far zone is

$$\mathbf{E} \times \mathbf{H} = \frac{k^2 M^2}{c\epsilon_0} \frac{\sin^2\theta}{r^2} \sin^2(kr - \omega t)\mathbf{1}_r. \tag{3}$$

We will now find the time average of this vector.

Let $f(t)$ be a periodic function of t with the period T, and write $\langle f(t) \rangle$ for its average value during a period. Next, let $f(t) = \sin^2\alpha$, where $\alpha = \omega(t - r/c)$ and r is kept constant. It is perhaps obvious that $\langle \sin^2\alpha \rangle = \langle \cos^2\alpha \rangle$. But $\sin^2\alpha + \cos^2\alpha = 1$, and therefore $\langle \sin^2\alpha \rangle = \langle \cos^2\alpha \rangle = \frac{1}{2}$. Consequently,

$$\left\langle \sin^2\omega\left(t - \frac{r}{c}\right) \right\rangle = \frac{1}{2} \tag{4}$$

and

$$\langle \mathbf{E} \times \mathbf{H} \rangle = \frac{k^2 M^2}{2c\epsilon_0} \frac{\sin^2\theta}{r^2} \mathbf{1}_r. \tag{5}$$

The coefficient of $\mathbf{1}_r$ in (3) is never negative, so, in the far zone, electromagnetic energy never flows back toward the antenna. The average density (watts/meter²) of the flow away from the antenna is equal to the coefficient of $\mathbf{1}_r$ in (5). This rate is an *inverse-square* function of the distance r.

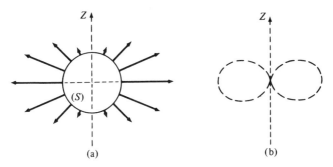

Fig. 237. Radiation pattern of a short antenna: (a) Poynting's vectors at points on a spherical surface centered on the antenna; (b) outline of the tips of these vectors when their tails are placed at the antenna.

To display the effect of the factor $\sin^2\theta$, we consider a spherical surface (S) of radius r, lying in the far zone and centered on the origin. The average Poynting vector at several points on (S) can then be pictured as in Fig. 237(a), which pertains to any plane containing the z axis. The standard way of picturing the *radiation pattern* of the antenna is shown in Fig. 237(b), in which (S) is shrunk to a point and only the locus of the tips of the Poynting vectors is plotted. The factor $\sin^2\theta$ causes the density (watts/meter²) of the energy flow to be largest in the xy plane and zero up and down the z axis.

Let $\langle w \rangle$ be the average power (watts) sent out by the antenna across the entire surface (S). Since

$$\langle w \rangle = \iint_{(S)} \langle \mathbf{E} \times \mathbf{H} \rangle_n \, da$$
$$= \frac{k^2 M^2}{2c\epsilon_0} \int_0^{2\pi} \int_0^{\pi} \left(\frac{\sin^2 \theta}{r^2}\right) r^2 \sin \theta \, d\theta \, d\phi, \tag{6}$$

we have

$$\langle w \rangle = \frac{4\pi k^2 M^2}{3c\epsilon_0}. \tag{7}$$

Equation (7) is not restricted to a surface in the far zone (Exercise 2), and the fact that it does not involve r implies that, in free space, the electromagnetic energy emitted by the antenna is not being converted into other forms of energy.

EXERCISES

1. Verify (4) by integration.

2. Compute Poynting's vector for the complete field [Equations (19^{108}) through (22^{108})] and show that (5) holds for that field.

3. Show that the distant fields (1) and (2) satisfy Maxwell's equations for empty space, except that div $\mathbf{E} \neq 0$. Evaluate div $\mathbf{E}$ explicitly and show that it is compatible with the pictorial argument given in the paragraph preceding (3).

4. Compute and compare the energy densities u_e and u_m at a point $P(r, \theta, \phi)$ in the far zone and plot $u_e + u_m$ as a function of t. Also plot, as a function of t, the divergence of Poynting's vector at P and verify that the two graphs are consistent. Finally, take time averages of $u_e + u_m$ and of div $\mathbf{E} \times \mathbf{H}$, show that *they* are consistent, and discuss them in the light of the results found in Exercise 2^{106} for *static* fields in vacuum.

5. If distances are measured from the origin, the distance between successive crests of the factor $\sin(kr - \omega t)$ in (1) and (2) at any instant, called the *wavelength* of the radiation, is

$$\lambda = \frac{2\pi}{k}. \tag{6}$$

Show that (6) is consistent with (18^3).

If r is large enough, the complete field reduces to the distant field. Express in terms of r and λ the condition for being "large enough."

6. Besides solutions involving the *retarded time*, $t - r/c$, Maxwell's equations have solutions involving an *advanced time*, $t + r/c$. For example, the following approximate solutions hold in the far zone of a short antenna:

$$\mathbf{E} = \frac{kM}{c\epsilon_0} \frac{\sin\theta}{r} \sin\omega\left(t + \frac{r}{c}\right) \mathbf{1}_\theta \tag{7}$$

and

$$\mathbf{H} = -kM \frac{\sin\theta}{r} \sin\omega\left(t + \frac{r}{c}\right) \mathbf{1}_\phi. \tag{8}$$

Verify that (7) and (8) satisfy Maxwell's equations, except that, as in (1), div $\mathbf{E} \neq 0$. Point out some curious physical features of (7) and (8).

CHAPTER 25

Plane Electromagnetic Waves

A line of **E** in Fig. 236[109] runs along the meridian of a spherical surface, say (S), and a line of **H** runs along a parallel of latitude. In this chapter we consider a region R in the far zone that subtends at the antenna small angles $d\theta$ and $d\phi$, as in Fig. 238. This region contains portions of spherical surfaces such as (S), but we assume that $d\theta$ and $d\phi$ are small enough to permit replacing these portions by planes. When this is done, we find that the distant fields (1^{109}) and (2^{109}) go over into *exact* solutions of Maxwell's equations, called *plane waves*.

Until §114 we deal with waves in vacuum.

110. PASSAGE TO PLANE WAVES

A point $P(x, y > 0, z)$ lies in the far zone in the region R of Fig. 238, and so close to the y axis that

$$|x| \ll y, \qquad |z| \ll y. \tag{1}$$

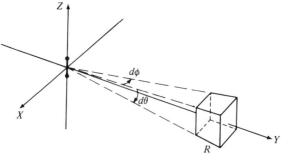

Fig. 238. A region R in the far zone of a short antenna.

The following approximations then hold at P:

$$r \approx y, \quad \sin \theta \approx 1, \tag{2}$$

and

$$\mathbf{1}_r \approx \mathbf{1}_y, \quad \mathbf{1}_\theta \approx -\mathbf{1}_z, \quad \mathbf{1}_\phi \approx -\mathbf{1}_x. \tag{3}$$

When we make these approximations in (1^{109}) and (2^{109}), and change the signs of $\mathbf{E}$ and $\mathbf{H}$ to simplify appearances, we get the approximate formulas

$$\mathbf{E} = \frac{kM}{c\epsilon_0} \frac{1}{y} \sin(ky - \omega t)\mathbf{1}_z, \quad \mathbf{H} = kM \frac{1}{y} \sin(ky - \omega t)\mathbf{1}_x. \tag{4}$$

Since $d(1/y)/dy = -1/y^2$, we conclude that, if y is large enough, the coefficients of the sine factors in (4) can be regarded as constants. We denote these coefficients by E_0 and H_0, and write

$$\eta_0 = \sqrt{\frac{\mu_0}{\epsilon_0}} \approx 377 \text{ ohms}. \tag{5}$$

In terms of the new symbols, the field (4) becomes

$$\mathbf{E} = E_0 \sin(ky - \omega t)\mathbf{1}_z, \quad \mathbf{H} = H_0 \sin(ky - \omega t)\mathbf{1}_x, \tag{6}$$

where the constants E_0 and H_0 satisfy the equation

$$E_0 = \eta_0 H_0. \tag{7}$$

The parameter η_0 is called the *intrinsic impedance* of free space.[1] We used its reciprocal in (13^{105}).

The electromagnetic field (6) can be pictured by a set of line maps, as in Fig. 239, or by curves that indicate the directions and magnitudes of $\mathbf{E}$ and $\mathbf{H}$, as in Fig. 240. Each of these figures pertains to a fixed instant of time. Comparison with (19^3) shows that the field (6) is "moving" in the $+y$ direction.

What we have done above to derive (6) is to make further approximations in the approximate solutions (1^{109}) and (2^{109}) for the far zone. Equation (6), however, proves to be an *exact* solution of Maxwell's equations—and not only far from the origin, but for *any* value of y (Exercise 1). In fact, it is usually convenient to picture the wave (6) as produced by an antenna located not at the origin, but on the negative part of the y axis and far from the origin.

A person familiar with partial differential equations can, of course, write the solution (6) without referring to any antennas. One of the advantages of leading up to (6) by first discussing the radiation from an antenna is that this

---

[1] S. A. Schelkunoff, *Bell System Technical Journal*, Vol. 3 (1938), p. 17.

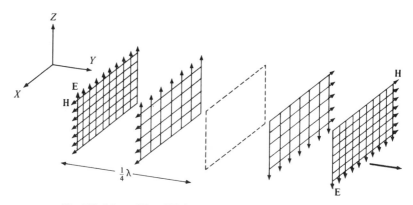

Fig. 239. Maps of **E** and **H** describing, at a fixed value of t, a plane wave moving in the $+y$ direction.

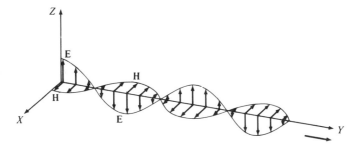

Fig. 240. A plane wave moving in the $+y$ direction, described at a fixed value of t by the vectors **E** and **H** at points on the y axis.

should help to make clear that, *physically*, electromagnetic waves can be regarded as plane only in limited regions. For example, the light radiated by an atom in the sun reaches the earth as a plane wave, but in the larger view it is, of course, a spherical wave.

EXERCISES

1. Show that, if E_0, H_0, and ϕ_0 are constants and $E_0 = \eta_0 H_0$, then the fields

$$\mathbf{E} = E_0 \cos(ky - \omega t + \phi_0)\mathbf{1}_z, \qquad \mathbf{H} = H_0 \cos(ky - \omega t + \phi_0)\mathbf{1}_x, \qquad (8)$$

satisfy all four of Maxwell's equations for free space, and that so do the fields

$$\mathbf{E} = E_0 \cos(-ky - \omega t + \phi_0)\mathbf{1}_z, \quad \mathbf{H} = -H_0 \cos(-ky - \omega t + \phi_0)\mathbf{1}_x. \quad (9)$$

Show also that both **E**'s satisfy (2^{105}) and that the **H**'s satisfy a similar wave equation.

2. Show that in the case (8) electromagnetic energy is flowing in the $+y$ direction, and that the time average of the rate of the flow of energy across any plane perpendicular to the y axis is

$$|\langle \mathbf{E} \times \mathbf{H} \rangle| = \frac{1}{2} E_0 H_0 \frac{\text{watts}}{\text{meter}^2}. \tag{10}$$

Analyze (9) in a similar way.

3. Transform (1^{109}) and (2^{109}) into plane waves near a point in the *far zone* that lies (a) near the negative part of the y axis, (b) near a line passing through the origin and having arbitrary direction cosines, say $\cos \alpha_x$, $\cos \alpha_y$, and $\cos \alpha_z$. Verify that your solutions satisfy Maxwell's free-space equations, as well as the wave equations for **E** and **H**.

4. As an aid to memory, verify that, by coincidence, the numerical value of η_0, expressed in ohms, is nearly equal to the value of ω for a 60-cycle circuit, expressed in radians per second.

5. With the scales used in Fig. 240, the amplitudes of **E** and **H** are about equal. How would they compare if 1 inch were used to represent 1 volt/meter on the **E** curve and 1 ampere/meter on the **H** curve?

111. RADIATION FROM A SHEET OF ALTERNATING CURRENT

To get another idealized example of plane waves, we imagine that the entire zx plane in Fig. 241 carries in the $+z$ direction a steady current, whose uniform linear density is constant and equal to J^* amperes per meter of width. The field **H** at a point $P(x, y, z)$ then depends only on the sign of the y coordinate of P; in fact,

(6^{88})
$$\mathbf{H} = \begin{cases} +\tfrac{1}{2} J^* \mathbf{1}_x & \text{if } y < 0 \\ -\tfrac{1}{2} J^* \mathbf{1}_x & \text{if } y > 0. \end{cases} \tag{1}$$

Next, suppose that the current flowing in the zx plane is not steady, but has the time-varying linear density

$$J_x^* = 0, \qquad J_y^* = 0, \qquad J_z^* = J_0^* \cos \omega t, \tag{2}$$

where J_0^* is a positive constant. We will compute the fields **H** and **E** for this case.

We begin with the lower line in (1), consider the point P_1 in Fig. 241, assume that the fields at P_1 do not depend on the z and x coordinates of P_1, and write c for the speed of propagation of electromagnetic effects in vacuum.

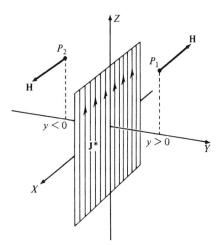

Fig. 241. A current of uniform and constant linear density **J*** is spread over an infinite plane sheet. It produces a field **H** that is perpendicular to **J*** and parallel to the sheet. The magnitude of **H** is the same everywhere outside the sheet, but the direction of **H** reverses as we cross the sheet.

The time required for such an effect to reach P_1 from the zx plane is then y/c, and hence we conclude that $\mathbf{H} = -\tfrac{1}{2}J_0^* \cos \omega(t - y/c)\mathbf{1}_x$ at P_1. When we write

$$H_0 = \tfrac{1}{2}J_0^* \tag{3}$$

and $k = \omega/c$, we get the lower line of (4). Next, we turn to the top line in (1), consider the point P_2 in the figure, note that (since y is now negative) the propagation time from the zx plane to P_2 is the positive time $-y/c$, and get complete tentative formulas for **H**:

$$H_x = \begin{cases} H_0 \cos(-ky - \omega t) & \text{if } y < 0 \\ -H_0 \cos(ky - \omega t) & \text{if } y > 0, \end{cases} \tag{4}$$

and

$$H_y = 0, \quad H_z = 0. \tag{5}$$

Equations (4) approach (1) for sufficiently small values of $|y|$, and hence are satisfactory so far as the conditions near the zx plane are concerned.

Inspection of (4) in the light of (8^{110}) and (9^{110}) suggests that

$$E_x = 0, \quad E_y = 0, \tag{6}$$

and, since $\cos(-ky - \omega t) = \cos(ky + \omega t)$,

$$E_z = \begin{cases} -E_0 \cos(ky + \omega t) & \text{if } y < 0 \\ -E_0 \cos(ky - \omega t) & \text{if } y > 0. \end{cases} \tag{7}$$

Here

$$E_0 = \eta_0 H_0 = \tfrac{1}{2}\eta_0 J_0^*. \tag{8}$$

The values of E_z are plotted in Fig. 242 for several values of t, expressed in terms of the period T of the oscillation of the current density (2), so that $T = 2\pi/\omega$. The arrows at a pair of crests in the figure show that the waves of **E** are moving away from the zx plane on both sides of this plane. The field **H** given by (4) and (5) has a discontinuity as we cross the zx plane. On the other hand, as we might expect intuitively, the field **E** given by (6) and (7) is everywhere continuous and is symmetric about the zx plane. Thus, both fields behave properly near the zx plane; since they satisfy Maxwell's equations, we accept them as correct.

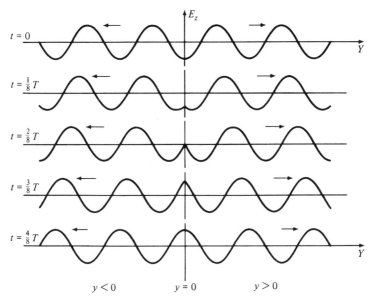

Fig. 242. If the current density in the sheet of Fig. 241 has the form $J_0^* \cos \omega t \mathbf{1}_z$, the sheet radiates plane waves in the $\pm y$ directions, as shown here by graphs of **E** for several values of t.

Our computation of the field caused by an infinite plane sheet of alternating current has proved to be much shorter than our computation of the field of a point-dipole antenna, even though we started in both cases with a formula for the current. The factors that account for this brevity are: a highly idealized current distribution, leading to simple geometry; the simple way of handling retardation; boundary conditions that are easy to visualize; and the availability of the typical fields (8^{110}) and (9^{110}) that satisfy Maxwell's equations.

We assumed in this section than a current of linear density $J_0^* \cos \omega t$ is somehow driven up and down the zx plane. Now that we have computed the electromagnetic field induced by this current, we notice that, according to (7), the electric field induced up and down the zx plane is $-E_0 \cos \omega t$. Consequently, this field *opposes* the flow of the current that generates it.

EXERCISES

1. Illustrate the behavior of H_x in (4) by a diagram similar to Fig. 242.

2. The imagined generator driving a current of density (2) up and down the zx plane must supply a certain amount of power per square meter of the plane. Show that the time average of the power required to drive the current against the field (7) is $\frac{1}{2} E_0 J_0^*$ watts/meter². Then consider planes parallel to the zx plane, and use Poynting's vector to show that $\frac{1}{2} E_0 J_0^*$ is precisely the average rate at which power is radiated away by the sheet of alternating current.

112. TERMINOLOGY AND NOTATION

The quantity $ky - \omega t + \phi_0$ in $\mathbf{E}$ in (8^{110}) is called the *phase* of $\mathbf{E}$ and is often denoted by ϕ; it is also the phase of $\mathbf{H}$ in (8^{110}). The surfaces on which the phase is constant at a fixed instant of time are called *surfaces of constant phase* pertaining to that instant. In the case of the field (8^{110}), these surfaces, pictured in Fig. 239, are planes perpendicular to the y axis; as time goes on, they move in the $+y$ direction.

Let the vector $\mathbf{1}$ point in the direction of propagation of a plane wave. The *propagation vector*, say $\mathbf{k}$, is then defined as

$$\mathbf{k} = k\mathbf{1} = \frac{\omega}{c} \mathbf{1}, \qquad (1)$$

where k is the propagation constant (2^{107}). Equation (20^3) reads $\omega = 2\pi/T$, where T is the period, and Equation (18^3) for the wavelength can be written in the present case as $\lambda = cT$. Consequently, $\lambda = 2\pi c/\omega$ and

$$k = \frac{2\pi}{\lambda}. \qquad (2)$$

In (8^{110}) we have $\mathbf{k} = (\omega/c)\mathbf{1}_y$. Now, $y = (\mathbf{1}_y \cdot \mathbf{1}_r)r = \mathbf{1}_y \cdot \mathbf{r}$, and hence (8^{110}) can be written as

$$\mathbf{E} = E_0 \cos(\mathbf{k} \cdot \mathbf{r} - \omega t + \phi_0)\mathbf{1}_z, \qquad \mathbf{H} = H_0 \cos(\mathbf{k} \cdot \mathbf{r} - \omega t + \phi_0)\mathbf{1}_x. \qquad (3)$$

According to Exercise 7^{10}, the locus of points for which

$$\mathbf{k} \cdot \mathbf{r} = \text{constant} \tag{4}$$

is indeed a plane perpendicular to the vector **k**.

At any point P, the vector **E** in (3) is perpendicular to the propagation vector **k**; therefore, the field (3) is called a *transverse electric* wave (a TE wave). Similarly, **H** in (3) is perpendicular to **k**, and therefore (3) is called a *transverse magnetic* wave (a TM wave). Since both **E** and **H** are perpendicular to **k**, this wave is a *transverse electromagnetic* wave (a TEM wave). In (3), the vector **E** at any point is parallel to a fixed line (in this case, the z axis); for this reason the wave (3) is said to be *linearly polarized*. At any instant, the fields **E** and **H** in (3) are constant over any plane of constant phase; therefore, this wave is called *uniform*. The frequency of (3), namely, $\omega/2\pi$, is a constant; light of a fixed frequency has a "pure" color and is called *monochromatic*; so is the wave (3). Note that (3) is *sinusoidal* in both time and space; that is, **E** and **H** vary sinusoidally not only with t at any point, but also with y at any instant.

A formula for linearly polarized plane monochromatic TEM waves moving in *any* direction can be constructed as follows: Write $E_0 \mathbf{1}_z = \mathbf{E}_0$ in (3) and note that, since $\mathbf{k} = (\omega/c)\mathbf{1}_y$ in (3), we have

$$\mathbf{k} \cdot \mathbf{E}_0 = 0, \qquad \mathbf{1}_y = \frac{c}{\omega}\mathbf{k}, \qquad \mathbf{1}_z = \frac{\mathbf{E}_0}{E_0}, \tag{5}$$

so

$$\mathbf{1}_x = \mathbf{1}_y \times \mathbf{1}_z = \frac{c}{\omega E_0}\mathbf{k} \times \mathbf{E}_0. \tag{6}$$

Consequently,

$$H_0 \mathbf{1}_x = \frac{E_0}{\eta_0}\mathbf{1}_x = \frac{c}{\omega \eta_0}\mathbf{k} \times \mathbf{E}_0 = \frac{1}{\omega \mu_0}\mathbf{k} \times \mathbf{E}_0. \tag{7}$$

Therefore, (3) can be written as

$$\mathbf{E} = \mathbf{E}_0 \cos(\mathbf{k} \cdot \mathbf{r} - \omega t + \phi_0), \qquad \mathbf{H} = \frac{1}{\omega \mu_0}\mathbf{k} \times \mathbf{E}_0 \cos(\mathbf{k} \cdot \mathbf{r} - \omega t + \phi_0), \tag{8}$$

where, to repeat, $\mathbf{E}_0$ is a constant vector and

$$\mathbf{k} \cdot \mathbf{E}_0 = 0. \tag{9}$$

The formulas (8) comprise an exact solution of Maxwell's free-space equations for any choice of the direction of the propagation vector **k**, provided that the condition (9) is met (Exercise 1). For example, to verify that div $\mathbf{E} = 0$, we write

$$\phi = \mathbf{k} \cdot \mathbf{r} - \omega t + \phi_0 = k_x x + k_y y + k_z z - \omega t + \phi_0, \tag{10}$$

and note that $\partial \phi / \partial x = k_x$, so that

$$\frac{\partial}{\partial x} E_x = \frac{\partial}{\partial x}(E_0)_x \cos\phi = (E_0)_x \frac{\partial}{\partial x}\cos\phi = -(E_0)_x k_x \sin\phi. \quad (11)$$

Treating the terms $\partial E_y/\partial y$ and $\partial E_z/\partial z$ in the same way, we find that div **E** has the factor $(E_0)_x k_x + (E_0)_y k_y + (E_0)_z k_z$, which is zero in view of (9).

Circular polarization. Imagine two identical short antennas located at the origin in Fig. 238; one is aligned with the z axis, the other with the x axis; they are activated by separate identical generators. To obtain a formula that corresponds to the **E** in (8^{110}) but pertains to the *second* antenna, we need only to replace the vector $\mathbf{1}_z$ by $\mathbf{1}_x$. Finally (for simplicity), we set $\phi_0 = 0$ for the first antenna and (to illustrate two kinds of circular polarization) set $\phi_0 = \pm 90$ deg for the second. Since

$$\cos(ky - \omega t \pm 90°) = \mp \sin(ky - \omega t), \quad (12)$$

the superposition of the two electric fields gives either

$$E_x = -E_0 \sin(ky - \omega t), \quad E_z = E_0 \cos(ky - \omega t), \quad (13)$$

or

$$E_x = E_0 \sin(ky - \omega t), \quad E_z = E_0 \cos(ky - \omega t), \quad (14)$$

depending on whether the second generator leads or lags behind the first one. We also have $E_y = 0$ in either case.

Consider now a fixed point $P(x_0, y_0, z_0)$ and the field (13). As time goes on, the tip of the vector **E** pertaining to P keeps moving uniformly in a circle that has the radius E_0 and lies in a plane parallel to the zx plane. In the case (14), the situation is similar, except that **E** turns in the opposite sense. In either case, the combination of the electric wave and the accompanying magnetic wave is called a *circularly polarized plane electromagnetic wave*. The definitions of the term *right* circular polarization differ from one branch of optics to another, and the definitions of *left* circular polarization differ accordingly.[2]

EXERCISES

1. Show that, under the condition (9), the field (8) satisfies all of Maxwell's free-space equations.

2. Show that the electromagnetic fields called for in parts (a) and (b) of Exercises 3^{110} can be written in the form (8).

3. Verify that **H** in (8) is expressed in amperes per meter.

---

[2] W. A. Shurcliff, *Polarized Light* (Cambridge, Mass.: Harvard University Press, 1966), pp. 2–7.

4. Compute the fields **H** associated, respectively, with (13) and with (14).

5. One short antenna, located at O in Fig. 238, lies along the z axis. Another one, also at O, lies in the zx plane. The two generators have the same frequency, but the magnitudes and phases of their emf's are adjustable. (That is, the angle between the antennas, and the values of the E_0's and ϕ_0's are all adjustable.) Choose the simplest combination of these parameters that you can think of, for which the wave is polarized elliptically—that is, the tip of the vector **E** at a fixed point P moves in an ellipse. Restrict yourself to the far zone.

113. MORE BOUNDARY CONDITIONS

To prepare for a discussion of reflection and transmission of electromagnetic waves at the interface of two mediums, we will now summarize the results of §70 and then work out the boundary conditions for magnetic fields.

In §70 we denoted the field **D** by **D'** in one medium and by **D''** in an adjoining medium. Using Fig. 174[70], we then found that, if the interface of these mediums is uncharged, the normal component of **D**, taken relative to the interface, does not change as we cross the interface; that is,

$$(2^{70}) \qquad D_n'' = D_n'. \qquad (1)$$

Now, since div **B** = 0, we may write B_n for D_n in (1^{70}), proceed as before, and conclude that

$$B_n'' = B_n'. \qquad (2)$$

Thus, the normal component of **B** relative to the interface is continuous. In

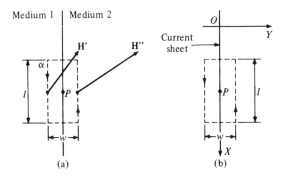

Fig. 243. Diagrams for deriving boundary conditions for **H**: (a) at the interface of two dielectric mediums and (b) in the presence of a surface current.

§70 we also found that the tangential component of **E** relative to the interface is continuous:

$$(4^{70}) \qquad E_\tau'' = E_\tau'. \qquad (3)$$

Figure 175^{70} is redrawn in Fig. 243(a) with symbols pertaining to **H**; as before, $w \ll l$. By Maxwell's mmf law, the integral of H_t around the contour α is equal to the current flowing through α; in a nonconductor this current would be zero, unless it is a displacement current. If the current density is everywhere finite, this current will approach zero when we let the contour shrink upon P, and the procedure will reduce to that of §70 and lead to the equation

$$H_\tau'' = H_\tau'. \qquad (4)$$

Surface charges and surface currents. If the boundary between two regions carries a surface charge of density q_a, other than a polarization charge, the condition (1) must be replaced by

$$(8^{70}) \qquad D_n'' - D_n' = q_a. \qquad (5)$$

Similarly, if the boundary carries a surface current, then (4) must be modified. We restrict ourselves to a current of linear density $J^* \mathbf{1}_z$ confined to the zx plane, use a contour α lying in the xy plane—as in Fig. 243(b)—and write H_x' and H_x'' for the instantaneous values of H_x just to the left and just to the right of the zx plane. The mmf in the left-hand edge of α is then lH_x', the mmf in the right-hand edge is $-lH_x''$, and the current linking α is lJ^* even when $w \to 0$. In the limit, Ampère's law then gives $lH_x' - lH_x'' = lJ^*$, and reduces to

$$H_x'' - H_x' = -J^*. \qquad (6)$$

For example, in the simple case of the steady current of Fig. 241, Equation (1^{111}) gives $H_x' = \tfrac{1}{2}J^*$ and $H_x'' = -\tfrac{1}{2}J^*$, and hence (6) is satisfied.

114. REFLECTION FROM NONCONDUCTORS; NORMAL INCIDENCE

The half-space to the left of the zx plane in Fig. 244 is filled with one medium and the half-space to the right with another. Both mediums are uncharged, homogeneous, isotropic, nonconducting, and nonmagnetic. Therefore, Maxwell's equations, the speed of electromagnetic waves, and the propagation constant for each medium can be obtained from the corresponding free-space formulas by replacing ϵ_0 by ϵ. Thus, in either medium,

Reflection from Nonconductors; Normal Incidence   § 114

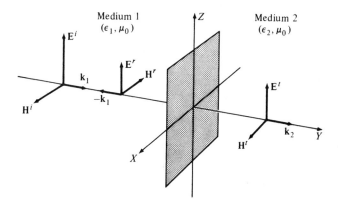

Fig. 244. A plane electromagnetic wave falling on the interface of two dielectric mediums; normal incidence.

(7[103]) $$\text{div } \mathbf{E} = 0, \quad \text{div } \mathbf{H} = 0, \tag{1}$$

and

(8[103]) $$\text{curl } \mathbf{E} = -\mu_0 \frac{\partial \mathbf{H}}{\partial t}, \quad \text{curl } \mathbf{H} = \epsilon \frac{\partial \mathbf{E}}{\partial t}, \tag{2}$$

where ϵ equals ϵ_1 in medium 1 and ϵ_2 in medium 2. The respective impedances and propagation speeds for the two mediums are

(5[110]) $$\eta_1 = \sqrt{\frac{\mu_0}{\epsilon_1}}, \quad \eta_2 = \sqrt{\frac{\mu_0}{\epsilon_2}}, \tag{3}$$

(10[108]) $$c_1 = \frac{1}{\sqrt{\epsilon_1 \mu_0}}, \quad c_2 = \frac{1}{\sqrt{\epsilon_2 \mu_0}}, \tag{4}$$

and the propagation constants for waves of angular frequency ω are

$$k_1 = \frac{\omega}{c_1}, \quad k_2 = \frac{\omega}{c_2}. \tag{5}$$

Another important quantity is the *index of refraction* of each medium. It is denoted by n and is defined as the ratio of the speed of light in vacuum to that in the medium. According to (4) and (10[108]),

$$n_1 = \sqrt{\frac{\epsilon_1}{\epsilon_0}}, \quad n_2 = \sqrt{\frac{\epsilon_2}{\epsilon_0}}. \tag{6}$$

A simple antenna is located in medium 1 on the negative part of the y

axis, so far from the origin that the waves it emits approach the zx plane as plane waves of the form

$$(8^{110}) \qquad \mathbf{E}^i = E_0^i \cos(k_1 y - \omega t)\mathbf{1}_z, \qquad \mathbf{H}^i = H_0^i \cos(k_1 y - \omega t)\mathbf{1}_x, \qquad (7)$$

where the superscript i stands for *incident*, and the constants E_0^i and H_0^i, which we take to be positive, satisfy the relation

$$(7^{110}) \qquad\qquad\qquad E_0^i = \eta_1 H_0^i. \qquad\qquad\qquad (8)$$

We assume outright that, as suggested in Fig. 244, the incident wave will be partly reflected at the interface back into medium 1, and partly transmitted into medium 2, and that these waves will have the following forms:

$$\mathbf{E}^r = E_0^r \cos(k_1 y + \omega t + e)\mathbf{1}_z, \qquad \mathbf{H}^r = H_0^r \cos(k_1 y + \omega t + h)\mathbf{1}_x \qquad (9)$$

and

$$\mathbf{E}^t = E_0^t \cos(k_2 y - \omega t)\mathbf{1}_z, \qquad \mathbf{H}^t = H_0^t \cos(k_2 y - \omega t)\mathbf{1}_x; \qquad (10)$$

here the superscript r stands for *reflected* and the superscript t for *transmitted*, and

$$E_0^r = \eta_1 H_0^r, \qquad E_0^t = \eta_2 H_0^t. \qquad (11)$$

We require all the constants E_0^r, H_0^r, E_0^t, and H_0^t to be *positive* if the two mediums are different, and we rely on the phase angles e and h to take care of any minus signs.

The components of our **E**'s and **H**'s normal to the boundary of the two mediums are zero, so the conditions (1^{113}) and (2^{113}) are met. The condition (3^{113}) requires that $E_z^i + E_z^r = E_z^t$ at $y = 0$. At the instant when $\omega t = 90°$, it reduces to $\cos(90° + e) = 0$, so that we can consider without loss of generality only two alternatives, namely

$$e = 0, \qquad e = 180°. \qquad (12)$$

The tangential components of the **H**'s lead to similar alternatives:

$$h = 0, \qquad h = 180°. \qquad (13)$$

When we let $t = 0$, the condition $E_z^i + E_z^r = E_z^t$ yields (14). Similarly, the condition $H_x^i + H_x^r = H_x^t$ leads to the equation $H_0^i + H_0^r \cos h = H_0^t$, so that, in view of (6), (8), and (11), we get (15). Thus

$$E_0^i + E_0^r \cos e = E_0^t, \qquad (14)$$

$$n_1(E_0^i + E_0^r \cos h) = n_2(E_0^t + E_0^r \cos e). \qquad (15)$$

Now consider the case when $n_1 > n_2$, the "denser-to-lighter" case. It is then apparent from (15) that $(E_0^i + E_0^r \cos h) < (E_0^i + E_0^r \cos e)$ and that, consequently, $E_0^r \cos h < E_0^r \cos e$ and $\cos h < \cos e$. Therefore, in view of (12) and (13), we have $e = 0$ and $h = 180°$. In this case, then, the direction of $\mathbf{E}^r$ at the interface is at every instant *the same* as that of $\mathbf{E}^i$, but the direction of $\mathbf{H}^r$ at the interface is at every instant *opposite* to that of $\mathbf{H}^i$. In short, in the denser-to-lighter case Equations (9) become

$$\mathbf{E}^r = E_0^r \cos(k_1 y + \omega t)\mathbf{1}_z, \qquad \mathbf{H}^r = -H_0^r \cos(k_1 y + \omega t)\mathbf{1}_x \qquad (16)$$

When we set $e = 0$ and $h = 180°$ in (14) and (15), we find that

$$E_0^r = \frac{n_1 - n_2}{n_1 + n_2} E_0^i, \qquad E_0^t = \frac{2n_1}{n_1 + n_2} E_0^i. \qquad (17)$$

Similar computations for the lighter-to-denser case show that then $e = 180°$ and $h = 0$.

The electromagnetic energy carried by the incident wave is flowing toward the interface; the average rate (watts/meter^2) at which it crosses from left to right any plane in Fig. 244 parallel to the interface and located to the left of the interface is

(10^{110})
$$\langle \mathbf{E}^i \times \mathbf{H}^i \rangle_y = \tfrac{1}{2} E_0^i H_0^i. \qquad (18)$$

The corresponding rates for the reflected and the transmitted wave are

$$\langle \mathbf{E}^r \times \mathbf{H}^r \rangle_y = -\frac{1}{2} \frac{(n_1 - n_2)^2}{(n_1 + n_2)^2} E_0^i H_0^i \qquad (19)$$

and

$$\langle \mathbf{E}^t \times \mathbf{H}^t \rangle_y = \frac{1}{2} \frac{4 n_1 n_2}{(n_1 + n_2)^2} E_0^i H_0^i; \qquad (20)$$

the minus sign in (19) implies that the reflected wave carries energy leftward, as in Fig. 244. Comparing the three equations, we find that all the energy delivered by the incident wave is carried away by the reflected and the transmitted waves. These formulas apply, of course, only if the mediums in question are transparent to the waves of frequency $\omega/2\pi$; that is, if no electromagnetic energy is converted into other forms. The ratio of the magnitudes of the right-hand sides of (19) and (18), say $R_{\text{intensity}}$, is called the *"intensity" reflection coefficient* or simply the *reflection coefficient* at normal incidence:

$$R_{\text{intensity}} = \frac{(n_1 - n_2)^2}{(n_1 + n_2)^2} \quad \text{(normal incidence)}. \qquad (21)$$

The term "reflection coefficient" is also used for the ratio of the *amplitudes* of E^r and E^i.

EXERCISES

1. Derive in detail the forms taken by the four equations in (9) and (17) in the lighter-to-denser case.

2. Let medium 2 in Fig. 244 be vacuum, so that $n_2 = 1$. Write the incident wave as in (7) and compare the values of E_0^r/E_0^i for the following values of n_1: 1.0 (vacuum), 1.5 (glass), and 2.4 (diamond). Make similar comparisons for E_0^t/E_0^i, H_0^r/H_0^i and H_0^t/H_0^i. Also, compute the reflection coefficient (20) for each value of n_1.

3. Suppose that mediums 1 and 2 in Fig. 244 are interchanged, but the incident wave still comes from the left and is described by (7), except that k_2 replaces k_1. How do the values of E_0^r, E_0^t, H_0^r, and H_0^t compare with those computed in the text? What happens to the values of the reflection coefficient (21)?

4. The total E and the total H in the left-hand medium are $E^i + E^r$ and $H^i + H^r$. Compute the Poynting vector using these total fields and verify that the right-hand side of (20) gives the average rate of flow of transmitted energy (watts/meter^2).

115. REFLECTION FROM PERFECT CONDUCTORS; NORMAL INCIDENCE

Let us now turn to Fig. 245 and assume that medium 1 is vacuum and medium 2 is an idealized nonmagnetic metal whose conductivity is infinite.

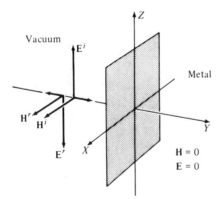

Fig. 245. A plane wave falling on a perfect reflector; normal incidence.

Since $g = \infty$ in the metal, any field **E** other than zero would produce infinite current in the metal, which is not acceptable even in our idealized case. Consequently, the transmitted field $\mathbf{E}^t$ must vanish, and (14^{114}) turns into

$$E_0^r = -E_0^i. \tag{1}$$

Furthermore, since $\mathbf{E}^t = 0$, the equation curl $\mathbf{E} = -\partial \mathbf{B}/\partial t$ implies that the transmitted **H** does not vary with time. Such an **H** could be produced, for example, by stationary permanent magnets, but we would certainly not expect the incident wave, which does involve t, to produce a nonzero field that is independent of t. Therefore, we set $\mathbf{H}^t = 0$.

When we combine (1) with the cosine factors taken from (7^{114}) and (9^{114}), we get

$$\mathbf{E}^i = E_0^i \cos(ky - \omega t)\mathbf{1}_z, \qquad \mathbf{E}^r = -E_0^i \cos(ky + \omega t)\mathbf{1}_z. \tag{2}$$

The corresponding **H**'s are

$$\mathbf{H}^i = H_0^i \cos(ky - \omega t)\mathbf{1}_x, \qquad \mathbf{H}^r = H_0^i \cos(ky + \omega t)\mathbf{1}_x. \tag{3}$$

Remembering that $\mathbf{E}^t = 0$ and $\mathbf{H}^t = 0$, and evaluating the fields $\mathbf{E}^i + \mathbf{E}^r$ and $\mathbf{H}^i + \mathbf{H}^r$, we get equations (4) through (7):

$$E_x = 0, \quad E_y = 0 \qquad \text{(everywhere)}, \tag{4}$$

$$E_z = \begin{cases} 2E_0^i \sin ky \sin \omega t & \text{(in vacuum)} \\ 0 & \text{(in metal)}, \end{cases} \tag{5}$$

$$H_x = \begin{cases} 2H_0^i \cos ky \cos \omega t & \text{(in vacuum)} \\ 0 & \text{(in metal)}, \end{cases} \tag{6}$$

$$H_y = 0, \quad H_z = 0 \qquad \text{(everywhere)}. \tag{7}$$

Figure 245 shows the orientation of the **E**'s and **H**'s at two points near the interface at an instant when $\mathbf{E}^i$ is directed upward. Figure 246, based on (5), illustrates the fact that, in the vacuum, E_z is a *standing wave*. This wave involves back-and-forth surges of electromagnetic energy, but no net flow (see Exercise 3).

Surface current. Using a prime for the vacuum and a double prime for the metal, we find from (6) that $H_x'' = 0$ and that, when $y \to 0$, we have $H_x' \to 2H_0^i \cos \omega t$. Consequently, the tangential component of **H** is *discontinuous*, and the simple boundary condition (4^{113}) is *not satisfied*. Therefore, we turn to the more general condition, namely,

(6^{113}) $$H_x'' - H_x' = -J^*, \qquad (8)$$

and substitute into it the values of H_x'' and H_x' found just above. The result is

$$0 - 2H_0^i \cos \omega t = -J^*. \qquad (9)$$

We are thus forced to conclude that an alternating surface current must be flowing on the exposed face of the metal and that its linear density is

$$J^* = 2H_0^i \cos \omega t. \qquad (10)$$

Inspection of Fig. 243(b) then shows that the surface current flows in the $+z$ direction when $\cos \omega t$ is positive.

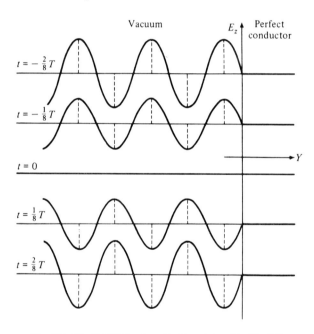

Fig. 246. The field **E** for the reflection pictured in Fig. 245.

Shielding. In our present problem, the electromagnetic field vanishes inside the metal, which somehow shields its interior from the electromagnetic effects that are taking place outside. By what mechanism does the metal do this? Does it prevent the incident wave from entering? Or does it let this wave go in, only to cancel it by a special wave of its own? We touched upon shielding in §72 and remarked that in the case of time-varying fields the second view is preferable. Let us then look from this standpoint at the field **E** given by (4) and (5). It is pictured in Fig. 246 and, for a single instant, in Fig. 247(c).

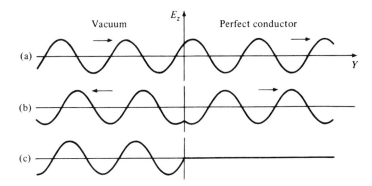

Fig. 247. Perfect reflection at normal incidence: (a) the field **E** of the incident wave at the instant $t = \frac{1}{8}T$; (b) the two-way wave caused by the surface current, taken from the curve marked $t = \frac{1}{8}T$ in Fig. 242; (c) the resultant of these waves, taken from Fig. 246.

The incident wave originates in some antenna, far away from the metal. We write E^{antenna} for the electric intensity of this wave and assume that the wave propagates without change into the metal, so that

$$E_z^{\text{antenna}} = E_0^i \cos(ky - \omega t) \tag{11}$$

for all values of y; that is

$$E_z^{\text{antenna}} = \begin{cases} E_0^i \cos(ky - \omega t) & \text{(in vacuum)} \\ E_0^i \cos(ky - \omega t) & \text{(in metal).} \end{cases} \tag{12}$$

The metal responds to the incident wave by a surface current of density given by (10), which can be written as $J^* = J_0^* \cos \omega t$, where

$$J_0^* = 2H_0^i = 2\eta_0^{-1} E_0^i. \tag{13}$$

This surface current produces the field (7^{111}), which we will denote by E_z^{current}. According to (8^{111}), the coefficient E_0 in that field is $\frac{1}{2}\eta_0 J_0^*$, so in view of (13), we have $E_0 = E_0^i$, and (7^{111}) becomes

$$E_z^{\text{current}} = \begin{cases} -E_0^i \cos(ky + \omega t) & \text{(in vacuum)} \\ -E_0^i \cos(ky - \omega t) & \text{(in metal).} \end{cases} \tag{14}$$

The sum of (12) and (14) is precisely the field (5), pictured in Fig. 246—a standing wave in the vacuum and a zero field in the metal. We may, therefore, say that the metal lets the incident wave go inside it without change, but it also develops a surface current whose field cancels the incident field inside the metal and provides a reflected field in the vacuum.

Remark on actual metals. Let the metal in Fig. 245 be an actual metal of conductivity g. A part of the incident plane wave streaming from the left will then be reflected, and a part will penetrate into the metal, causing currents and a conversion of electromagnetic energy into heat. As a result, the wave in the metal will be *attenuated*—its amplitude will decrease with the depth of penetration. More precisely, as one can deduce from (15^{105}), the amplitude will include the factor $e^{-y/\delta}$, where δ is a constant. Consequently, the electromagnetic disturbance in the metal will be essentially confined to a depth δ, called *skin depth*. The general formula for δ is complicated, but if the metal is a good conductor—say copper or silver—and if the frequency of the incident wave satisfies the condition $\omega \ll g/\epsilon_0$, we have the approximate formula

$$\delta = \sqrt{\frac{2}{\omega g \mu_0}}. \qquad (15)$$

If we let $g \to \infty$, the skin depth approaches zero, the current in the metal becomes confined to the surface, and we are led to the idealized case worked out above. In practice, δ may be so small that we may regard the current as a surface current.

EXERCISES

1. Compute the fields H^{antenna} and H^{current} that go with (12) and (14), and verify that their sum is (6).

2. A microwave whose wavelength in vacuum is 3 centimeters falls on a silver plate. Compute the skin depth.

3. Consider a region R, shaped as a right circular cylinder, parallel to the y axis in Fig. 245, and extending from $y = y_1$ to $y = y_2$, where $y_1 < y_2 < 0$. Write a for the cross-sectional area of R and use (5) and (6) to compute, as functions of t: (a) the density u of electromagnetic energy in R; (b) the total electromagnetic energy U in R; (c) the time rate of change of U; (d) the respective time rates at which energy flows into R at $y = y_1$ and $y = y_2$. Check the consistency of your results for (c) and (d). Compute the time averages of u and of Poynting's vector. Find convenient values of y_1 and y_2 for which Poynting's vector vanishes at both flat ends of R, and describe the behavior of U for this case.

116. REFLECTION FROM PERFECT CONDUCTORS; OBLIQUE INCIDENCE

The surface of the idealized, perfectly conducting, nonmagnetic metal is the zx plane, as before, but the propagation vector of the incident plane wave makes

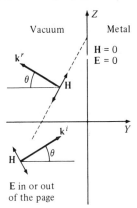

Fig. 248. Reflection from a perfect conductor; oblique incidence.

the angle θ with the y axis, as in Fig. 248. We will consider an incident wave of the form (8^{112}) with ϕ_0 put equal to zero:

$$\mathbf{E}^i = \mathbf{E}_0^i \cos(\mathbf{k}^i \cdot \mathbf{r} - \omega t), \qquad \mathbf{H}^i = \frac{1}{\omega\mu_0} \mathbf{k}^i \times \mathbf{E}_0^i \cos(\mathbf{k}^i \cdot \mathbf{r} - \omega t). \tag{1}$$

We also assume, for simplicity, that

$$E_x^i = E_0^i \cos(\mathbf{k}^i \cdot \mathbf{r} - \omega t), \qquad E_y^i = 0, \qquad E_z^i = 0, \tag{2}$$

$$k_x^i = 0, \qquad k_y^i = k\cos\theta, \qquad k_z^i = k\sin\theta; \tag{3}$$

as usual, $k = \omega/c$ and $\mathbf{r} = x\mathbf{1}_x + y\mathbf{1}_y + z\mathbf{1}_z$. The formula for $\mathbf{H}^i$ in (1) then gives

$$\mathbf{H}^i = H_0^i (\sin\theta \mathbf{1}_y - \cos\theta \mathbf{1}_z) \cos(\mathbf{k}^i \cdot \mathbf{r} - \omega t), \tag{4}$$

where

(7^{110})
$$H_0^i = \eta_0^{-1} E_0^i. \tag{5}$$

In this case, $\mathbf{k}^i$ and $\mathbf{H}^i$ are both parallel to the yz plane, but $\mathbf{E}^i$ is perpendicular to it. Note that

$$\mathbf{k}^i \cdot \mathbf{r} = ky\cos\theta + kz\sin\theta. \tag{6}$$

The tangential component of $\mathbf{E}^i$ at the interface is its x component evaluated at $y = 0$:

$$E_\tau^i = E_0^i \cos(kz \sin\theta - \omega t). \tag{7}$$

The electromagnetic field vanishes inside the metal, so, to satisfy the continuity condition (3^{113}), the right-hand side of (7) must be set equal to zero for all values of z and t, which is possible only if $E_0^i = 0$. We conclude that the wave marked $\mathbf{k}^i$ in Fig. 248 cannot propagate *alone* in the vacuum to the left of the metal. Consequently, there must also be a reflected wave, say $\mathbf{E}^r$, that cancels (7) at the interface.

The angle of reflection. To cancel (7) at the interface, we must have

$$E_x^r = -E_0^i \cos(kz \sin\theta - \omega t) \tag{8}$$

for all values of x and z, and for $y = 0$. Therefore, for other values of y,

$$E_x^r = -E_0^i \cos(Ay + kz \sin\theta - \omega t), \tag{9}$$

where the constant A is as yet undetermined. Equation (9) can be put into the general form (1) for plane waves by introducing a propagation vector $\mathbf{k}^r$ for which

$$\mathbf{k}^r \cdot \mathbf{r} = Ay + kz \sin\theta, \tag{10}$$

so that

$$k_x^r = 0, \qquad k_y^r = A, \qquad k_z^r = k \sin\theta. \tag{11}$$

Now, the magnitude of the propagation vector in the vacuum should be ω/c, or simply k. Therefore, $A^2 + k^2 \sin^2\theta = k^2$, and hence $A = \pm k \cos\theta$. If we take the plus sign, (9) will cancel (2) everywhere, so we must use the minus sign. Consequently, the proper components of k^r are

$$k_x^r = 0, \qquad k_y^r = -k \cos\theta, \qquad k_z^r = k \sin\theta, \tag{12}$$

as in Fig. 248. Thus the angle of reflection is equal to the angle of incidence.

EXERCISES

1. The dotted line in Fig. 248 marks an instantaneous position of a plane of constant phase of the reflected magnetic field. Show that the line of intersection of this plane with the surface of the metal travels with the speed

$$v_p = \frac{c}{\sin\theta}, \tag{13}$$

which is greater than the speed of light in vacuum.

2. Show that the formulas for the reflected electromagnetic field in Fig. 248 are

$$\mathbf{E}^r = -E_0^i \cos(\mathbf{k}^r \cdot \mathbf{r} - \omega t)\mathbf{1}_x, \tag{14}$$

$$\mathbf{H}^r = -\eta_0^{-1} E_0^i (\sin\theta \mathbf{1}_y + \cos\theta \mathbf{1}_z) \cos(\mathbf{k}^r \cdot \mathbf{r} - \omega t), \tag{15}$$

and that the tangential component of $\mathbf{H}^r$ at the surface of the metal is equal to that of $\mathbf{H}^i$.

3. Show that the linear density of the surface current in the zx plane in Fig. 248 is

$$\mathbf{J}^* = 2\eta_0^{-1} E_0^i \cos\theta \cos(kz\sin\theta - \omega t)\mathbf{1}_x. \tag{16}$$

4. In the text we took it for granted that the reflected wave in Fig. 248 is a *plane* wave. Justify this assumption.

5. Extend the discussion of §114 to oblique incidence. In particular, show that, if θ_1 and θ_2, respectively, are the angles of incidence and refraction, then $n_1 \sin\theta_1 = n_2 \sin\theta_2$ (Snell's law[3]).

---

[3] Willebrord Snellius (1591-1626), who worked with a corpuscular theory of light.

CHAPTER 26

Rectangular Waveguides with Perfectly Conducting Walls

The most common device for transmitting electric power is a pair of wires, as in a house circuit. Another is the coaxial cable, in which one wire is replaced by a metal sheath surrounding the other wire, and radiation losses are reduced. Still another way is to let electrical power flow inside a hollow metal pipe. Each possibility can be studied from either of two viewpoints: (a) as a problem in generalized circuit theory, involving currents in conductors of particular shapes or (b) as a problem in the theory of electromagnetic waves guided by conductors. In the case of low frequencies and transmission distances at which retardation can be ignored, the circuit method is simpler. But for hollow pipes, the wave method is best in helping to visualize the physical phenomena; however, for practical computations concerning combinations of wave guides, still another procedure—the method of "equivalent circuits," not included in this book because it is highly specialized—is more manageable. The reason for the interest in transmission through pipes is that in many modern devices electric power is produced at frequencies of some 3×10^9 or more hertz (3×10^9 or more cycles per second, which means wavelengths of 10 centimeters or less); at such frequencies, transmission by wires would involve prohibitive radiation losses.

We will consider only an idealized pipe or "guide," oriented as in Fig. 249. It has perfectly conducting walls, is infinitely long, and is activated by a Hertzian dipole oscillator, pictured in Fig. 249 as a short vertical rod; this oscillator is presumably located at $z = -\infty$ and operates at a single frequency, which we write as $\omega/2\pi$. We also assume that all the transient effects have subsided—the effects that arise in an actual guide when the oscillator is first turned on.

The simplest waves that a rectangular guide can transmit are transverse

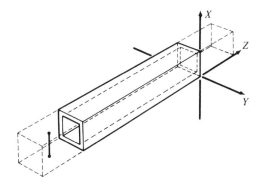

Fig. 249. A long square waveguide, activated by a Hertzian dipole (the vertical rod in the figure) located on the center line of the guide, far from the origin.

electric (TE) waves of various "modes," and transverse magnetic (TM) waves. We will consider the simplest varieties of TE waves in some detail, but will touch upon TM waves only in the exercises.

As an introduction to rectangular guides, we consider propagation between parallel metal plates.

117. TE WAVES BETWEEN PARALLEL PERFECTLY CONDUCTING PLATES

An infinite portion of space is bounded by two perfectly conducting plane metal walls located at $y = 0$ and $y = b$, as in Fig. 250. If a plane wave of the type

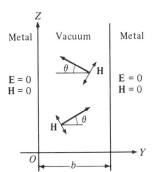

Fig. 250. Plane waves, reflected back and forth between metal plates.

indicated in the figure is somehow launched in this space by a properly oriented distant antenna, the wave will be repeatedly reflected back and forth between the walls as it makes its zigzag way in the $+z$ direction, and the space will become pervaded by a superposition of waves whose electric portions are given in (1¹¹⁶) and (14¹¹⁶). When we drop the superscripts, the superposition becomes

$$\mathbf{E} = E_0 \cos(ky \cos\theta + kz \sin\theta - \omega t)\mathbf{1}_x$$
$$- E_0 \cos(-ky \cos\theta + kz \sin\theta - \omega t)\mathbf{1}_x. \quad (1)$$

Trigonometric formulas then lead to the following field, in which we have omitted the factor -2:

$$\mathbf{E} = E_0 \sin(ky \cos\theta) \sin(kz \sin\theta - \omega t)\mathbf{1}_x. \quad (2)$$

The boundary condition (3¹¹³), which in the present case reads $E_x = 0$, must hold at both metal walls. It is already satisfied at $y = 0$, and to satisfy it at $y = b$ we must set $\sin(kb \cos\theta) = 0$. This means that $kb \cos\theta = n\pi$, where $n = 1, 2, 3, \ldots$, and consequently the following condition must hold:

$$\omega = \frac{n\pi c}{b \cos\theta}, \quad n = 1, 2, 3, \ldots. \quad (3)$$

The integer n is called a *mode index*. The dependence of $\mathbf{E}$ on n is illustrated in Fig. 251.

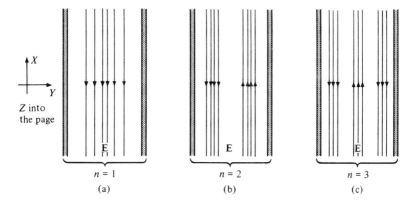

Fig. 251. Examples of lines of $\mathbf{E}$ in transverse electric waves between metal plates. These lines lie in planes that are moving away from the reader and satisfy the equation $kz \sin\theta - \omega t = -\tfrac{1}{2}\pi$.

Equation (3) gives ω as a function of the angle θ. In practice, the important parameter is the frequency of the waves, and consequently it is sometimes more suggestive to write (3) in the form (1¹¹⁸), which expresses $\cos\theta$ in terms of ω.

At this point it is convenient to rewrite (2) as

$$\mathbf{E} = E_0 \sin\left(\frac{2\pi y}{\lambda_c}\right) \sin\left(\frac{2\pi z}{\lambda_z} - \omega t\right) \mathbf{1}_x, \quad (4)$$

where

$$\lambda_0 = \frac{2\pi}{k}, \quad \lambda_c = \frac{\lambda_0}{\cos\theta}, \quad \lambda_z = \frac{\lambda_0}{\sin\theta}; \quad (5)$$

here λ_0 is the free-space wavelength of radiation whose frequency is $\omega/2\pi$, and $k = 2\pi/\lambda_0$. Since (4) vanishes at $y = b$, we have $2\pi b/\lambda_c = n\pi$, and (3) yields the formula

$$\lambda_c = \frac{2b}{n}, \quad n = 1, 2, 3, \ldots. \quad (6)$$

The components of $\mathbf{H}$ that go with (4) are

$$H_x = 0, \quad (7)$$

$$H_y = H_0 \sin\theta \sin\left(\frac{2\pi y}{\lambda_c}\right) \sin\left(\frac{2\pi z}{\lambda_z} - \omega t\right), \quad (8)$$

$$H_z = H_0 \cos\theta \cos\left(\frac{2\pi y}{\lambda_c}\right) \cos\left(\frac{2\pi z}{\lambda_z} - \omega t\right), \quad (9)$$

where

(5^{110})
$$H_0 = \frac{E_0}{\eta_0}. \quad (10)$$

To illustrate the shapes of the direction lines of $\mathbf{H}$ for a simple case, we set $t = 0$, $\lambda_c = 2b$, and $\theta = 45$ deg; that is, we let

$$\theta = 45°, \quad \lambda_0 = \sqrt{2}\, b, \quad \lambda_c = 2b, \quad \lambda_z = 2b. \quad (11)$$

We then proceed as in (4^{17}), write $dz/dy = H_z/H_y$, and find that

$$\frac{\sin(\pi z/b)}{\cos(\pi z/b)} dz = \frac{\cos(\pi y/b)}{\sin(\pi y/b)} dy. \quad (12)$$

Each side of (12) integrates to a logarithm, and the final result is

$$\cos\frac{\pi z}{b} \sin\frac{\pi y}{b} = \text{constant} = C. \quad (13)$$

When we set C equal in turn to 0.25, 0.50, and 0.75, we get the loops shown in Fig. 252. The space period of the pattern formed by the direction lines is λ_z, as one can infer from the last factors in (8) and (9). To determine the speed with which the pattern moves in the $+z$ direction, we set the phase $(2\pi z/\lambda_z) - \omega t$ equal to a constant and differentiate with respect to t. Writing v_p for

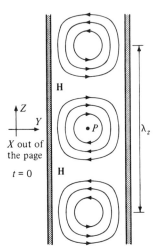

Fig. 252. Examples of lines of **H** in a transverse electric wave between metal plates. The loop-pattern moves in the $+z$ direction, up the page.

$\partial z/\partial t$, we then get $(2\pi/\lambda_z)v_p = \omega$, a formula that reduces to

$$v_p = \frac{c}{\sin\theta}. \tag{14}$$

This speed is called the *phase speed* or the *pattern speed* of our waves.

The flow of energy. The density u_e of electric energy associated with the wave (4) is $\tfrac{1}{2}\mathbf{E}\cdot\mathbf{D}$, where $\mathbf{D} = \epsilon_0\mathbf{E}$; consequently, from (4),

$$u_e = \frac{1}{2}\,\epsilon_0 E_0^2 \sin^2\!\left(\frac{2\pi y}{\lambda_c}\right) \sin^2\!\left[\left(\frac{2\pi z}{\lambda_z}\right) - \omega t\right]. \tag{15}$$

Its time average, say $\langle u_e\rangle_t$, is $\tfrac{1}{4}\epsilon_0 E_0^2 \sin^2(2\pi y/\lambda_c)$, and its average with respect to both t and y is

$$\langle u_e\rangle_{t,y} = \tfrac{1}{8}\epsilon_0 E_0^2. \tag{16}$$

The density u_m of magnetic energy associated with the wave is $\tfrac{1}{2}\mathbf{H}\cdot\mathbf{B}$, where $\mathbf{B} = \mu_0\mathbf{H}$; therefore, since $\mu_0\eta_0^{-2} = \epsilon_0$, we have

$$\begin{aligned}u_m = &\frac{1}{2}\,\epsilon_0 E_0^2 \sin^2\theta \sin^2\!\left(\frac{2\pi y}{\lambda_c}\right) \sin^2\!\left[\left(\frac{2\pi z}{\lambda_z}\right) - \omega t\right] \\ &+ \frac{1}{2}\,\epsilon_0 E_0^2 \cos^2\theta \cos^2\!\left(\frac{2\pi y}{\lambda_c}\right) \cos^2\!\left[\left(\frac{2\pi z}{\lambda_z}\right) - \omega t\right].\end{aligned} \tag{17}$$

Accordingly, $\langle u_m\rangle_{t,y} = \tfrac{1}{8}\epsilon_0 E_0^2 = \langle u_e\rangle_{t,y}$, and

$$\langle u \rangle_{t,y} = \langle u_e + u_m \rangle_{t,y} = \frac{1}{4} \epsilon_0 E_0^2 \frac{\text{joules}}{\text{meter}^3}. \tag{18}$$

This average holds for any plane parallel to the xy plane and reaching from one metal wall to the other.

Since $E_y = 0$ and $E_z = 0$, Poynting's vector has only two nonvanishing components:

$$(\mathbf{E} \times \mathbf{H})_x = 0, \tag{19}$$

$$(\mathbf{E} \times \mathbf{H})_y = -\frac{1}{4} E_0 H_0 \cos\theta \sin\left(\frac{4\pi y}{\lambda_c}\right) \sin\left[\left(\frac{4\pi z}{\lambda_z}\right) - 2\omega t\right], \tag{20}$$

$$(\mathbf{E} \times \mathbf{H})_z = E_0 H_0 \sin\theta \sin^2\left(\frac{2\pi y}{\lambda_c}\right) \sin^2\left[\left(\frac{2\pi z}{\lambda_z}\right) - \omega t\right]. \tag{21}$$

The z component of $\mathbf{E} \times \mathbf{H}$ is never negative and implies a pulsating flow of energy in the $+z$ direction. By contrast, at any point where $\sin(4\pi y/\lambda_c) \neq 0$, the y component keeps changing sign at twice the frequency of the wave—roughly speaking, the electromagnetic energy flows in Fig. 250 not only in the $+z$ direction, but also bounces from left to right and from right to left.

When the components of $\mathbf{E} \times \mathbf{H}$ are averaged with respect to t and y, they become

$$\langle (\mathbf{E} \times \mathbf{H})_x \rangle_{t,y} = 0, \langle (\mathbf{E} \times \mathbf{H})_y \rangle_{t,y} = 0, \quad \langle (\mathbf{E} \times \mathbf{H})_z \rangle_{t,y} = \tfrac{1}{4} E_0 H_0 \sin\theta.$$

The one nonzero term can be written as

$$\langle (\mathbf{E} \times \mathbf{H})_z \rangle_{t,y} = \frac{1}{4\eta_0} E_0^2 \sin\theta \frac{\text{watts}}{\text{meter}^2}. \tag{22}$$

Let us now write v_g for the speed of propagation of energy in the $+z$ direction. To compute v_g, we claim that the average time rate of the flow of energy is equal to the average energy density multiplied by v_g; that is, using (18) and (22), we set $\eta_0^{-1} E_0^2 \sin\theta$ equal to $\epsilon_0 E_0^2 v_g$ and find that

$$v_g = c \sin\theta. \tag{23}$$

Pattern speed and group speed. The field pattern shown in Fig. 252 consists of square portions of edge-length $\tfrac{1}{2}\lambda_z$. In adjoining squares, the fields $\mathbf{E}$ and $\mathbf{H}$ differ only in sign, so the energy distributions in all the squares are the same. A chain of these squares, which presumably extends infinitely far in the $\pm z$ directions, is sketched in Fig. 253(a) for two instants of time, one period apart. The pointers show which square moved into which position during a period. For the case (11), the pattern speed is given by (14) as

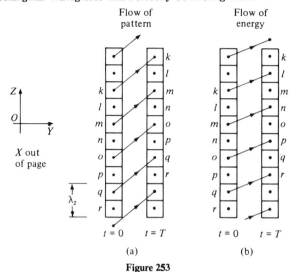

Figure 253

$$v_p = \sqrt{2}\, c. \tag{24}$$

One might think offhand that the electromagnetic energy associated with the field pattern should move in the $+z$ direction together with the pattern, and thus have the speed (14), which is greater than c. Such a speed, however, is not compatible with the theory of relativity insofar as propagation of *energy* is concerned. Furthermore, between the instants $t = 0$ and $t = T$, the energy not only drifts in a pulsating fashion in the $+z$ direction, but also surges back and forth between the metal walls. Consequently, the argument that "energy should move with the pattern" is not altogether convincing. In fact, we already know that, at least on the average, the energy should move with the speed v_g, which in this example is $c/\sqrt{2}$, so

$$v_g = \tfrac{1}{2} v_p. \tag{25}$$

In our particularly simple example, one can picture the situation as in Fig. 253(b), where the energy contained in the square marked k when $t = 0$ has moved into the square l when $t = T$. Taken together, the two parts of Fig. 253 display the fact that, on the average, energy travels at half the speed of the wave pattern, as required by (25). If $\theta \neq 45$ deg., the situation is more complicated; in fact, one may have to imagine that the energy stored in a particular square or rectangular portion of the pattern at time zero has spread into more than one such portion when $t = T$ (Exercise 6).

The symbol v_g stands for *group speed* or *group velocity*, which is the speed of propagation of a transitory electric pulse consisting of a narrow range of frequencies. In a wave guide, this speed proves to be the speed of propagation of energy by waves that go on forever; that is why we used the symbol v_g in (23). One must be sure not to confuse the *group speed* v_g with the *pattern speed* v_p, whose usual name is *phase speed*. Note that, according to (14) and (23),

$$v_p v_g = c^2 \qquad (26)$$

for any value of θ. Incidentally, the length that we denoted by λ_z is often written as λ_g and called the "guide wavelength."

The waves studied in this section are called transverse electric (TE) waves because their energy flows on the average in the $+z$ direction, at right angles to E, which has only an x component. But they are not transverse electromagnetic (TEM) waves, because H does have a component in the direction of the flow.

The tangential components of H in the vacuum, just next to the metal walls, can be computed from (9). These components vanish in the (perfectly conducting) metal. The discontinuity implies surface currents.

EXERCISES

1. Why would one expect the agreement between (14) and (13[116])?

2. Draw a point Q in the plane of Fig. 252 and let its (y, z) coordinates be $(\tfrac{1}{2}b + \xi, \eta)$, so that the square of its distance from the point P in the figure is $r_0^2 = \xi^2 + \eta^2$. Show that if r_0 is small enough compared to b, the line of H passing through Q is a circle. Take the counterclockwise sense of this circle as positive, and show that the component of H tangential to this circle is $-H_0 \pi r_0/(\sqrt{2}\, b)$, and that

$$\text{mmf}_{\text{small circle}} = -\frac{\sqrt{2}\, \pi^2 r_0^2}{b} H_0. \qquad (27)$$

3. To construct a TM wave, let

$$\mathbf{H} = H_0 f(y) \sin\left(\frac{2\pi z}{\lambda_z} - \omega t\right) \mathbf{1}_x, \qquad (28)$$

where H_0, ω, and λ_z are constants, and determine the corresponding E that satisfies the necessary boundary conditions at the conducting walls. In this work, restrict yourself to some simple form of $f(y)$ that leads to a satisfactory E.

4. Show that, in the case (11), the linear density of the surface current on each metal wall is

$$\mathbf{J}^* = \frac{1}{\sqrt{2}} H_0 \cos\left[\left(\frac{\pi z}{b}\right) - \omega t\right] \mathbf{1}_x. \qquad (29)$$

5. Verify that the magnetic field described by (7), (8), and (9) satisfies the boundary condition (2^{113}) for **B**.

6. Suppose that the value of the frequency in such equations as (4) is increased by the factor $\sqrt{3/2}$. How will this change affect the general λ's in (5), the angle θ and the particular λ's in (11), and the speeds v_p and v_g? Using the increased value of the frequency, draw diagrams similar to Fig. 253. (To get a "complete" set, patterns should be shown for $t = 0, T,$ and $2T$.)

118. THE "CUTOFF" PHENOMENON

The condition (3^{117}) can be written as

$$\cos \theta = \frac{n\pi c}{b\omega}, \quad n = 1, 2, 3, \ldots. \tag{1}$$

When b, n, and ω are given, Equation (1) leads to one or the other of two possibilities: either

$$\omega \geq \frac{n\pi c}{b} \quad \text{and hence } \cos \theta \leq 1, \tag{2}$$

or

$$\omega < \frac{n\pi c}{b} \quad \text{and hence } \cos \theta > 1. \tag{3}$$

In the first case the solutions of §117 are valid as they stand, and the waves and their energy propagate in the space between the perfectly conducting walls without attenuation. In the second case, when $\cos \theta > 1$, a sketch as unsophisticated as Fig. 250 no longer applies, and our solutions must be revised. We will illustrate below that the result is an attenuated wave, which keeps receiving energy and sending it back in the $-z$ direction. This type of wave is said to be "beyond cutoff." The value of ω that separates the possibilities (2) and (3) is the "cutoff omega" for the given values of b and n:

$$\omega_c = \frac{n\pi c}{b}. \tag{4}$$

The *cutoff frequency* is

$$f_c = \frac{\omega_c}{2\pi} = \frac{nc}{2b}, \tag{5}$$

and the *cutoff wavelength* is

$$\lambda_c = \frac{c}{f_c} = \frac{2b}{n}. \tag{6}$$

Thus, it follows from (6^{117}) that the cutoff wavelength is the λ_c defined in (5^{117}).

This section involves complex numbers, which are also used for a different purpose and in a different notation in Appendix C.

To illustrate cutoff by a numerical example, we let

$$\omega = \frac{1}{2}\omega_c = \frac{n\pi c}{2b} \tag{7}$$

so that
$$\cos\theta = 2 \tag{8}$$

and
$$\sin\theta = \pm\sqrt{1-\cos^2\theta} = \pm i\sqrt{3}, \tag{9}$$

where $i^2 = -1$. We also have

$$\lambda_z = \frac{\lambda_0}{\sin\theta} = \mp i\frac{\lambda_0}{\sqrt{3}}. \tag{10}$$

To abbreviate further formulas, we write

$$\alpha = \frac{2\sqrt{3}\,\pi}{\lambda_0}. \tag{11}$$

As we will see, α is an *attenuation constant*.

The last factor in (4^{117}) now becomes

$$\sin(\pm i\alpha z - \omega t) = \sin(\pm i\alpha z)\cos\omega t - \cos(\pm i\alpha z)\sin\omega t. \tag{12}$$

Whether we take the upper or the lower sign, this factor is complex rather than real, and consequently the corresponding solution of Maxwell's equations is formal rather than physical. However, since these equations do not contain the imaginary unit i, the real and the imaginary parts of (12) lead to solutions when taken separately. Also, since the equations are linear, linear combinations of these parts lead to further solutions. Now, as the reader should verify (Exercise 5),

$$\sin(i\alpha z) = \tfrac{1}{2}i(e^{\alpha z} - e^{-\alpha z}), \qquad \cos(i\alpha z) = \tfrac{1}{2}(e^{\alpha z} + e^{-\alpha z}), \tag{13}$$

and consequently the simplest possibilities to consider as replacements for the last factor of (4^{117}) are

$$e^{-\alpha z}\cos\omega t, \qquad e^{-\alpha z}\sin\omega t, \qquad e^{\alpha z}\cos\omega t, \qquad e^{\alpha z}\sin\omega t. \tag{14}$$

If we adopt the first possibility listed in (14), Equation (4^{117}) changes to

$$\mathbf{E} = E_0 \sin\left(\frac{2\pi y}{\lambda_c}\right) e^{-\alpha z} \cos\omega t \mathbf{1}_x, \tag{15}$$

where $e^{-\alpha z}$ is the *attenuation factor*. The components of $\mathbf{H}$ now are

$$H_x = 0, \tag{16}$$

$$H_y = \sqrt{3}\, H_0 \sin\left(\frac{2\pi y}{\lambda_c}\right) e^{-\alpha z} \sin \omega t, \tag{17}$$

$$H_z = 2 H_0 \cos\left(\frac{2\pi y}{\lambda_c}\right) e^{-\alpha z} \sin \omega t. \tag{18}$$

The components of Poynting's vector for this electromagnetic field are

$$(\mathbf{E} \times \mathbf{H})_x = 0, \tag{19}$$

$$(\mathbf{E} \times \mathbf{H})_y = -\frac{1}{2} E_0 H_0 \sin\left(\frac{4\pi y}{\lambda_c}\right) e^{-2\alpha z} \sin 2\omega t, \tag{20}$$

$$(\mathbf{E} \times \mathbf{H})_z = \frac{\sqrt{3}}{2} E_0 H_0 \sin^2\left(\frac{2\pi y}{\lambda_c}\right) e^{-2\alpha z} \sin 2\omega t. \tag{21}$$

Note that the time average of each component is zero—the waves are "cut off." Actual metal walls are not perfect conductors, and consequently the attenuation is even more pronounced than indicated above; besides, there is a gradual change of phase as a function of z.[1]

EXERCISES

1. Verify that the fields (15) through (18) satisfy Maxwell's equations.
2. Replace the last factor of (4^{117}) by $e^{-\alpha z} \sin \omega t$ and compute the corresponding **H**.
3. Suppose that the last factor of (4^{117}) is replaced by $e^{\alpha z} \cos \omega t$ and describe the resulting situation without going into details.
4. Show that

$$\frac{1}{\lambda_c^2} + \frac{1}{\lambda_z^2} = \frac{1}{\lambda_0^2} \tag{22}$$

and restate the cutoff condition (3) in terms of n, b, and λ_0.
5. Recall Exercise 5^6 and verify the formulas (13).

119. TE WAVES IN RECTANGULAR GUIDES; THE DOMINANT MODE

We now turn to the rectangular guide of Fig. 254; the distance between its side walls is b; that between its top and bottom is a. The field **E** in §117 is perpendicular to the top and the bottom, and vanishes at the side walls. The field **H**

---

[1] See, for example, R. M. Whitmer, *Electromagnetics*, 2nd Edition (Englewood Cliffs, N. J.: Prentice-Hall, Inc., 1962), pp. 314-320. The complex-number notation used in this reference is taken up in our Appendix C.

of §117 is parallel to the top and the bottom; near the side walls it is parallel to them. Consequently, these fields satisfy the boundary conditions of our new problem, and constitute a solution. In these solutions E_x does not depend on x, and they prove to comprise only a limited set of solutions for the guide. This set is denoted by TE_{0n}, where n is the integer appearing in (6^{117}); if $n = 1$, we have what is called the "dominant" TE mode.

When we restrict ourselves to the special case (11^{117}), the fields of §117 become

$$E_x = E_0 \sin\left(\frac{\pi y}{b}\right) \sin\left(\frac{\pi z}{b} - \omega t\right), \tag{1}$$

$E_y = E_z = 0$; and $H_x = 0$,

$$H_y = \frac{1}{\sqrt{2}} H_0 \sin\left(\frac{\pi y}{b}\right) \sin\left(\frac{\pi z}{b} - \omega t\right), \tag{2}$$

$$H_z = \frac{1}{\sqrt{2}} H_0 \cos\left(\frac{\pi y}{b}\right) \cos\left(\frac{\pi z}{b} - \omega t\right). \tag{3}$$

Furthermore, since $n = 1$ and $\theta = 45$ deg., we have

(3^{117})
$$\omega = \frac{\sqrt{2}\,\pi c}{b}. \tag{4}$$

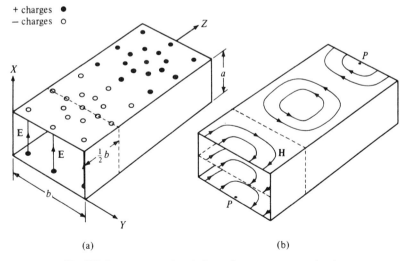

Fig. 254. Instantaneous descriptions of a TE_{01} wave moving in a rectangular waveguide: (a) the field **E** and the charges on the inner surfaces of the top and bottom of the guide; (b) the field **H**.

In regions where E_x is positive, the lines of E in Fig. 254 run from the lower to the upper plate; that is, the charges that develop on the upper plate in such regions are negative. According to (1^{72}), we have $q_a = \epsilon_0 E_x$, so the general formula for the charge density on the *upper* plate is

$$q_a = -\epsilon_0 E_0 \sin\left(\frac{\pi y}{b}\right) \sin\left(\frac{\pi z}{b} - \omega t\right); \tag{5}$$

on the lower plate q_a is the negative of (5). A charge distribution of this kind is suggested in Fig. 254(a).

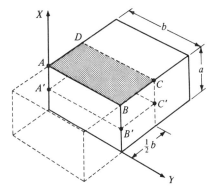

Figure 255

To illustrate how the charge, say q, located on a fixed portion of the upper plate varies with time, we consider the shaded strip in Fig. 255, where

$$0 \leq y \leq b, \quad 0 \leq z \leq \tfrac{1}{2} b. \tag{6}$$

Integration of (5) over this strip gives

$$q = -\frac{2}{\pi^2} \epsilon_0 E_0 b^2 (\cos \omega t - \sin \omega t). \tag{7}$$

Direction lines of H are sketched in Fig. 254(b) for three horizontal planes: just above the bottom of the guide, in the midplane, and just below the top; they are patterned after Fig. 252, which pertains to any plane parallel to the yz plane.

The boundary condition (6^{113}), with a similar condition for H_y, implies surface currents in the guide, flowing at right angles to the lines of H, as suggested in Fig. 256. To make sure that the solution we adopted from the two-plate case of §117 is satisfactory, one should show that these surface currents are

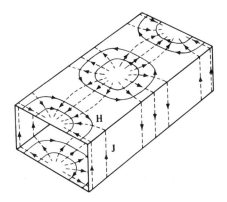

Fig. 256. The linear density of currents on the inner surfaces of the guide of Fig. 254.

consistent with the time rate of change of the surface charge on the top and bottom of the guide (Exercise 3).

One can compute the electromagnetic fields in a wave guide using the wave equations for **E** and **H** and the proper boundary conditions; one can do this in a straightforward mathematical way and without any reference to surface charges and surface currents. But the physical picture becomes complete only when the surface phenomena are included, for electric and magnetic fields go hand in hand with electric charges and currents.

A superposition of the simplest TE waves.   In view of (5^{117}) and (6^{117}), the field (4^{117}) can be written as

$$\mathbf{E} = E_0 \sin \frac{n\pi y}{b} f(z, t) \mathbf{1}_x, \tag{8}$$

where $n = 1, 2, 3, \ldots$, and

$$f(z, t) = \sin\left(\frac{2\pi z}{\lambda_z} - \omega t\right). \tag{9}$$

Let the oscillator activating the guide of Fig. 254 be turned through 90 deg. in a plane parallel to the xy plane. It is then perhaps obvious that a TE solution can be obtained by replacing y by x and $n\pi/b$ by $m\pi/a$, where $m = 1, 2, 3, \ldots$. Hence the guide will also accommodate the field

$$\mathbf{E} = E_0 \sin \frac{m\pi x}{a} f(z, t) \mathbf{1}_y, \tag{10}$$

if ω is large enough.

Linear combinations of (8) and (10) form solutions of Maxwell's equations for the guide when combined with the corresponding **H**'s.

EXERCISES

1. Use (29^{117}) to show that the conduction current flowing at the instant t into the shaded rectangle in Fig. 255 across the line BC is

$$i_{BC} = \frac{1}{\sqrt{2}\pi} bH_0(\cos \omega t + \sin \omega t). \tag{11}$$

2. Consider the top of the guide under the conditions (11^{117}), and use (2) and (3) to show that, at points on the line AB in Fig. 255, the linear density of the surface current is

$$\mathbf{J}^* = \frac{1}{\sqrt{2}} H_0 \left[\mathbf{1}_y \cos\left(\frac{\pi y}{b}\right) \cos \omega t + \mathbf{1}_z \sin\left(\frac{\pi y}{b}\right) \sin \omega t \right]. \tag{12}$$

Also, compute $\mathbf{J}^*$ at points on the line CD in the figure.

3. The total charge q on the shaded rectangle in Fig. 255 is (7). Use the results of Exercises 1 and 2 to verify that the surface current flowing into the rectangle across its four edges is equal to dq/dt, as required by the law of conservation of charge.

4. A charged insulating strip of width b is parallel to the yz plane and, as in Fig. 257, is moving with the speed v in the $+z$ direction relative to a fixed coordinate frame; it drags the charges with it, so that v is also the speed of the charge pattern. The line AB in the figure lies along the y axis. (a) Write $q_a(y, z, t)$

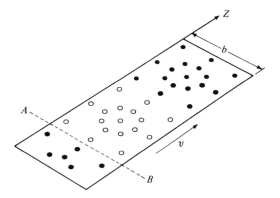

Fig. 257. A moving nonconducting strip that drags with it a charge distribution similar to that of Fig. 254(a).

for the charge density on the strip and show that the density of the current flowing across AB is $\mathbf{J}^* = vq_a(y, 0, t)\mathbf{1}_z$. (b) Explain why the method of part (a) cannot be applied to the charges on the top of the guide in Fig. 254. (c) What *is* the speed by which one should multiply the charge density (5) in order to get the z component of the current density (12)?

5. Suppose that the ω in (4) is increased by the factor $\sqrt{3/2}$, as in Exercise 6^{117}, and investigate the effects of this change on the fields (1), (2), and (3), and on the charge density (5).

6. (a) Use (1) to find the *density* of the displacement current flowing inside the guide. (b) Find the displacement *current* flowing in the guide through a rectangle $A'B'C'D'A'$, which lies inside the guide as in Fig. 255. (c) Verify that the net conduction current flowing into the shaded rectangle across its edges is equal to the net displacement current leaving this rectangle at the same instant.

7. Use (2) and (3) to find the mmf in the path $A'B'C'D'A'$ in Fig. 255. Then show that this mmf and the displacement current computed in Exercise 6 satisfy Maxwell's mmf law.

8. Imagine a small circle lying in a plane parallel to the top plate in Fig. 254(b); its center lies just below the point P, whose position relative to the wave pattern is the same as that of the point P in Fig. 252. The radius r_0 of the circle is small enough to justify the approximations made in Exercise (2^{117}). Find the displacement current flowing through this circle and show that it is consistent with the mmf given in (27^{117}).

9. Use the boundary conditions for **E** and **H** to show that an idealized perfectly conducting guide of infinite length shields the space outside it from electromagnetic phenomena that take place inside it; base the proof essentially on general integral formulas, such as Gauss's form of Coulomb's law and the integral form of Maxwell's mmf law. (One can say that the waves generated inside the guide go out through the metal, and that the waves generated outside the guide by the surface currents cancel the escaping waves. We have proved this in §115 for shielding by a flat metal plate of infinite extent, but a proof along these lines for a guide of finite cross section would be complicated.)

120. OTHER TYPES OF TE WAVES

To obtain further TE waves for a rectangular guide, we must look for electric fields that satisfy the proper boundary conditions, satisfy the wave equation

$$(2^{105}) \qquad \nabla^2 \mathbf{E} = \frac{1}{c^2} \frac{\partial^2 \mathbf{E}}{\partial t^2}, \qquad (1)$$

and also the equation

$$\text{div } \mathbf{E} = 0 \qquad (2)$$

inside the evacuated guide. We must also look for magnetic fields that satisfy *their* wave equation, satisfy *their* boundary conditions, and, together with **E**, satisfy Maxwell's equations. One standard way of doing all this is first to separate the variables, as we did on a smaller scale in the steps from (3^{68}) to (12^{68}) and in Exercise 3^{68}. We can, however, take a short cut, because we already know some of the E's for the guide.

The first step in the method of separation of variables is to assume, in our case, that $E_x = X(x)Y(y)Z(z)T(t)$, with similar expressions for E_y and the three components of **H**. The final step is to take linear combinations of solutions of this form that satisfy all the necessary requirements.

The fields E and H. Since we are looking for TE waves moving in the z direction, we may set $E_z = 0$ right away and replace the factor $Z(z)T(t)$ by $f(z, t) = \sin[(2\pi z/\lambda_z) - \omega t]$, which is a superposition of the functions $\sin(2\pi z/\lambda_z)\cos \omega t$ and $\cos(2\pi z/\lambda_z)\sin \omega t$, each of which has the form $Z(z)T(t)$; the constant λ_z will remain undetermined until later. Accordingly, we multiply (8^{119}) by a function $X_1(x)$ of x alone, multiply (10^{119}) by a function $Y_2(y)$ of y alone, and begin with an **E** of the form

$$E_x = E_0 X_1(x) \sin \frac{n\pi y}{b} f(z, t), \tag{3}$$

$$E_y = E_0 \sin \frac{m\pi x}{a} Y_2(y) f(z, t), \tag{4}$$

and $E_z = 0$. Equation (2), which reads $\partial E_x/\partial x + \partial E_y/\partial y + \partial E_z/\partial z = 0$, then gives

$$E_0 \frac{dX_1}{dx} \sin \frac{n\pi y}{b} f(z, t) + E_0 \sin \frac{m\pi x}{a} \frac{dY_2}{dy} f(z, t) = 0. \tag{5}$$

Consequently,

$$\frac{dX_1/dx}{\sin(m\pi x/a)} = -\frac{dY_2/dy}{\sin(n\pi y/b)}. \tag{6}$$

The first term in (6) depends only on x, the second only on y, hence both are equal to the same constant. In order to have the final results in a convenient form, we write $dX_1/dx = -(C_0 m n \pi/abE_0)\sin(m\pi x/a)$, and ignore the constant of integration in X_1. After we treat the right-hand side of (6) in a similar way, the fields (3) and (4) become

$$E_x = C_0 \frac{n}{b} \cos \frac{m\pi x}{a} \sin \frac{n\pi y}{b} \sin \left(\frac{2\pi z}{\lambda_z} - \omega t\right), \tag{7}$$

$$E_y = -C_0 \frac{m}{a} \sin \frac{m\pi x}{a} \cos \frac{n\pi y}{b} \sin \left(\frac{2\pi z}{\lambda_z} - \omega t\right), \tag{8}$$

where C_0 is a constant, expressed in volts.

To determine the relation between λ_z and the other parameters, we substitute (7) into the wave equation for E_x, obtained from (1) by writing E_x for E. The factors involving the variables cancel, and we are left with the formula

$$\frac{1}{\lambda_z^2} = \frac{1}{\lambda_0^2} - \frac{1}{4}\left(\frac{m^2}{a^2} + \frac{n^2}{b^2}\right), \tag{9}$$

where λ_0 is the free-space wavelength of a wave of frequency $\omega/2\pi$. Note that if $m = 0$ and $n = 1$, then (9) reduces to (22^{118}).

Since $E_z = 0$, the equation curl $\mathbf{E} = -\partial \mathbf{B}/\partial t$ leads from (7) and (8) to the following magnetic field:

$$H_x = -\frac{\lambda_0}{\eta_0 \lambda_z} E_y, \tag{10}$$

$$H_y = \frac{\lambda_0}{\eta_0 \lambda_z} E_x, \tag{11}$$

$$H_z = \frac{\lambda_0 C_0}{2\eta_0}\left(\frac{m^2}{a^2} + \frac{n^2}{b^2}\right) \cos\frac{m\pi x}{a} \cos\frac{n\pi y}{b} \cos\left(\frac{2\pi z}{\lambda_z} - \omega t\right). \tag{12}$$

Here η_0 is the intrinsic impedance of free space, introduced in (5^{110}) and equal to $\sqrt{\mu_0/\epsilon_0}$.

The waves consisting of the fields $\mathbf{E}$ and $\mathbf{H}$ computed above are called TE_{mn} waves; the case when $n = m = 0$ is, of course, trivial.

Lines of E and H in TE_{11} waves. To illustrate the shapes of field lines, we consider the TE_{11} mode and a moving plane on which

$$\frac{2\pi z}{\lambda_z} - \omega t = \frac{\pi}{2}. \tag{13}$$

If we also set $x = 0$, Equation (7) gives

$$E_x = \frac{C_0}{b} \sin\frac{\pi y}{b}. \tag{14}$$

Let us now turn back for a moment to the TE_{01} mode and to a plane on which the last factor of (1^{119}) is unity, so that

$$E_x = E_0 \sin\frac{\pi y}{b}. \tag{15}$$

Comparison with (14) shows that, in the moving planes under consideration, the lines of $\mathbf{E}$ begin on the bottom of the guide in the TE_{11} mode in the same way as they do in the TE_{01} mode, illustrated in Fig. 254(a). They do not continue straight up, however, because now E_y is not zero. In fact, if we set $y = 0$ in (8) under the condition (13), we get

$$E_y = -\frac{C_0}{a} \sin \frac{\pi x}{a}. \tag{16}$$

Therefore, in the TE$_{11}$ mode and in the plane under consideration, as many lines of **E** end on the left-hand wall as begin on the bottom of the guide. More detailed analysis shows that these lines are shaped as in Fig. 258(a).

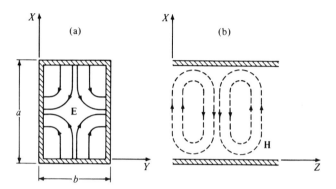

Fig. 258. A TE$_{11}$ wave moving in a rectangular guide in the $+z$ direction: (a) typical lines of **E** and (b) *projections* upon a sidewall of typical lines of **H**.

The lines of **H** are more difficult to picture, because they do not lie in planes. We will merely set $t = 0$ and consider the *projections* of some of these lines on a side wall of the guide, say the zx plane. When we let $m = 1, n = 1, t = 0$, ignore H_y, and write $dx/dz = H_x/H_z$, we get an equation that has the form (12^{117}) and leads to Fig. 258(b). More detailed diagrams are given in several books.[2]

EXERCISES

1. Compute the pattern speed and the group speed for the TE$_{11}$ mode and note that they satisfy 26^{117}.

2. Compute the cutoff frequency for the TE$_{11}$ mode. If $a = 3$ cm and $b = 4$ cm, what is the maximum free-space wavelength that the guide will transmit in the TE$_{11}$ mode? In the TE$_{10}$ mode? In the TE$_{01}$ mode?

---

[2] S. A. Schelkunoff, *Electromagnetic Waves* (Princeton, N. J.: D. Van Nostrand Co., Inc., 1943) pp. 395 and 396. Simon Ramo and John R. Whinnery, *Fields and Waves in Modern Radio* (New York: Wiley & Sons, Inc., and London, Chapman & Hall, Ltd., 2nd Ed., 1953), Table 9.03. Also *American Institute of Physics Handbook* (New York: McGraw-Hill Book Co., 1957) p. **5**-61.

CHAPTER 27

Magnetic Vector Potential

The potential V is a very useful *scalar* field, even though it does not appear in Maxwell's equations. We will now introduce a *vector* field—the magnetic vector potential **A**—that also does not appear in Maxwell's equations but is nevertheless of great interest. The field **A** simplifies many computations, as we illustrate for the case of a short antenna. And its time derivative, taken with a minus sign, proves to be the intensity of the induced electric field of the Faraday type.

The definition of **A** reads curl **A** = **B**. In §121 we illustrate the forms of **A** pertaining to a straight filamentary current, a short segment of such a current, and a plane electromagnetic wave. Next, we express Faraday's law of induction in terms of **A**, consider a short antenna, compute the effects of retardation on **A** and V, and in §123 recompute by straightforward differentiation the fields **E** and **H**, found by trial and error in §108. The chapter ends with remarks on currents in extended regions.

If **B** is given, the single equation curl **A** = **B** does not specify **A** uniquely. In §121 we simply discard the ambiguous term in **A** and find in §125 that, as a result, our **A** satisfies the equation called the *Lorentz condition*. We then mention the fact that, if the arbitrary features of **A** are removed so as to satisfy the Lorentz condition, then the three cartesian components of **A** and a constant multiple of V form together an important "four-vector" of the theory of relativity. The usefulness of the Lorentz condition in computations is illustrated more thoroughly in Appendix C.

As before, we confine ourselves to nonmagnetic mediums and set $\mu = \mu_0$.

121. FILAMENTARY CURRENTS

In §82 we remarked without proof that if a vector field **F** satisfies the equation

$$\operatorname{curl} \mathbf{F} = 0, \qquad (1)$$

then there are scalar fields f, such that

$$\mathbf{F} = \operatorname{grad} f. \qquad (2)$$

For example, let the z axis be uniformly charged and let q_l be the linear density of the charge. The resulting electric field is electrostatic and hence conservative, so that curl $\mathbf{E} = 0$. Therefore, we can conclude from the theorem quoted above that $\mathbf{E}$ must be the gradient of a scalar field. In other words, the field $\mathbf{E}$, whatever it may be, is "derivable" from a scalar potential, say V. In the present example, we already know both $\mathbf{E}$ and V:

(11^{60}) $$\mathbf{E} = \frac{q_l}{2\pi\epsilon_0} \frac{1}{\rho} \mathbf{1}_\rho, \qquad (3)$$

(31^{60}) $$V = -\frac{q_l}{2\pi\epsilon_0} \ln \frac{\rho}{\rho_0}. \qquad (4)$$

These expressions satisfy the equation $\mathbf{E} = -\operatorname{grad} V$ for any arbitrary choice of the nonzero length ρ_0.

We also remarked in §79 that if

$$\operatorname{div} \mathbf{G} = 0, \qquad (5)$$

then there are fields $\mathbf{F}$, such that

$$\mathbf{G} = \operatorname{curl} \mathbf{F}. \qquad (6)$$

Consequently, Equation (4^{103}), namely div $\mathbf{B} = 0$, implies that the magnetic flux density $\mathbf{B}$ is the curl of another vector field. This field is denoted by $\mathbf{A}$ and is called the *magnetic vector potential*. In symbols,

$$\mathbf{B} = \operatorname{curl} \mathbf{A}. \qquad (7)$$

To illustrate, let a steady current i flow along the z axis in the $+z$ direction, so that, for $\rho > 0$,

(7^{87}) $$\mathbf{B} = \frac{\mu_0 i}{2\pi} \frac{1}{\rho} \mathbf{1}_\phi. \qquad (8)$$

The theorem stated just above implies that the field (8) must be the curl of a vector field. Indeed,

(10^{81}) $$\frac{1}{\rho} \mathbf{1}_\phi = -\operatorname{curl}\left(\ln \frac{\rho}{\rho_0} \mathbf{1}_z\right) \qquad (9)$$

and, therefore, (7) will hold if we set

$$\mathbf{A} = -\frac{\mu_0 i}{2\pi} \ln \frac{\rho}{\rho_0} \mathbf{1}_z + \mathbf{A}_0, \qquad (10)$$

where the field $\mathbf{A}_0$ satisfies the condition

$$\text{curl } \mathbf{A}_0 = 0 \qquad (11)$$

but is otherwise arbitrary. For simplicity, we let

$$\mathbf{A}_0 = 0. \qquad (12)$$

For the computation of the $\mathbf{A}$ pertaining to a filamentary current of *any* shape, we need a formula for the contribution, say $d\mathbf{A}$, made to $\mathbf{A}$ by a short segment of a filamentary current. To find such a formula, we first consider a uniformly charged z axis in Fig. 259(a), write V for the electrostatic potential

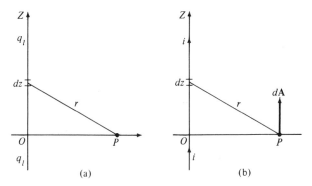

Fig. 259. A filamentary charge and a filamentary current. Comparison of their respective fields V and $\mathbf{B}$ suggests the formula $d\mathbf{A} = \mathbf{1}_z(\mu_0 i/4\pi r) \, dz$ for the contribution to the vector potential $\mathbf{A}$ at P, made by the segment dz of the current i.

at the point P, and write dV for the contribution to V made by the charge located on the element dz shown in the figure. This charge is $q_l \, dz$, its distance from P is r, and, therefore, according to (3^{59}),

$$dV = \frac{q_l}{4\pi\epsilon_0} \frac{dz}{r}. \qquad (13)$$

Next, we note that, in view of (12), the $\mathbf{A}$ in (10) can be obtained from the V in (4) by replacing the factor q_l/ϵ_0 by $\mu_0 i$ and multiplying the result by $\mathbf{1}_z$. We may expect that the same changes will convert (13) into $d\mathbf{A}$ and, therefore, we write

$$d\mathbf{A} = \frac{\mu_0 i\, dz}{4\pi\, r} \mathbf{1}_z. \tag{14}$$

The local field* B *of a short antenna. To check (14) in a small way, we return to the antenna of Fig. 233[107] and restrict ourselves to the near zone, so that retardation can be ignored. As in §107, we have $i = I\cos\omega t$ and $dz = l$. Since only one current element is involved, we write $\mathbf{A}$ for $d\mathbf{A}$ and find from (14) that, at the point P in the figure,

$$\mathbf{A} = \frac{\mu_0 lI \cos \omega t}{4\pi} \frac{1}{r}\mathbf{1}_z. \tag{15}$$

Consequently,

$$\mathbf{B} = \operatorname{curl} \mathbf{A} = \frac{\mu_0 lI \cos \omega t}{4\pi} \frac{\sin \theta}{r^2}\mathbf{1}_\phi, \tag{16}$$

in agreement with (10^{107}).

This computation of $\mathbf{B}$ from (14) is perhaps a little shorter than that based on the $d\mathbf{H}$ formula, as in Exercise 4^{88}. However, the utility of the potential $\mathbf{A}$ is revealed in a much more striking way when this potential is introduced into Faraday's emf law.

Plane waves in free space. Let

$$(6^{110}) \qquad \mathbf{E} = E_0 \sin(ky - \omega t)\mathbf{1}_z, \qquad \mathbf{H} = H_0 \sin(ky - \omega t)\mathbf{1}_x, \tag{17}$$

where $E_0 = \sqrt{\mu_0/\epsilon_0}\, H_0 = \eta_0 H_0$, as in ($7^{110}$). To identify a vector potential $\mathbf{A}$ whose curl is μ_0 times the $\mathbf{H}$ in (17), we begin with Faraday's law, namely curl $\mathbf{E} = -\mu_0\, \partial \mathbf{H}/\partial t$. Combining this law with the equation curl $\mathbf{A} = \mu_0 \mathbf{H}$, we find that curl $(\mathbf{E} + \partial \mathbf{A}/\partial t) = 0$. Therefore, we expect that some appropriate $\mathbf{A}$ will resemble $\mathbf{E}$, except for a constant factor and a change from $\sin(ky - \omega t)$ to $\cos(ky - \omega t)$. When we set $\mathbf{A} = C\cos(ky - \omega t)\mathbf{1}_z$, compute curl $\mathbf{A}$, and compare with the $\mathbf{H}$ in (17), we find that one possibility for $\mathbf{A}$ is

$$\mathbf{A} = -\frac{1}{\omega} E_0 \cos(ky - \omega t)\mathbf{1}_z. \tag{18}$$

This formula is generalized in Exercise 3^{125}.

EXERCISES

1. Check the computation of curl $\mathbf{A}$ in (16).
2. Verify that (18) is consistent with the $\mathbf{H}$ in (17).
3. A solenoid of infinite length is oriented as in Fig. 204^{88}. Its radius is ρ_0, it

has n turns per meter of length, and it carries the current i. According to Exercise 6⁹³, we then have

$$\mathbf{B} = \begin{cases} \mu_0 ni \mathbf{1}_z & \text{inside} \\ 0 & \text{outside.} \end{cases} \quad (19)$$

Use the equation $\mathbf{B} = \text{curl } \mathbf{A}$ to verify that a suitable vector potential for this case is

$$\mathbf{A} = \begin{cases} \dfrac{\rho}{\rho_0} A_0 \mathbf{1}_\phi & \text{inside} \\ \dfrac{\rho_0}{\rho} A_0 \mathbf{1}_\phi & \text{outside,} \end{cases} \quad (20)$$

where $A_0 = \tfrac{1}{2}\mu_0 ni\rho_0$. Note that (20) is patterned after (6⁹³), and explain why we can expect (6⁹³) to provide a useful clue.

4. Two long solenoids (radii ρ_1 and ρ_2) are coaxial with the z axis, as in Fig. 260; the magnitudes of ni are the same in both, but the currents have

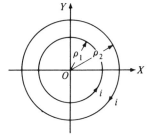

Fig. 260. Cross sections of two coaxial solenoids.

the opposite senses shown in the figure, so that $\mathbf{B} = 0$ inside the inner solenoid. Verify that in this case a suitable vector potential is $\mathbf{A} = A_\rho \mathbf{1}_\phi$, where

$$A_\rho = \begin{cases} 0 & \text{for } \rho \leq \rho_1 \\ \tfrac{1}{2}\mu_0 ni(\rho_1^2 - \rho^2)/\rho & \text{for } \rho_1 \leq \rho \leq \rho_2 \\ \tfrac{1}{2}\mu_0 ni(\rho_1^2 - \rho_2^2)/\rho & \text{for } \rho \geq \rho_2. \end{cases} \quad (21)$$

5. Show that if the field $\mathbf{B}$ is uniform and constant, and has an arbitrarily fixed direction, then a suitable choice of $\mathbf{A}$ is $\tfrac{1}{2}\mathbf{B} \times \mathbf{r}$, where $\mathbf{r} = x\mathbf{1}_x + y\mathbf{1}_y + z\mathbf{1}_z$. Next consider the special case when $\mathbf{B} = B_0 \mathbf{1}_z$, and correlate the result with the top line of (20).

122. FARADAY'S LAW

Since $\mathbf{B} = \operatorname{curl} \mathbf{A}$, the equation

$$(1^{103}) \qquad \operatorname{curl} \mathbf{E} = -\frac{\partial \mathbf{B}}{\partial t} \qquad (1)$$

can be written as

$$\operatorname{curl}\left(\mathbf{E} + \frac{\partial \mathbf{A}}{\partial t}\right) = 0. \qquad (2)$$

Equation (2) has the form (1^{121}) and implies that the field $\mathbf{E} + \partial \mathbf{A}/\partial t$ is the gradient of a scalar field, which we denote by $-V$, so that in statics the equations will reduce to familiar formulas. That is, we write $\mathbf{E} + \partial \mathbf{A}/\partial t = -\operatorname{grad} V$ and conclude that

$$\mathbf{E} = -\operatorname{grad} V - \frac{\partial \mathbf{A}}{\partial t}. \qquad (3)$$

This result has the form

$$(11^{98}) \qquad \mathbf{E} = -\operatorname{grad} V + \mathbf{E}^f, \qquad (4)$$

where $\mathbf{E}^f$ is the intensity of an electric field of the Faraday type.

Equation (3) is one of the most mystifying equations of Maxwell's theory, at least until one looks at the theory from the relativistic standpoint. For example, according to (3), an electric field may exist at a point P at an instant t_0, even if $\operatorname{grad} V = 0$, $\mathbf{B} = 0$, and $\mathbf{A} = 0$ at P at the instant t_0, provided that $\partial \mathbf{A}/\partial t \neq 0$ at that instant. We will not illustrate this possibility but will be content with a simpler case.

Example. Let us return to §97 and to the long slender solenoid lying along the z axis, write Φ for the magnetic flux in the solenoid, write

$$(2^{97}) \qquad i_m = \frac{d\Phi}{dt}, \qquad (5)$$

call i_m the magnetic displacement current flowing in the solenoid, and assume, as in §97, that i_m is a constant. The intensity of the Faraday field induced in the space outside the solenoid is

$$(5^{97}) \qquad \mathbf{E} = -\frac{i_m}{2\pi} \frac{1}{\rho} \mathbf{1}_\phi. \qquad (6)$$

In §97, Equation (6) was derived from the $d\mathbf{E}$ formula. It can also be derived from the integral form of Faraday's law and the assumption of axial symmetry (Exercise 6[98]). In this section we will derive it in a still different way, based on (3).

In the light of the assumption stated in footnote 5[97], Equation (3) reduces outside the windings of the solenoid to

$$\mathbf{E} = -\frac{\partial \mathbf{A}}{\partial t}, \tag{7}$$

where $\mathbf{E}$ can be written as $\mathbf{E}^f$. Inside the solenoid, $B = \mu_0 n i$ and $\Phi = B\pi\rho_0^2 = \pi\mu_0 n i \rho_0^2$, where ρ_0 is the cross-sectional radius of the solenoid. The formula for $\mathbf{A}$ outside the solenoid is

(20[121]) $$\mathbf{A} = \frac{1}{2}\mu_0 n i \rho_0^2 \frac{1}{\rho} \mathbf{1}_\phi, \tag{8}$$

so that

$$\mathbf{A} = \frac{\Phi}{2\pi} \frac{1}{\rho} \mathbf{1}_\phi. \tag{9}$$

Now, Φ does not depend on spatial coordinates, so that $\partial\Phi/\partial t$ is simply $d\Phi/dt$. Consequently, (7) leads to the formula

$$\mathbf{E} = -\frac{d\Phi/dt}{2\pi} \frac{1}{\rho} \mathbf{1}_\phi = -\frac{i_m}{2\pi} \frac{1}{\rho} \mathbf{1}_\phi, \tag{10}$$

and we have arrived at (6) once again.

123. RETARDED POTENTIALS FOR A SHORT ANTENNA

We will now prepare the ground for the use of the potential $\mathbf{A}$ for computing the radiation from the short antenna pictured in Fig. 233[107]. As before, we let

$$i = I\cos\omega t, \qquad q = \frac{I}{\omega}\sin\omega t, \tag{1}$$

and use the symbols $k = \omega/c$ and $M = Il/4\pi$, where l is the length of the antenna.

The scalar potential. Let V_1 and V_2 be the respective potentials produced at the point P in Fig. 152[60] by the charges q and $-q$, located on the upper and lower ends of the antenna, whose distances from P are s_1 and s_2. The potential felt at P at the instant t corresponds to the value that q had at the earlier instant $t - s_1/c$, when q was equal to $(I/\omega) \sin \omega(t - s_1/c)$. Therefore, from Coulomb's law,

$$V_1 = \frac{I}{4\pi\epsilon_0 \omega} \frac{1}{s_1} \sin \omega\left(t - \frac{s_1}{c}\right). \tag{2}$$

Since l is small, Equations (24[7]) and (25[7]) convert (2) into

$$V_1 = \frac{I}{4\pi\epsilon_0 \omega}\left(\frac{1}{r} + \frac{l}{2r^2}\cos\theta\right) \sin\left[\omega\left(t - \frac{r}{c}\right) + \frac{\omega l}{2c}\cos\theta\right]. \tag{3}$$

The second term in the argument of the sine is small by our standards. Therefore, we let

$$\sin\left(\frac{\omega l}{2c}\cos\theta\right) = \frac{\omega l}{2c}\cos\theta, \qquad \cos\left(\frac{\omega l}{2c}\cos\theta\right) = 1. \tag{4}$$

Ignoring in (3) the term that involves l^2, we then get

$$V_1 = \frac{I}{4\pi\epsilon_0 \omega}\left(\frac{1}{r}\sin\omega\tau + \frac{\omega l \cos\theta}{2cr}\cos\omega\tau + \frac{l \cos\theta}{2r^2}\sin\omega\tau\right), \tag{5}$$

where $\tau = t - r/c$, as in (5[107]).

When we treat in a similar way the charge on the lower end of the antenna, we find that

$$V_2 = -\frac{I}{4\pi\epsilon_0 \omega}\left(\frac{1}{r}\sin\omega\tau - \frac{\omega l \cos\theta}{2cr}\cos\omega\tau - \frac{l \cos\theta}{2r^2}\sin\omega\tau\right). \tag{6}$$

Adding V_1 and V_2, we finally get the total retarded scalar potential, namely,

$$V = \frac{M \cos\theta}{\epsilon_0 \omega}\left(\frac{k}{r}\cos\omega\tau + \frac{1}{r^2}\sin\omega\tau\right). \tag{7}$$

The vector potential. If retardation is ignored, the potential **A** for a short antenna is given by (15[121]). To generalize that expression to larger distances from the antenna, we need only to replace t by τ, getting

$$\mathbf{A} = \frac{\mu_0 M}{r}\cos\omega\tau \mathbf{1}_z \tag{8}$$

or, since $\mathbf{1}_z = \mathbf{1}_r \cos\theta - \mathbf{1}_\theta \sin\theta$,

$$\mathbf{A} = \mu_0 M \frac{\cos\theta}{r}\cos\omega\tau \mathbf{1}_r - \mu_0 M \frac{\sin\theta}{r}\cos\omega\tau \mathbf{1}_\theta. \tag{9}$$

124. THE ELECTROMAGNETIC FIELD OF A SHORT ANTENNA

Now that the potentials V and $\mathbf{A}$ have been found, we will compute the fields $\mathbf{E}$ and $\mathbf{H}$ from (3^{122}) and (7^{121}), namely,

$$\mathbf{E} = -\operatorname{grad} V - \frac{\partial \mathbf{A}}{\partial t}, \qquad \mathbf{H} = \frac{1}{\mu_0} \operatorname{curl} \mathbf{A}, \tag{1}$$

where $\mathbf{H}$ is written for free space.

The electric field. According to (7^{123}) and the relations

$$(3^{108}) \qquad \frac{\partial}{\partial r} \sin \omega \tau = -k \cos \omega \tau, \qquad \frac{\partial}{\partial r} \cos \omega \tau = k \sin \omega \tau, \tag{2}$$

we have

$$\operatorname{grad}_r V = -\frac{2M \cos \theta}{\omega \epsilon_0} \frac{1}{r^3} \sin \omega \tau - \frac{2M \cos \theta}{c \epsilon_0} \frac{1}{r^2} \cos \omega \tau + \mu_0 \omega M \frac{\cos \theta}{r} \sin \omega \tau, \tag{3}$$

$$\operatorname{grad}_\theta V = -\frac{M \sin \theta}{\omega \epsilon_0} \frac{1}{r^3} \sin \omega \tau - \frac{M \sin \theta}{c \epsilon_0} \frac{1}{r^2} \cos \omega \tau, \tag{4}$$

and $\operatorname{grad}_\phi V = 0$. Also, according to ($9^{123}$) and the relation $\epsilon_0 \mu_0 = 1/c^2$, we have

$$\frac{\partial A_r}{\partial t} = -\mu_0 M \omega \frac{\cos \theta}{r} \sin \omega \tau, \tag{5}$$

$$\frac{\partial A_\theta}{\partial t} = \frac{kM \sin \theta}{c \epsilon_0} \frac{1}{r} \sin \omega \tau, \tag{6}$$

and $\partial A_\phi / \partial t = 0$. When we substitute these expressions into the first equation in (1), we get the field $\mathbf{E}$ given in (19^{108}), (20^{108}), and (21^{108}), which is the field that we obtained earlier by trial, error, and more trial.

The magnetic field. Finally, we turn to (9^{123}) and to the second equation in (1). The resulting formulas are $H_r = \mu_0^{-1} \operatorname{curl}_r \mathbf{A} = 0$, $H_\theta = \mu_0^{-1} \operatorname{curl}_\theta \mathbf{A} = 0$, and

$$H_\phi = \frac{1}{\mu_0} \operatorname{curl}_\phi \mathbf{A} = M \frac{\sin \theta}{r^2} \cos \omega \tau - kM \frac{\sin \theta}{r} \sin \omega \tau, \tag{7}$$

in agreement with (22^{108}).

Once the retarded potentials V and $\mathbf{A}$ for the antenna are known, the computation of $\mathbf{E}$ and $\mathbf{H}$, presented in this section, is straightforward. The computation of the retarded V in §123 is also straightforward, although a bit tedious; but it can be simplified, as in Exercise 1^{125}

125. THE LORENTZ CONDITION

The potentials V and $\mathbf{A}$ given in (7^{123}) and (9^{123}) are generated by the same antenna, and hence we may expect them to be closely related. Indeed,

$$\frac{\partial}{\partial t} V = \frac{M \cos \theta}{\epsilon_0} \left(+\frac{1}{r^2} \cos \omega\tau - \frac{k}{r} \sin \omega\tau \right), \tag{1}$$

and

$$\operatorname{div} \mathbf{A} = \mu_0 M \cos \theta \left(-\frac{1}{r^2} \cos \omega\tau + \frac{k}{r} \sin \omega\tau \right), \tag{2}$$

so that $\operatorname{div} \mathbf{A} = -\epsilon_0 \mu_0 \, \partial V/\partial t$ and

$$\operatorname{div} \mathbf{A} + \frac{1}{c^2} \frac{\partial V}{\partial t} = 0. \tag{3}$$

Equation (3) is called the *Lorentz condition*. The reason why the fields $\mathbf{A}$ and V computed above satisfy this important condition can be traced back to the fact that we chose the field $\mathbf{A}_0$ in (10^{121}) to be zero. In §121, the equation curl $\mathbf{A}_0 = 0$ was necessary; but the equation $\mathbf{A}_0 = 0$ was not necessary, and we adopted it only for simplicity. We could have chosen a different $\mathbf{A}_0$, and then the resulting $\mathbf{A}$ would not necessarily satisfy the Lorentz condition.

As we have seen, if $\mathbf{B}$ is given, the equation curl $\mathbf{A} = \mathbf{B}$ does not determine $\mathbf{A}$ completely. A theorem of vector analysis states that to determine $\mathbf{A}$ in full detail, we must be given not only its curl, but also its divergence; and if only its curl is given, then, with due regard for the proper boundary conditions, we are at liberty to specify the divergence of $\mathbf{A}$ as we please. Consequently, instead of computing the fields $\mathbf{A}$ and V as in §123 and then discovering that they satisfy the Lorentz condition, we could have assumed from the start that this condition holds. This procedure would have simplified the computation of the retarded potential V (Exercise 1 and Appendix C).

One of the advantages of setting div $\mathbf{A}$ equal to $-c^{-2} \partial V/\partial t$, rather than to some other scalar field, is that this simplifies the wave equation for $\mathbf{A}$. If we write $\mathbf{D} = \epsilon_0 \mathbf{E}$ and $\mathbf{B} = \mu_0 \mathbf{H}$, Equations ($7^{121}$) and ($3^{122}$), respectively, become $\mathbf{H} = \mu_0^{-1}$ curl $\mathbf{A}$ and $\mathbf{D} = -\epsilon_0$ grad $V - \epsilon_0 \partial \mathbf{A}/\partial t$. The equation

$$(2^{103}) \qquad\qquad \operatorname{curl} \mathbf{H} = \mathbf{J} + \frac{\partial \mathbf{D}}{\partial t} \tag{4}$$

can then be written as

$$\operatorname{curl} \operatorname{curl} \mathbf{A} = \mu_0 \mathbf{J} - \frac{1}{c^2} \operatorname{grad} \frac{\partial V}{\partial t} - \frac{1}{c^2} \frac{\partial^2 \mathbf{A}}{\partial t^2}. \tag{5}$$

Therefore, in view of (47^A),

$$\nabla^2 \mathbf{A} - \frac{1}{c^2} \frac{\partial^2 \mathbf{A}}{\partial t^2} = -\mu_0 \mathbf{J} + \operatorname{grad}\left(\operatorname{div} \mathbf{A} + \frac{1}{c^2} \frac{\partial V}{\partial t}\right). \tag{6}$$

It is now apparent that, as long as we are free to choose the divergence of **A**, a convenient choice is provided by the Lorentz condition, which simplifies (6) to

$$\nabla^2 \mathbf{A} - \frac{1}{c^2} \frac{\partial^2 \mathbf{A}}{\partial t^2} = -\mu_0 \mathbf{J}. \tag{7}$$

Another reason for the usefulness of the Lorentz condition comes from relativity: when coordinate frames in relative motion are involved, this relation remains invariant in form and permits interpreting A_x, A_y, A_z and a multiple of V as the components of a single "four-vector." Note, incidentally, that Equation (8[40]) for the conservation of charge has the same form as (3).[3]

In uncharged vacuum, (7) reduces to the simple wave equation

$$\nabla^2 \mathbf{A} = \frac{1}{c^2} \frac{\partial^2 \mathbf{A}}{\partial t^2}. \tag{8}$$

An equation of this type also holds in uncharged vacuum for both **E** and **H** —recall Equation (2[105]) and Exercise 3[105].

If none of the quantities relevant to a given problem varies with the time, Equation (7) becomes

$$\nabla^2 \mathbf{A} = -\mu_0 \mathbf{J}, \tag{9}$$

so that each cartesian component of **A** satisfies a Poisson equation of type (7[66]).

Currents in extended regions. The appearance of the current density **J** in (7) and (9) indicates that these equations are not restricted only to filamentary currents. To generalize the formula (14[121]) for $d\mathbf{A}$, we consider Fig. 261, which is similar to Fig. 259(b), except that the current i is not confined to the z axis. The cross-sectional area of the wire in the figure is a, and the magnitude of the current density in the wire is a constant J, so that $i = aJ$. If the wire is thin enough, (14[121]) can be replaced by

---

[3] The structural similarity of (8[40]) and (3) is even clearer when these equations are written in the relativistic four-dimensional notation. See, for example, W. M. Schwarz, *Intermediate Electromagnetic Theory* (New York: John Wiley & Sons, 1964), Equations (7-2) and (8-4) on p. 387.

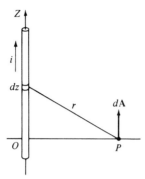

Fig. 261. Computation of the vector potential **A** caused by a thick current-carrying wire.

$$d\mathbf{A} = \frac{\mu_0 J}{4\pi} \frac{dv}{r} \mathbf{1}_z, \tag{10}$$

where dv, equal to $a\,dz$, is the volume of the element of the wire marked in Fig. 261. Furthermore, since in this example $\mathbf{J} = J\mathbf{1}_z$, we can write (10) in the form

$$d\mathbf{A} = \frac{\mu_0}{4\pi} \frac{\mathbf{J}}{r} dv, \tag{11}$$

which applies to any orientation of **J**.

Suppose that a region R is free from any magnetic materials and is pervaded by steady currents of density **J**, which may vary from point to point in R. Divide R into differential elements of volume dv and write r for the distance from an element to a fixed point P. It then follows from (11) that the vector potential produced at P by the currents flowing in R is

$$\mathbf{A} = \frac{\mu_0}{4\pi} \iiint_R \frac{\mathbf{J}(x,y,z)}{r} dv. \tag{12}$$

If **J** depends on t, (12) should be corrected for retardation (Exercise 6).

EXERCISES

1. Use the Lorentz condition and the formula (9^{123}) for **A** to determine the retarded V, and discuss the arbitrary features of this V. Compare this V with (7^{123}).

2. Show that the plane-wave potential **A** in (18^{121}) is consistent with the **E** in (17^{121}). In view of the result of Exercise 2^{121}, the single formula (18^{121}) accounts for both **E** and **H** in (17^{121}).

3. Let

$$\mathbf{A} = \frac{1}{\omega}\mathbf{E}_0 \sin(\mathbf{k} \cdot \mathbf{r} - \omega t), \tag{13}$$

where $\mathbf{E}_0$ and $\mathbf{k}$ are constant vectors, such that $\mathbf{E}_0 \cdot \mathbf{k} = 0$. Show that the single equation (13) describes both parts of (8^{112}) with $\phi_0 = 0$, namely,

$$\mathbf{E} = \mathbf{E}_0 \cos(\mathbf{k} \cdot \mathbf{r} - \omega t), \qquad \mathbf{H} = \frac{1}{\omega \mu_0} \mathbf{k} \times \mathbf{E}_0 \cos(\mathbf{k} \cdot \mathbf{r} - \omega t). \tag{14}$$

4. Recall the hints that led us to (18^{121}) and write a tentative formula for the vector potential $\mathbf{A}$ for the TE_{11} mode in a rectangular guide. Then adjust this formula, if necessary, so as to make it consistent with the fields $\mathbf{E}$ and $\mathbf{H}$ given in equations (7^{120}) through (12^{120}), other than (9^{120}).

5. The vector potential describing a certain mode of electromagnetic waves in a rectangular guide is

$$\mathbf{A} = \mathbf{1}_z A_0 \sin\frac{2\pi x}{a} \sin\frac{2\pi y}{b} \sin\left(\frac{2\pi z}{\lambda_z} - \omega t\right), \tag{15}$$

where A_0 is a constant. (a) Without computation, describe some features that you would expect $\mathbf{E}$ and $\mathbf{H}$ to have in this case. (b) Compute $\mathbf{E}$ and $\mathbf{H}$ in detail and illustrate them by rough field maps. (c) Is this a TE, or TM, or a TEM mode, or is it a superposition of such modes? (d) What is the cutoff frequency for this mode?

6. Assume that, throughout the region R, the current density $\mathbf{J}$ includes the factor $\cos \omega t$ and correct (12) for retardation.

7. Consider the hypothetical case of a static charge of density q_v somehow held fixed in a region R in vacuum, and show that the potential V at a point P can then be written as

$$V = \frac{1}{4\pi\epsilon_0} \iiint_R \frac{q_v(x, y, z)}{r} \, dv, \tag{16}$$

where r has the same meaning as in (12).

8. If $\mathbf{B}$ is given and all physical boundary conditions are ignored, the equation curl $\mathbf{A} = \mathbf{B}$ and the Lorentz condition do not determine $\mathbf{A}$ uniquely. Describe the remaining indeterminacy of $\mathbf{A}$.

APPENDIX A

Mathematical Formulas

I. COORDINATE VARIABLES

Relations among coordinates. The cartesian, cylindrical, and spherical coordinates of a point P, described in Fig. 8^5, are related as follows:

Cartesian and cylindrical

$$x = \rho \cos \phi, \qquad \rho = \sqrt{x^2 + y^2}, \tag{1}$$

$$y = \rho \sin \phi, \qquad \phi = \tan^{-1} \frac{y}{x}, \tag{2}$$

$$z = z. \qquad z = z. \tag{3}$$

Cartesian and spherical

$$x = r \sin \theta \cos \phi, \qquad r = \sqrt{x^2 + y^2 + z^2}, \tag{4}$$

$$y = r \sin \theta \sin \phi, \qquad \theta = \cos^{-1}\left(\frac{z}{\sqrt{x^2 + y^2 + z^2}}\right), \tag{5}$$

$$z = r \cos \theta. \qquad \phi = \tan^{-1} \frac{y}{x}. \tag{6}$$

Cylindrical and spherical

$$\rho = r \sin \theta, \qquad r = \sqrt{\rho^2 + z^2}, \tag{7}$$

$$\phi = \phi, \qquad \phi = \phi, \tag{8}$$

$$z = r \cos \theta. \qquad \theta = \cos^{-1}\left(\frac{z}{\sqrt{\rho^2 + z^2}}\right). \tag{9}$$

Neighboring points. We denote the vector step from a point P to a neighboring point Q by dl, and the length of this step by dl. Some special cases are

Cartesian

Step from $P(x, y, z)$ to $Q(x + dx, y + dy, z + dz)$

$$dl = \mathbf{1}_x\, dx + \mathbf{1}_y\, dy + \mathbf{1}_z\, dz, \tag{10}$$
$$(dl)^2 = (dx)^2 + (dy)^2 + (dz)^2. \tag{11}$$

Cylindrical

Step from $P(\rho, \phi, z)$ to $Q(\rho + d\rho, \phi + d\phi, z + dz)$

$$dl = \mathbf{1}_\rho\, d\rho + \mathbf{1}_\phi\, \rho\, d\phi + \mathbf{1}_z\, dz, \tag{12}$$
$$(dl)^2 = (d\rho)^2 + \rho^2 (d\phi)^2 + (dz)^2. \tag{13}$$

Spherical

Step from $P(r, \theta, \phi)$ to $Q(r + dr, \theta + d\theta, \phi + d\phi)$

$$dl = \mathbf{1}_r\, dr + \mathbf{1}_\theta\, r\, d\theta + \mathbf{1}_\phi\, r \sin\theta\, d\phi, \tag{14}$$
$$(dl)^2 = (dr)^2 + r^2 (d\theta)^2 + r^2 \sin^2\theta (d\phi)^2. \tag{15}$$

Elementary cells. The edge-lengths (shown below in parentheses) and the volumes of the elementary cells pictured in Fig. 262 are

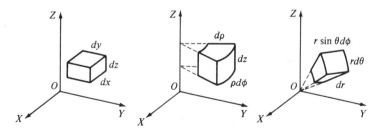

Fig. 262. Elementary cells of the cartesian, cylindrical, and spherical coordinate systems.

Cartesian

$$dv = (dx)(dy)(dz) = dx\, dy\, dz. \tag{16}$$

Cylindrical

$$dv = (d\rho)(\rho\,d\phi)(dz) = \rho\,d\rho\,d\phi\,dz. \tag{17}$$

Spherical

$$dv = (dr)(r\,d\theta)(r\sin\theta\,d\phi) = r^2\sin\theta\,dr\,d\theta\,d\phi. \tag{18}$$

II. BASE-VECTORS

Relations among vectors. The relations among the base-vectors shown in Figs. 58[16], 93[33], and 104[36] can be written as follows:

Cartesian and cylindrical

$$\mathbf{1}_x = \frac{x}{\rho}\mathbf{1}_\rho - \frac{y}{\rho}\mathbf{1}_\phi, \qquad \mathbf{1}_\rho = \frac{x}{\rho}\mathbf{1}_x + \frac{y}{\rho}\mathbf{1}_y, \tag{19}$$

$$\mathbf{1}_y = \frac{y}{\rho}\mathbf{1}_\rho + \frac{x}{\rho}\mathbf{1}_\phi, \qquad \mathbf{1}_\phi = -\frac{y}{\rho}\mathbf{1}_x + \frac{x}{\rho}\mathbf{1}_y, \tag{20}$$

$$\mathbf{1}_z = \mathbf{1}_z. \qquad \mathbf{1}_z = \mathbf{1}_z. \tag{21}$$

Cartesian and spherical

$$\mathbf{1}_x = \frac{x}{r}\mathbf{1}_r + \frac{xz}{\rho r}\mathbf{1}_\theta - \frac{y}{\rho}\mathbf{1}_\phi, \qquad \mathbf{1}_r = \frac{x}{r}\mathbf{1}_x + \frac{y}{r}\mathbf{1}_y + \frac{z}{r}\mathbf{1}_z, \tag{22}$$

$$\mathbf{1}_y = \frac{y}{r}\mathbf{1}_r + \frac{yz}{\rho r}\mathbf{1}_\theta + \frac{x}{\rho}\mathbf{1}_\phi, \qquad \mathbf{1}_\theta = \frac{xz}{\rho r}\mathbf{1}_x + \frac{yz}{\rho r}\mathbf{1}_y - \frac{\rho}{r}\mathbf{1}_z, \tag{23}$$

$$\mathbf{1}_z = \frac{z}{r}\mathbf{1}_r - \frac{\rho}{r}\mathbf{1}_\theta. \qquad \mathbf{1}_\phi = -\frac{y}{\rho}\mathbf{1}_x + \frac{x}{\rho}\mathbf{1}_y. \tag{24}$$

Cylindrical and spherical

$$\mathbf{1}_\rho = \frac{\rho}{r}\mathbf{1}_r + \frac{z}{r}\mathbf{1}_\theta, \qquad \mathbf{1}_r = \frac{\rho}{r}\mathbf{1}_\rho + \frac{z}{r}\mathbf{1}_z, \tag{25}$$

$$\mathbf{1}_z = \frac{z}{r}\mathbf{1}_r - \frac{\rho}{r}\mathbf{1}_\theta, \qquad \mathbf{1}_\theta = \frac{z}{\rho}\mathbf{1}_\rho - \frac{\rho}{r}\mathbf{1}_z, \tag{26}$$

$$\mathbf{1}_\phi = \mathbf{1}_\phi. \qquad \mathbf{1}_\phi = \mathbf{1}_\phi. \tag{27}$$

Derivatives of base-vectors.

Cartesian

$$\frac{\partial}{\partial x}\mathbf{1}_x = 0, \qquad \frac{\partial}{\partial y}\mathbf{1}_x = 0, \qquad \frac{\partial}{\partial z}\mathbf{1}_x = 0, \tag{28}$$

$$\frac{\partial}{\partial x}\mathbf{1}_y = 0, \quad \frac{\partial}{\partial y}\mathbf{1}_y = 0, \quad \frac{\partial}{\partial z}\mathbf{1}_y = 0, \tag{29}$$

$$\frac{\partial}{\partial x}\mathbf{1}_z = 0, \quad \frac{\partial}{\partial y}\mathbf{1}_z = 0, \quad \frac{\partial}{\partial z}\mathbf{1}_z = 0. \tag{30}$$

Cylindrical[1]

$$\frac{\partial}{\partial \rho}\mathbf{1}_\rho = 0, \quad \frac{\partial}{\partial \phi}\mathbf{1}_\rho = \mathbf{1}_\phi, \quad \frac{\partial}{\partial z}\mathbf{1}_\rho = 0, \tag{31}$$

$$\frac{\partial}{\partial \rho}\mathbf{1}_\phi = 0, \quad \frac{\partial}{\partial \phi}\mathbf{1}_\phi = -\mathbf{1}_\rho, \quad \frac{\partial}{\partial z}\mathbf{1}_\phi = 0, \tag{32}$$

$$\frac{\partial}{\partial \rho}\mathbf{1}_z = 0, \quad \frac{\partial}{\partial \phi}\mathbf{1}_z = 0, \quad \frac{\partial}{\partial z}\mathbf{1}_z = 0. \tag{33}$$

Spherical[1]

$$\frac{\partial}{\partial r}\mathbf{1}_r = 0, \quad \frac{\partial}{\partial \theta}\mathbf{1}_r = \mathbf{1}_\theta, \quad \frac{\partial}{\partial \phi}\mathbf{1}_r = \frac{\rho}{r}\mathbf{1}_\phi, \tag{34}$$

$$\frac{\partial}{\partial r}\mathbf{1}_\theta = 0, \quad \frac{\partial}{\partial \theta}\mathbf{1}_\theta = -\mathbf{1}_r, \quad \frac{\partial}{\partial \phi}\mathbf{1}_\theta = \frac{z}{r}\mathbf{1}_\phi, \tag{35}$$

$$\frac{\partial}{\partial r}\mathbf{1}_\phi = 0, \quad \frac{\partial}{\partial \theta}\mathbf{1}_\phi = 0, \quad \frac{\partial}{\partial \phi}\mathbf{1}_\phi = -\mathbf{1}_\rho. \tag{36}$$

III. VECTOR IDENTITIES

If the letters **A**, **B**, and **C** denote either individual vectors or vector fields, we have

$$\mathbf{A} \cdot \mathbf{B} = \mathbf{B} \cdot \mathbf{A}, \tag{37}$$

$$\mathbf{A} \times \mathbf{B} = -\mathbf{B} \times \mathbf{A}, \tag{38}$$

$$\mathbf{A} \cdot (\mathbf{B} + \mathbf{C}) = \mathbf{A} \cdot \mathbf{B} + \mathbf{A} \cdot \mathbf{C}, \tag{39}$$

$$\mathbf{A} \times (\mathbf{B} + \mathbf{C}) = \mathbf{A} \times \mathbf{B} + \mathbf{A} \times \mathbf{C}, \tag{40}$$

$$\mathbf{A} \cdot \mathbf{B} \times \mathbf{C} = \mathbf{B} \cdot \mathbf{C} \times \mathbf{A} = \mathbf{C} \cdot \mathbf{A} \times \mathbf{B}$$
$$= -\mathbf{A} \cdot \mathbf{C} \times \mathbf{B} = -\mathbf{B} \cdot \mathbf{A} \times \mathbf{C} = -\mathbf{C} \cdot \mathbf{B} \times \mathbf{A}, \tag{41}$$

$$\mathbf{A} \times (\mathbf{B} \times \mathbf{C}) = \mathbf{B}(\mathbf{A} \cdot \mathbf{C}) - \mathbf{C}(\mathbf{A} \cdot \mathbf{B}). \tag{42}$$

---

[1] Except at singular points. For example, the equation $\partial \mathbf{1}_\rho / \partial \phi = \mathbf{1}_\phi$ is meaningless at a point on the z axis.

IV. GRADIENT, DIVERGENCE, CURL, AND THE LAPLACIANS

Coordinate-free definitions pertaining to a point P are[2]

$$\operatorname{grad} f = \lim_{v \to 0} \frac{1}{v} \left(\iint_{(S)} \mathbf{1}_n f \, da \right), \tag{43}$$

$$\operatorname{div} \mathbf{F} = \lim_{v \to 0} \frac{1}{v} \left(\iint_{(S)} \mathbf{1}_n \cdot \mathbf{F} \, da \right), \tag{44}$$

$$\operatorname{curl} \mathbf{F} = \lim_{v \to 0} \frac{1}{v} \left(\iint_{(S)} \mathbf{1}_n \times \mathbf{F} \, da \right). \tag{45}$$

Here (S) is the closed surface bounding a region R that includes the point P and has the volume v. The condition $v \to 0$ implies that every spatial dimension of R tends to zero, so that every point on (S) approaches P. The letters f and $\mathbf{F}$ stand, respectively, for scalar and vector fields.

The two kinds of Laplacians are defined as follows:

$$\nabla^2 f = \operatorname{div} \operatorname{grad} f, \tag{46}$$

$$\nabla^2 \mathbf{F} = \operatorname{grad} \operatorname{div} \mathbf{F} - \operatorname{curl} \operatorname{curl} \mathbf{F}. \tag{47}$$

The coordinate formulas that follow from the definitions given above are

Cartesian

$$\operatorname{grad} f = \mathbf{1}_x \frac{\partial f}{\partial x} + \mathbf{1}_y \frac{\partial f}{\partial y} + \mathbf{1}_z \frac{\partial f}{\partial z}, \tag{48}$$

$$\operatorname{div} \mathbf{F} = \frac{\partial}{\partial x} F_x + \frac{\partial}{\partial y} F_y + \frac{\partial}{\partial z} F_z, \tag{49}$$

$$\nabla^2 f = \frac{\partial^2 f}{\partial x^2} + \frac{\partial^2 f}{\partial y^2} + \frac{\partial^2 f}{\partial z^2}, \tag{50}$$

$$\nabla^2 \mathbf{F} = \mathbf{1}_x \nabla^2 F_x + \mathbf{1}_y \nabla^2 F_y + \mathbf{1}_z \nabla^2 F_z, \tag{51}$$

$$\operatorname{curl} \mathbf{F} = \mathbf{1}_x \operatorname{curl}_x \mathbf{F} + \mathbf{1}_y \operatorname{curl}_y \mathbf{F} + \mathbf{1}_z \operatorname{curl}_z \mathbf{F}, \tag{52}$$

where

$$\operatorname{curl}_x \mathbf{F} = \frac{\partial}{\partial y} F_z - \frac{\partial}{\partial z} F_y, \tag{53}$$

$$\operatorname{curl}_y \mathbf{F} = \frac{\partial}{\partial z} F_x - \frac{\partial}{\partial x} F_z, \tag{54}$$

$$\operatorname{curl}_z \mathbf{F} = \frac{\partial}{\partial x} F_y - \frac{\partial}{\partial y} F_x. \tag{55}$$

[2] Another coordinate-free definition of gradients is given in §57.

Cylindrical

$$\operatorname{grad} f = \mathbf{1}_\rho \frac{\partial f}{\partial \rho} + \mathbf{1}_\phi \frac{1}{\rho} \frac{\partial f}{\partial \phi} + \mathbf{1}_z \frac{\partial f}{\partial z}, \tag{56}$$

$$\operatorname{div} \mathbf{F} = \frac{1}{\rho} \frac{\partial}{\partial \rho}(\rho F_\rho) + \frac{1}{\rho} \frac{\partial}{\partial \phi} F_\phi + \frac{\partial}{\partial z} F_z, \tag{57}$$

$$\nabla^2 f = \frac{1}{\rho} \frac{\partial}{\partial \rho}\left(\rho \frac{\partial f}{\partial \rho}\right) + \frac{1}{\rho^2} \frac{\partial^2 f}{\partial \phi^2} + \frac{\partial^2 f}{\partial z^2}, \tag{58}$$

$\nabla^2 \mathbf{F}$: See definition (47), (59)

$$\operatorname{curl} \mathbf{F} = \mathbf{1}_\rho \operatorname{curl}_\rho \mathbf{F} + \mathbf{1}_\phi \operatorname{curl}_\phi \mathbf{F} + \mathbf{1}_z \operatorname{curl}_z \mathbf{F}, \tag{60}$$

where

$$\operatorname{curl}_\rho \mathbf{F} = \frac{1}{\rho} \frac{\partial F_z}{\partial \phi} - \frac{\partial F_\phi}{\partial z}, \tag{61}$$

$$\operatorname{curl}_\phi \mathbf{F} = \frac{\partial F_\rho}{\partial z} - \frac{\partial F_z}{\partial \rho}, \tag{62}$$

$$\operatorname{curl}_z \mathbf{F} = \frac{1}{\rho} \frac{\partial}{\partial \rho}(\rho F_\phi) - \frac{1}{\rho} \frac{\partial F_\rho}{\partial \phi}. \tag{63}$$

Spherical

$$\operatorname{grad} f = \mathbf{1}_r \frac{\partial f}{\partial r} + \mathbf{1}_\theta \frac{1}{r} \frac{\partial f}{\partial \theta} + \mathbf{1}_\phi \frac{1}{r \sin \theta} \frac{\partial f}{\partial \phi}, \tag{64}$$

$$\operatorname{div} \mathbf{F} = \frac{1}{r^2} \frac{\partial}{\partial r}(r^2 F_r) + \frac{1}{r \sin \theta} \frac{\partial}{\partial \theta}(F_\theta \sin \theta) + \frac{1}{r \sin \theta} \frac{\partial F_\phi}{\partial \phi}, \tag{65}$$

$$\nabla^2 f = \frac{1}{r^2} \frac{\partial}{\partial r}\left(r^2 \frac{\partial f}{\partial r}\right) + \frac{1}{r^2 \sin \theta} \frac{\partial}{\partial \theta}\left(\frac{\partial f}{\partial \theta} \sin \theta\right) + \frac{1}{r^2 \sin^2 \theta} \frac{\partial^2 f}{\partial \phi^2}, \tag{66}$$

$\nabla^2 \mathbf{F}$: See definition (47), (67)

$$\operatorname{curl} \mathbf{F} = \mathbf{1}_r \operatorname{curl}_r \mathbf{F} + \mathbf{1}_\theta \operatorname{curl}_\theta \mathbf{F} + \mathbf{1}_\phi \operatorname{curl}_\phi \mathbf{F}, \tag{68}$$

where

$$\operatorname{curl}_r \mathbf{F} = \frac{1}{r \sin \theta} \frac{\partial}{\partial \theta}(F_\phi \sin \theta) - \frac{1}{r \sin \theta} \frac{\partial F_\theta}{\partial \phi}, \tag{69}$$

$$\operatorname{curl}_\theta \mathbf{F} = \frac{1}{r \sin \theta} \frac{\partial F_r}{\partial \phi} - \frac{1}{r} \frac{\partial}{\partial r}(rF_\phi), \tag{70}$$

$$\operatorname{curl}_\phi \mathbf{F} = \frac{1}{r} \frac{\partial}{\partial r}(rF_\theta) - \frac{1}{r} \frac{\partial F_r}{\partial \theta}. \tag{71}$$

In the following formulas, $\mathbf{F}$ and $\mathbf{G}$ are vector fields and f is a scalar field:

$$\operatorname{div}(f\mathbf{F}) = f \operatorname{div} \mathbf{F} + \mathbf{F} \cdot \operatorname{grad} f, \tag{72}$$

$$\operatorname{div}(\mathbf{F} \times \mathbf{G}) = \mathbf{G} \cdot \operatorname{curl} \mathbf{F} - \mathbf{F} \cdot \operatorname{curl} \mathbf{G}, \tag{73}$$

$$\text{div grad } f = \nabla^2 f, \tag{74}$$

$$\text{div curl } \mathbf{F} = 0, \tag{75}$$

$$\text{curl grad } f = 0. \tag{76}$$

V. FOUR THEOREMS

(a) Gauss's divergence theorem (§31):

$$\iiint_R \text{div } \mathbf{F} \, dv = \iint_{(S)} F_n \, da. \tag{77}$$

(b) Stokes's circulation theorem (§81):

$$\iint_S (\text{curl } \mathbf{F})_n \, da = \oint_C F_t \, dl. \tag{78}$$

(c) If div $\mathbf{G} = 0$, then there are vector fields $\mathbf{F}$ such that $\mathbf{G} = \text{curl } \mathbf{F}$; see §79.

(d) If curl $\mathbf{F} = 0$, then there are scalar fields f such that $\mathbf{F} = \text{grad } f$; see §82.

APPENDIX B

The MKS-Giorgi Units of Measurement

> "In the present state of science the most universal standard of length which we could assume would be the wave length in vacuum of a particular kind of light, emitted by some widely diffused substance such as sodium, which has well-defined lines in its spectrum. Such a standard would be independent of any changes in the dimensions of the earth, and should be adopted by those who expect their writings to be more permanent than that body."
>
> J. C. MAXWELL, after defining the meter in terms of a meridian of the earth.[1]

The system of units of measurement used in this book is a combination of metric units and "practical" electrical and magnetic units, first advocated in 1901 by Giovanni Giorgi (1871–1950). The older system in general use in physics—the centimeter-gram-second (cgs) system—is based on *three* fundamental units; in that system, the units of measurement for electric and magnetic quantities are expressed in terms of the units for length, mass, and time. The Giorgi system differs from the cgs system in two distinct ways. First, the meter and the kilogram replace the centimeter and the gram. Second, the Giorgi system takes one of the electrical or magnetic units of measurement as fundamental, and hence is based on *four* fundamental units. The name "MKS-Giorgi" was suggested for this system in 1935 by the International Electrotechnical Commission.[2] If the fourth unit is chosen to be the ampere, we have the MKSA

---

[1] *A Treatise on Electricity and Magnetism*, 3rd ed., Vol. 1, p. 3 (Oxford, 1873). Quoted by permission of the Oxford University Press.

[2] F. B. Silsbee, *Establishment and Maintenance of the Electrical Units*, National Bureau of Standards Circular 475, p. 20 (1949).

system; if the coulomb is chosen, we have the MKSC system, and so on. In fact, "any of the following practical units, ohm, ampere, volt, henry, farad, coulomb, weber already in use may equally serve as the fourth fundamental unit, because it is possible to derive each unit and its dimensions from any four others mutually independent."[3] For the purposes of this book it is not important to specify the fourth basic unit, and hence we use the general term "MKS-Giorgi." The MKS-Giorgi system is often called simply the MKS system, but one should then remember that it is, nevertheless, a four-unit rather than a three-unit system.

The units of the MKS-Giorgi system, other than the **meter** (unit of length), the **kilogram** (unit of mass), and the **second** (unit of time) can be defined as follows:[4]

The **newton** (unit of force). The newton is the force which gives to a mass of 1 kilogram an acceleration of 1 meter per second per second.

The **joule** (unit of energy or work). The joule is the work done when the point of application of a force equal to 1 newton is displaced a distance of 1 meter in the direction of the force.

The **watt** (unit of power). The watt is the power which gives rise to the production of energy at the rate of 1 joule per second.

The **ampere** (unit of electric current). The ampere is the constant current which, if maintained in each of two straight parallel conductors of infinite length, of negligible circular sections, and placed 1 meter apart in a vacuum, will produce between these conductors a force equal to 2×10^{-7} newton per meter of length.

The **coulomb** (unit of quantity of electricity). The coulomb is the quantity of electricity transported in 1 second by a current of 1 ampere.

The **volt** (unit of difference of potential and of electromotive force). The volt is the difference of electric potential between two points when the work done in transferring a small electric charge from the one point to the other is equal to 1 joule per coulomb of transferred charge.

The **ohm** (unit of electric resistance). The ohm is the electric resistance between two points of a conductor when a constant difference of potential of 1 volt, applied between

---

[3] See Silsbee, ibid., p. 21.

[4] Our definitions are equivalent to those recommended by the International Committee on Weights and Measures at its meeting in Paris on October 29, 1946 (see Silsbee, ibid, Appendix 6); and with a few exceptions they are worded in the same way.[5] Thus, they also agree with the latest legal U. S. definitions, which are, however, expressed in terms of the cgs electromagnetic units and do not include the newton, the weber, and the mho (Public Law 617, 81st Congress, 1950). The aim of the Committee was not to phrase the definitions in their most complete forms, but rather to express them "in as simple and easily understood language as is possible."

[5] The exceptions are: We write "newton" for "MKS unit of force." We inserted the words "each of" into the definition of the ampere. The definition of the volt, which in the Committee's wording is slanted toward circuit problems, has been somewhat generalized. In the definition of the ohm, we replaced "seat" by "source." The definition of the mho has been added.

these points, produces in this conductor a current of 1 ampere, this conductor not being the source of any electromotive force.

The **farad** (unit of electric capacitance). The farad is the capacitance of a capacitor between the plates of which there appears a difference of potential of 1 volt when it is charged by a quantity of electricity equal to 1 coulomb.

The **henry** (unit of electric inductance). The henry is the inductance of a closed circuit in which an electromotive force of 1 volt is produced when the electric current in the circuit varies uniformly at the rate of 1 ampere per second.

The **weber** (unit of magnetic flux). The weber is the magnetic flux which, linking a circuit of 1 turn, produces in it an electromotive force of 1 volt as it is reduced to zero at a uniform rate in 1 second.[6]

The **mho** (unit of electric conductance). The mho is the electric conductance between two points of a conductor whose electric resistance between these points is 1 ohm.

These definitions imply that the two sides of any one of the following "equivalences" can be used interchangeably:

$$1 \text{ newton} = 1 \text{ (kilogram} \cdot \text{meter)/second}^2,$$
$$1 \text{ joule} = 1 \text{ newton} \cdot \text{meter},$$
$$1 \text{ watt} = 1 \text{ joule/second},$$
$$1 \text{ coulomb} = 1 \text{ ampere} \cdot \text{second},$$
$$1 \text{ volt} = 1 \text{ joule/coulomb},$$
$$1 \text{ ohm} = 1 \text{ volt/ampere},$$
$$1 \text{ farad} = 1 \text{ coulomb/volt},$$
$$1 \text{ henry} = 1 \text{ (volt} \cdot \text{second)/ampere},$$
$$1 \text{ weber} = 1 \text{ volt} \cdot \text{second},$$

and

$$1 \text{ mho} = 1 \text{ ohm}^{-1}.$$

Some further equivalences are listed on cover two.

If the unit for measuring a quantity is a compound unit having the same numerator and denominator, this quantity is called a "pure number." The radian measure of an angle has the form (length of arc)/(length of radius), so that its unit of measurement is, say, the "meter/meter." Therefore, the radian measure of an angle is a pure number.

Equivalences should not be confused with definitions. For example, the words "kilogram·meter/second^2" do not define the newton, just as the words "meter/meter" do not define the radian.

The first electrical unit on our list, the ampere, is defined in terms of me-

---

[6] In this book, we take the weber to be also the unit for measuring magnetic pole strength. See the table on cover three.

chanical effects. Consequently, the ampere, and with it all the electrical and magnetic units of measurement, can be expressed in terms of the meter, kilogram, and second; but this proceduce is open to objections, which are avoided in the MKS-Giorgi system by introducing a fourth basic unit of measurement.

A sequence of definitions of electrical units that starts with the ampere fits in well with the procedures for establishing electrical standards in the laboratory. But it has a pedagogical drawback: we tell the reader in §1 that the "ampere" of which we speak is the ampere specified in the definition, but it is only in §89 that we prove this.

Rationalization. The process of rationalization adopted in §59 does not affect the MKS-Giorgi units of measurement; thus, the definition of the ampere given above remains the same (and hence the ampere remains the same) whether or not the equations are rationalized. What rationalization *does* affect is the appearance of some of the equations and the numerical values of certain parameters, such as the capacitivity (permittivity) ϵ_0 and the inductivity (magnetic permeability) μ_0 of free space.

APPENDIX C

The Complex-Number Shorthand

Formulas for waves include time factors such as $\sin \omega t$ and $\cos \omega t$. This notation is explicit, but in some computations it it inconvenient because differentiation changes sines into cosines, and vice versa. We will now describe another notation, which employs complex exponentials instead of sines and cosines.

The notation for the imaginary unit is not uniform in physics books. Some authors use i, as we did in the mathematical work of §118. Others replace i by j or by $-j$. In this appendix we will use j and a form of time factors that gives the familiar plus and minus signs in Maxwell's free-space curl equations:

$$\text{curl } \mathbf{H} = +\epsilon_0 \, \partial \mathbf{E}/\partial t, \qquad \text{curl } \mathbf{E} = -\mu_0 \, \partial \mathbf{H}/\partial t. \tag{1}$$

Euler's formula. Let j be the imaginary unit, so that $j^2 = -1$, and let c be the complex number

$$c = a + jb, \tag{2}$$

where a and b are real numbers; they are called, respectively, the *real part* and the *imaginary part* of c, even though b is real. The number c can be represented by a point in the *complex plane*, pictured in Fig. 263. The figure suggests that another form of (2) is

$$c = r(\cos \phi + j \sin \phi). \tag{3}$$

The positive number r is the *magnitude* or *absolute value* of c; the angle ϕ is the *phase* of c.

A theorem due to Leonhard Euler (1707–1783) states that

$$\cos \phi + j \sin \phi = e^{j\phi} \tag{4}$$

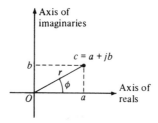

Fig. 263. A point $a + jb$ in the complex plane.

and permits us to write (2) in a still different way, namely,

$$c = r e^{j\phi}. \tag{5}$$

Here e is the base of the natural logarithms. It follows that $\cos \phi =$ real part of $e^{j\phi}$, and $\sin \phi =$ imaginary part of $e^{j\phi}$. If we multiply (4) by j, we get $j \cos \phi - \sin \phi = je^{j\phi}$, so that $\sin \phi =$ real part of $-je^{j\phi}$. We now adopt the following convention: If the number of an equation appears at the margin of the page not in parentheses but in *square brackets*, then the left-hand side of the equation is intended to be equal to the *real part* of the right-hand side, rather than to the entire right-hand side. For example, the equation

$$\cos \phi = \text{real part of } e^{j\phi} \tag{6}$$

is written in this notation as

$$\cos \phi = e^{j\phi}. \tag{7}$$

Similarly,

$$\sin \phi = -je^{j\phi}. \tag{8}$$

Now, let $f = f_0 \cos \omega t$, where f_0 is real and does not depend on t. The equations

$$f = f_0 \cos \omega t, \qquad \frac{\partial f}{\partial t} = -\omega f_0 \sin \omega t \tag{9}$$

can then be written as

$$f = f_0 e^{j\omega t}, \qquad \frac{\partial f}{\partial t} = -\omega f_0(-je^{j\omega t}). \tag{10}$$

That is, $\partial f/\partial t = j\omega f_0 e^{j\omega t}$, and hence

$$\frac{\partial f}{\partial t} = j\omega f. \tag{11}$$

If $f = f_0 \sin \omega t$, we get [11] once again. It follows that, in the notation adopted above, multiplication of a function f by $j\omega$ replaces differentiation of f with respect to time, *provided* that f has the form $f_1 \cos \omega t + f_2 \sin \omega t$, where f_1

and f_2 are real and independent of t. If f does not have this form, *the complex-number notation may lead into error* (Exercise 3).

According to [11], Maxwell's free-space equations (8^{103}) can be written in the single-frequency case as

$$\text{curl } \mathbf{E} = -j\omega\mu_0 \mathbf{H}, \qquad \text{curl } \mathbf{H} = j\omega\epsilon_0 \mathbf{E}, \qquad [12]$$

while Faraday's law (3^{122}) turns into

$$\mathbf{E} = -\text{grad } V - j\omega\mathbf{A}, \qquad [13]$$

and the Lorentz condition (3^{125}) becomes

$$V = j\frac{c^2}{\omega} \text{ div } \mathbf{A}. \qquad [14]$$

We also note that, since $\tau = t - r/c$,

$$\frac{\partial}{\partial r} e^{j\omega\tau} = -jk e^{j\omega\tau}. \qquad (15)$$

Radiation from a short antenna. To illustrate the use of the complex-number notation, we return once again to the short antenna of Fig. 233^{107}. The current (1^{107}) is

$$i = I e^{j\omega t}. \qquad [16]$$

Consequently, the retarded vector potential, given in (8^{123}) as $\mathbf{1}_z(\mu_0 M/r) \cos \omega\tau$, can be written as $\mathbf{1}_z(\mu_0 M/r) e^{j\omega\tau}$, so that, in spherical coordinates,

$$\mathbf{A} = \mathbf{1}_r \mu_0 M \frac{\cos \theta}{r} e^{j\omega\tau} - \mathbf{1}_\theta \mu_0 M \frac{\sin \theta}{r} e^{j\omega\tau}, \qquad [17]$$

in agreement with (9^{123}). Here $M = lI/4\pi$, where l is the length of the antenna, and τ is the retarded time, equal to $t - r/c$.

Since $A_\phi = 0$ and [17] does not involve ϕ, equations (69^A), (70^A), and (71^A) give $\text{curl}_r \mathbf{A} = \text{curl}_\theta \mathbf{A} = 0$, and

$$\text{curl}_\phi \mathbf{A} = \frac{1}{r} \frac{\partial}{\partial r}(rA_\theta) - \frac{1}{r}\frac{\partial}{\partial \theta} A_r$$
$$= -\mu_0 M \frac{\sin \theta}{r} \frac{\partial}{\partial r} e^{j\omega\tau} - \mu_0 M \frac{1}{r^2} e^{j\omega\tau} \frac{\partial}{\partial \theta} \cos \theta. \qquad [18]$$

Therefore, in view of (15),

$$\text{curl}_\phi \mathbf{A} = jk\mu_0 M \frac{\sin \theta}{r} e^{j\omega\tau} + \mu_0 M \frac{\sin \theta}{r^2} e^{j\omega\tau}. \qquad [19]$$

Now, $\text{curl } \mathbf{A} = \mathbf{B} = \mu_0 \mathbf{H}$. To evaluate the one nonvanishing component of $\mathbf{H}$, we divide the right-hand side of [19] by μ_0 and take its real part. The result agrees with (22^{108}).

To compute E, we first use (65^A) to evaluate the divergence of the right-hand side of [17] and find that

$$\text{div } \mathbf{A} = -\mu_0 M \cos\theta \left(\frac{1}{r^2} + j\frac{k}{r} \right) e^{j\omega\tau}. \qquad [20]$$

Substitution into the Lorentz condition [14] then yields the following scalar potential:

$$V = \frac{\mu_0 M c^2 \cos\theta}{\omega} \left(\frac{k}{r} - j\frac{1}{r^2} \right) e^{j\omega\tau}. \qquad [21]$$

Let us now use [13] to find the radial component of E. Since $\mu_0 c^2 = 1/\epsilon_0$, we have

$$\text{grad}_r V = -\frac{M \cos\theta}{\omega\epsilon_0} \left[\frac{2k}{r^2} + j\left(\frac{k^2}{r} - \frac{2}{r^3} \right) \right] e^{j\omega\tau}. \qquad [22]$$

Also, in view of [17],

$$j\omega A_r = j\omega\mu_0 M \frac{\cos\theta}{r} e^{j\omega\tau}. \qquad (23)$$

Consequently, since $(k^2/\omega\epsilon_0) - \omega\mu_0 = 0$, we have

$$E_r = \frac{M \cos\theta}{\omega\epsilon_0} \left(\frac{2k}{r^2} - j\frac{2}{r^3} \right) e^{j\omega\tau}. \qquad [24]$$

The real part of the right-hand side of [24] agrees with (19^{108}). The value of E_θ can be found in a similar way (Exercise 5).

EXERCISES

1. Derive Euler's formula (4) from (13^{118}).
2. Use (4) to derive the trigonometric formulas for the sine and the cosine of $\phi_1 + \phi_2$.
3. Suppose that E is the real part of $\mathbf{E}_0 e^{j\omega t}$, where $\mathbf{E}_0$ does not depend on t, and show that the electric energy density is then *not* equal to the real part of $\frac{1}{2}\epsilon(E_0 e^{j\omega t})^2$.
4. Show that (15) includes both equations in (3^{108}).
5. Use [13], [17], and [21] to check the fields E_θ and E_ϕ in (20^{108}) and (21^{108}).
6. Check the r components of the two sides of the equation $\text{curl } \mathbf{H} = j\omega\epsilon_0 \mathbf{E}$ for the case of a short antenna.

APPENDIX D

The Del Notation

Vector formulas are often abbreviated with the help of the following mathematical operators: the operator ∇, called *gradient* or *del* (an alternative name is *nabla*); the operator $\nabla \cdot$, called *divergence* or *del dot*; and the operator $\nabla \times$, called *curl* or *del cross*. The definitions read:

$$\nabla f = \text{grad } f, \tag{1}$$

$$\nabla \cdot \mathbf{F} = \text{div } \mathbf{F}, \tag{2}$$

$$\nabla \times \mathbf{F} = \text{curl } \mathbf{F}, \tag{3}$$

for every f and $\mathbf{F}$ that satisfy the respective continuity conditions required for the existence of the fields grad f, div $\mathbf{F}$, and curl $\mathbf{F}$. To get more explicit definitions, we replace the right-hand sides of (1), (2), and (3) by the right-hand sides of (43^A), (44^A), and (45^A). In tensor analysis, the symbol ∇ may be used in still other ways.

Comparison of (1), (2), and (3) shows that ∇, $\nabla \cdot$, and $\nabla \times$ are indeed different operators. Another viewpoint is that we have introduced not three but only one operator, denoted by ∇, which can operate on its operands in different ways. If we adopt this viewpoint, the reason for the symbols $\nabla \cdot$ and $\nabla \times$ can be brought out by considering the cartesian case. Since grad $f = \mathbf{1}_x \, \partial f/\partial x + \mathbf{1}_y \, \partial f/\partial y + \mathbf{1}_z \, \partial f/\partial z$, it follows from (1) that $\nabla f = \mathbf{1}_x \, \partial f/\partial x + \mathbf{1}_y \, \partial f/\partial y + \mathbf{1}_z \, \partial f/\partial z$. This equation can be abbreviated to

$$\nabla f = \left(\mathbf{1}_x \frac{\partial}{\partial x} + \mathbf{1}_y \frac{\partial}{\partial y} + \mathbf{1}_z \frac{\partial}{\partial z} \right) f, \tag{4}$$

and suggests writing

$$\nabla = \mathbf{1}_x \frac{\partial}{\partial x} + \mathbf{1}_y \frac{\partial}{\partial y} + \mathbf{1}_z \frac{\partial}{\partial z}. \tag{5}$$

To test whether or not (5) gives a versatile explicit expression for $\mathbf{V}$ in the cartesian case, we consider the formula

$$\mathbf{V} \cdot \mathbf{F} = \left(\mathbf{1}_x \frac{\partial}{\partial x} + \mathbf{1}_y \frac{\partial}{\partial y} + \mathbf{1}_z \frac{\partial}{\partial z}\right) \cdot (\mathbf{1}_x F_x + \mathbf{1}_y F_y + \mathbf{1}_z F_z). \tag{6}$$

The cartesian base-vectors do not vary from point to point. Therefore, since $\mathbf{1}_x \cdot \mathbf{1}_x = 1, \mathbf{1}_x \cdot \mathbf{1}_y = 0$, and so on, we have

$$\mathbf{1}_x \frac{\partial}{\partial x} \cdot \mathbf{1}_x F_x = (\mathbf{1}_x \cdot \mathbf{1}_x) \frac{\partial}{\partial x} F_x = \frac{\partial}{\partial x} F_x, \tag{7}$$

and so on. All in all,

$$\mathbf{V} \cdot \mathbf{F} = \frac{\partial F_x}{\partial x} + \frac{\partial F_y}{\partial y} + \frac{\partial F_z}{\partial z} = \text{div } \mathbf{F}, \tag{8}$$

in agreement with (2). If we interpret the symbol $\mathbf{V} \times \mathbf{F}$ as the cross product of the right-hand side of (5) and $\mathbf{F}$, we find that (3) is also satisfied. Again, if we interpret the symbol $\mathbf{V}^2$ as $\mathbf{V} \cdot \mathbf{V}$, we find, without using (5), that

$$\mathbf{V}^2 f = \mathbf{V} \cdot \mathbf{V} f = \mathbf{V} \cdot (\mathbf{V} f) = \text{div grad } f, \tag{9}$$

in agreement with definition (4^{65}) of the Laplacian.

The steps leading from (6) to (8) are simple, because the base-vectors $\mathbf{1}_x$, $\mathbf{1}_y$, and $\mathbf{1}_z$ do not change from point to point. By contrast, the vectors $\mathbf{1}_\rho$, $\mathbf{1}_r$, $\mathbf{1}_\theta$, and $\mathbf{1}_\phi$ are not constants, as seen from such equations as (31^A). As a result, it is not practicable, in cylindrical or spherical coordinates, to identify $\mathbf{V}$ with a simple and sufficiently versatile explicit operator. Therefore, in formulating the del notation, it is important to stress the definitions (1), (2), and (3)—which hold in any coordinate frame—rather than to stress the special case (5).

Which notation is better—symbols such as grad f, curl $\mathbf{F}$, and div curl $\mathbf{F}$, or the equivalent symbols $\mathbf{V} f$, $\mathbf{V} \times \mathbf{F}$, and $\mathbf{V} \cdot \mathbf{V} \times \mathbf{F}$? Opinions differ, because each notation has merits of its own. For example, to verify the equation $\mathbf{V} \cdot \mathbf{V} \times \mathbf{F} = 0$ in a formal way, we need only to interpret the term $\mathbf{V} \cdot \mathbf{V} \times \mathbf{F}$ as a triple scalar product with a repeated factor and note that the determinant (19^{77}) then has two identical rows. But when the equation $\mathbf{V} \cdot \mathbf{V} \times \mathbf{F} = 0$ is written as div curl $\mathbf{F} = 0$, it somehow brings out more vividly the striking fact that the direction lines of a curl have no beginnings and no ends.

Index

Section numbers for some topics are given in the chart in the preface. Some alternative names of physical quantities are listed on the inside of the back cover.

A

Absolute value, 15, 23
Across *vs.* through, 45
Advanced time, 386
Allen, J. R., 347 n.
Ampère, A. M., 1
Ampere (unit), 320, 450
Ampère's mmf law, 328
 differential form, 334
 integral (circuital) form, 329
 Maxwell's modification of, 335
Ampèrian currents, 306, 329
Analogies among fields, 366
Antenna (*see* Short antenna)
Associative property, 59

B

Balloon, charged, 224
Base-vectors, 66, 134, 144, 444
 derivatives of, 444, 445
Betatron, 165
Bewley, L. V., 356 n.
Biot, J. B., 313
Biot-Savart law, 315
Boundary conditions, 258, 271
 on **B** and **H**, 397, 398
 on **D** and **E**, 261, 262, 398
Bragg, W. H., 338 n.

C

Cable, leaky, 77, 151, 208
Campbell, G. A., 218 n.
Capacitance, 217, 275
 distributed, 169
 of coaxial cylinders, 278
 of concentric spheres, 276
 of parallel plates, 275
Capacitivity (permittivity), 217, 232
 of free space, 217
 of metals, 277
Capacitor:
 cylindrical, 278
 parallel-plate, 217, 222, 275
  leaky, 337
 spherical, 276
Cavity in conductor, 94, 146, 260, 266, 273

Celsius, A., 19 n.
Charge:
 conservation of, 154, 156
 density, 11
  linear, 14, 218
  surface, 80, 218, 265, 270, 356
  volume, 12, 218
 electrostatic unit of, 218
 image, 271
 on straight line, 219
 on waveguide walls, 422
 polarization, electric, 232, 240
  magnetic, 305
 static, 2
 test, 160
Circular:
 fields, 105, 142, 198
 polarization of wave, 396
Circulation, 185, 299, 448
Commutative property, 60
Component, 61 (*see also* Resolvent)
 azimuthal, 136
 cartesian (rectangular), 62, 64
 circular, 136, 145
 cylindrical, 136
 forward, 92
 normal, 68, 92
 radial, 136, 145
 spherical, 145
 tangential, 68, 92
Conductance, 4
Conduction current density, 74, 285
 in leaky cable, 151
 near cylindrical hole, 79, 82, 259
 near spherical cavity, 94
Conduction electrons, 2
 drift speed, 7
Conductivity, 9
Conservation:
 of charge, 154, 156
 of energy, 370
Conservative fields, 180, 301
 differential test for, 193, 195
 integral test for, 192
Constitutive relations, 365
Contour (terminology), 43–46
Coordinate-free equations, 23, 70
Coordinate variables, 23, 442

459

Coulomb, C. A., 1
Coulomb (unit), 11, 450
Coulomb's law, 163, 216, 234
  Gauss's form, 239
  Laplace's form, 251
  Maxwell's form, 252
  Poisson's form, 251
Cross products, 287
Curl, 293, 446, 457
  direction lines of, 295, 300
  flux lines of, 300
  torque on pinwheel, 197, 302
Current, conduction, 2
  along straight line, 315, 324, 329, 430
  Ampèrian, 306, 329
  branch, 5
  definition of, 73, 85
  density of, 74
  eddy, 194
  estimation of, 115
  image, 274
  in a plate, 318
  in a rod, 330
  in a sheet, 317, 391
  in walls of waveguide, 423
  leakage, 77, 151
  loop (mesh), 6, 347
  true, 329
Current, convection, 169, 365
Current, displacement, 284
  density of, 284
Current loop, 311
Curve (terminology), 42–47
Cutoff (see Waveguide)
Cyclotron, 308
Cylindrical hole in conductor, 78
  current density, 79, 82, 259
  electric fields, 206, 265
  surface charge, 80, 265

D

$d\mathbf{B}$ formula, 313
$d\mathbf{E}$ formula, 341
Deaver, B. S., 96 n.
Del, 249, 457
Delta function, 244
Density (see Charge, Current, Energy, Field line, Source)
$d\mathbf{H}$ formula, 316
Dielectric constant, 233
Dielectric in uniform field:
  plate, 233
  rod, 267
  sphere, 268
Dielectrics, 12, 231
  linear, 232
Difference of potential, 186, 201
Differential antenna, 378
Differential forms:
  Ampère's mmf law, 334
  Coulomb's law, 250–252
  Faraday's emf law, 355
  Maxwell's mmf law, 335

Dipole, dipole moment:
  electric, 163, 312
  Hertzian, 378
  magnetic, 304, 312
  point, 378
Dirac, P. A. M., 96 n., 245 n., 304
Direction lines, 99
Directional derivative, 28, 37, 210
Displacement current:
  electric, 283
  magnetic, 342
Distributive property, 61
Divergence (div), 125, 128, 446, 447
  theorem, 129, 448
Dot product, 59, 90
Drift speed of electrons, 3, 7
Duhamel, J. M. C., 226 n.

E

Electret, 235, 305
Electric displacement density, **D**, 230, 234, 302
Electric field, properties of:
  charge-displacing, **D**, 229, 302
  force-exerting, **E**, 160, 302
  torque-exerting, curl, **E**, 194, 302
Electric flux density, **D**, 230
Electric intensity, **E**, 160, 161 (see also Faraday field)
Electric intensity:
  of point charge, 217
  near charged conductor, 270
Electric moment, 163, 312
Electric pinwheel, 194, 302
Electromotance, 186
Electromotive force (emf), 5, 186, 200 (see also Faraday emf)
Electron, 164 (see also Conduction electron)
Electrostatic:
  force field, 2, 159, 181
  potential field, 199
Energy density:
  electric, 281
  magnetic, 353
Energy, flow of, 370, 385, 401, 414
Equipotentials, 201
Euler, L., 453

F

Fairbank, W. M., 96 n.
Farad, 451
Faraday, M., 1
Faraday:
  emf, 346
  field, $\mathbf{E}'$, 159, 342
  force, 344
  ice-pail experiment, 274
Faraday's law of induction (Faraday's emf law), 339
  differential form, 355, 434
  integral form, 344
Fermi (unit), 279

Index

Field:
 circular, 105, 142, 198
 conservative, 192, 301
 electric, 159
 electrostatic, 159, 181
 Faraday, 159, 342
 invariant, 18, 101, 439
 magnetic, 306, 308
 radial, 138, 140, 147
 scalar, 18
 solenoidal, 131
 vector, 89
Field line, 95, 98, 100 (see also under Direction, Force, and Flux)
 equation of, 99, 413
 refraction of, 237, 263
 unit for, 100
Filamentary:
 current, 313, 429
 solenoid, 342
Flux:
 definition of:
  for contours, 131, 308
  for surfaces, 109, 117
 density, 110
  electric, 230
  magnetic, 306
 graphical estimates of, 115
 integral, 109
 line, 89, 98, 111, 132
 magnetic, 308
 theorem (Gauss), 129, 448
Force:
 charge-deflecting, 165, 307
 charge-driving, 164
 electromotive, 186
 Faraday, 344
 line of, 98
 Lorentz, 187, 307, 374
 magnetomotive, 327
 on current-carrying wires, 310
Four-vector, 439

G

Gauss, K. F., 48 n.
Gaussian surface, 239
Gauss's divergence theorem, 129, 448
Gauss's form of Coulomb's law, 239
Giorgi, G., 449
Gradient:
 coordinate formulas, 213, 214, 446, 447
 definitions, 209, 446
Group speed (velocity), 416, 417
Guard ring, 97
Gutman, F., 235 n.

H

Harmonic functions (harmonics), 253–258
Heaviside, O., 218 n.
Henricksen, R. N., 347 n.
Henry (unit), 451
Hertz, H. R., 378 n., 382 n.

Hertz (unit), 410
Hertzian dipole, 378
Hole (see Cylindrical hole)
Homogeneous, 8

I

Image charge, 271
Image current, 274
Impedance, intrinsic, 367, 389
Inductivity (magnetic permeability), 309
 of free space, 321
Inequalities ($<$ vs. $\ll$), 32
Integral forms:
 Ampère's mmf law, 328
 Coulomb's law, 239
 Faraday's emf law, 344
 Maxwell's mmf law, 335
Integrals, notation for, 29
Intensity:
 electric, $\mathbf{E}$, 161
 magnetic, $\mathbf{H}$, 308
 of magnetization, $\mathbf{M}$, 309, 367
 of polarization, $\mathbf{P}$, 235, 366
Intrinsic impedance, 367, 389
Iona, M., 133 n.
Isotropic, 8

J

Jefimenko, O. D., 230 n., 234
Joule, J. P., 170
Joule (unit), 450
Joule's law:
 circuit form, 170
 field form, 171

K

Kirchhoff, G. R., 1
Kirchhoff's laws:
 current (junction) law: 5, 115, 149
  field form, 151
 voltage (loop) law, 347

L

Laplace, P. S., 249 n.
Laplace's equation, 251, 253
 form of Coulomb's law, 250
Laplacian, 249, 446
Leakage current, 77, 151, 208
Lenz, H. F. E., 339 n.
Level surface, 20
Line (see also under Direction, Field, Flux, and Force):
 density, 95, 98
 map, 95
Line vs. length, 8
Linear:
 combination of functions, 251, 256, 423
 dielectric, 232
 polarization, 395
 term, 34
Linkage, 131, 308, 329

Index

Loop (mesh) current, 6, 347
Lorentz, H. A., 307
Lorentz condition, 438, 455
  force, 187, 347, 374

M

Macroscopic phenomena, 3, 366
Magnetic:
  dipole, 304, 312
  displacement current, 342
  energy, energy density, 353
  flux, 110, 132, 308
  flux density, **B**, 110, 132, 306
  induction, **B**, 110, 132
  inductivity (permeability), 309
    of free space, 321
  intensity, **H**, 308
  moment, 312
  monopole, 304
  permeability (inductivity), 309
  polarization charge, 305
  scalar potential, 325
  vector potential, **A**, 429
Magnetization, intensity of, **M**, 367
Magnetomotive force (mmf), 327
Magnetostatics, 105, 322, 331
  rôle of solid angles, 322
Map, line, 95
  pointer, 79
Matzek, M. A., 374 n.
Maxwell, J. C., 1, 449
Maxwell's equations, 364
  form of Coulomb's law, 252, 365
  mmf law, 335, 365
Measure-numbers of vectors, 66
Mho, 451
Microscopic phenomena, 3, 366
MKS-Giorgi units, 449
Mmf law, Ampère's, 328, 333
  Maxwell's, 335
Möbius strip, 45 n.
Molecules, polar, nonpolar, 231, 232
Moment (see Dipole)
Monopole, magnetic, 304
Moorcroft, D. R., 356 n.

N

Nabla, 457
Neighboring points, 24, 443
Neumann, F. E., 339
Newton (unit), 450
Nonhomogeneous conductor, 81
Nonnegative number, 19
Normal (positive) to a surface, 44

O

Odell, T. W., 126
Oersted, H. C., 313
Ohm, G. S., 4
Ohm (unit), 450
Ohm's law, circuit form, 4
  field form, 166

Ohmic (nonohmic) conductor, 4

P

Paradox, 362
Parker, S., 356 n.
Path, 42
Pattern speed (velocity), 415
Permeability, magnetic (inductivity), 3
  of free space, 321
Permittivity (capacitivity), 217, 232
  of free space, 217
  of metals, 277
Phase (see Wave terminology)
Phase speed (velocity), 17, 417
Pinwheel, electric, 194, 302
Plane waves, 388, 394, 396
Pohl, R. W., 230 n.
Point quantity, 8
Pointer, 56
  map, 79
Poisson, S. D., 251
Poisson's equation, 251
  form of Coulomb's law, 250
Polar (nonpolar) molecules, 231, 232
Polarization charges, 232, 239, 240
Polarization of a dielectric, 232
  intensity of, **P**, 235, 366
Polarized wave, 395, 396
Positive normal, 44
Potential:
  Coulomb, 203, 217
  electric, V, 199
  magnetic, scalar, 325
  vector, **A**, 429
Potential drop (rise), 5, 200, 360
Poynting, J. H., 371 n.
Poynting's vector, 371
Probe, two-coin, 227, 302
Product, cross (vector), 288, 289
  dot (scalar), 59, 90, 289
  triple scalar, 291
Projection, 47, 60
Propagation constant, 17
  vector, 394

Q

"Q" of a medium, 336

R

Radial component, 136, 145
Radial vector fields:
  three-dimensional, 147
  two-dimensional, 138, 140
Radiation (see Short antenna)
Ramanathan, K. G., 340 n.
Ramo, S., 428 n.
Rationalization, 218, 452
Reflection, angle of, 408
Reflection from:
  nonconductors:
    normal incidence, 398

Reflection from (*cont.*):
  perfect conductors:
    normal incidence, 402
    oblique incidence, 407
Refraction, index of, 399
Refraction of field lines, 237, 263
Region, 8
  shrinking, 30
  simply connected, 47
Region *vs.* volume, 8
Relaxation time, 273, 336
Resistance, 3
  leakage, 208
  specific, 9
Resistivity, 9
Resolvent ("vector component"), 61, 65, 69, 90
Retarded:
  time, 377, 386
  potentials, 435
Right-hand rule, 44, 45, 47
Right-handed frame, 23
Rod, carrying a current, 330
Rod in external field, 266, 267
Rosser, W. G. V., 360 n.
Russell, B. R., 374 n.

S

S vs. (S), 8
Savart, F., 313
Scalar:
  field (function), 18, 21
  invariant, 18
  potential:
    electric, 199
    magnetic, 325
  product, 59, 90
Scanlon, P. J., 347 n.
Schelkunoff, S. A., 230 n., 389 n., 428 n.
Schwarz, W. M., 439 n.
Sears, F. W., 340 n.
Self-inductance, 169
Separation of variables, 255, 426
Sheet of current, a. c., 391
  d.c., 317
Shielding, 272, 404
Short antenna:
  complete field, 381, 437, 455
    maps of **E**, 382, 383
    vector potential, 436
  distant field, 384
    radiation pattern, 385
    vectors **E** and **H**, 384
  far zone, 384
  local field, 378, 432
    vector potential, 432
  near zone, 378
  retarded potentials, 435, 436
Side *vs.* edge, 43, 46
Signature, "signed" (*see* "*terminology*" *under* Contour, Curve, and Surface)
Silsbee, F. B., 449 n.

Sinks (sources) of vector fields, 105, 121
Skilling, H. H., 63 n., 383 n.
Skin depth, 406
Snellius, W., 409 n.
Snell's law, 409
Solenoid, 319, 342–344, 352
Solid angles and magnetic fields, 322
Solid angles, subtended by:
  contours, 326
  surfaces, 48
Source density, definition of:
  mathematical, 124, 125
  pictorial, 124
Sources (sinks) of vector fields, 105, 121
Space, 8
Speed of electron drift, 3, 7
  of light, 336, 366, 376, 380
Speed (velocity), group, 415, 416
  pattern, 414, 415, 417
  phase, 414, 417
Stokes, G. G., 298 n.
Stokes's circulation theorem, 298, 448
Superposition, 251, 423
Surface (terminology) 42–47 (*see also* Level and Equipotential)
Surface charges, 80, 189, 218, 265, 270, 273, 359, 398, 421
Surface currents, 398, 404, 423
Surface *vs.* area, 8

T

Taylor, B., 34 n.
Taylor's equation, series, etc., 33, 37
TE, TEM, TM waves, 395, 425
Tesla, Nikola, 307
Tesla (unit), 307
Test charge, 160, 230, 302
  surface, 74
Torque, 287
  and curl, 302
  on current loop, 311
  on pinwheel, 194
  on small pinwheel, 195
Two-coin probe, 227, 302

U

Unit vectors, 63
Units of measurement:
  esu of charge, 218
  for field lines, 100
  gauss, 307
  MKS-Giorgi, 449
  Tesla, 307

V

Van de Graaff machine, 169
Vector:
  algebra, 60, 90
  base, 66, 134, 144, 444
  component, 61

Vector (*cont.*):
  cross product, 287
  dot product, 59, 90
  identities, 445
  potential, 436
  step, 68, 443
Vector field, 89
  radial, 138, 140, 147
  time-varying, 282
  uniform, 100, 202
Velocity (speed): (*see* Group, Pattern, and Phase)
Volt, 450
Voltage, 186
Voltmeter, a.c., 349
  d.c., 187
Volume charge density, 12

W

Watt (unit), 450
Wave:
  attenuated, 419
  plane, 388
  polarized, circularly, 396
    linearly, 395

Wave (*cont.*):
  transverse (electric, electromagnetic, magnetic), 395, 425
Wave equations, 33, 368, 370, 439
Wave terminology, 16–17, 394–396, 415, 416
Waves between parallel plates, 411
  cutoff phenomenon, 418
  energy flow, 414
  group and phase (pattern) speeds, 415, 416
  surface currents, 417
  TE type, 411
  TM type, 417
Waveguide, rectangular, 410
  dominant TE mode, 420
    lines of **E** and **H,** 421
    surface charges, 422
    surface currents, 423
  other TE waves, 423–428
Weber, W. E., 306 n.
Weber (unit), 451
Whinnery, J. R., 428 n.
Whitmer, R. M., 420 n.
Wire circuits in time-varying magnetic fields, 347, 356
  effects of surface charges, 360
Work, 174